Herbicides and Plant Physiology

T0261688

Herbicides and Plant Physiology

Third Edition

Professor Dr Andrew H. Cobb

Formerly Emeritus Professor of Plant Science,
Dean of Academic Affairs and Director of
Research at Harper Adams University,
Shropshire,
UK

WILEY Blackwell

This edition first published 2022
© 2022 John Wiley & Sons Ltd

Edition History
(1e, 1992) Andrew Cobb, published by Chapman and Hall
(2e, 2010) A.H. Cobb and J.P.H. Reade, published by Wiley-Blackwell

All rights reserved. No part of this publication may be reproduced, stored in a retrieval system, or transmitted, in any form or by any means, electronic, mechanical, photocopying, recording or otherwise, except as permitted by law. Advice on how to obtain permission to reuse material from this title is available at http://www.wiley.com/go/permissions.

The right of Andrew H. Cobb to be identified as the author of this work has been asserted in accordance with law.

Registered Offices
John Wiley & Sons, Inc., 111 River Street, Hoboken, NJ 07030, USA
John Wiley & Sons Ltd, The Atrium, Southern Gate, Chichester, West Sussex, PO19 8SQ, UK

Editorial Office
The Atrium, Southern Gate, Chichester, West Sussex, PO19 8SQ, UK

For details of our global editorial offices, customer services, and more information about Wiley products visit us at www.wiley.com.

Wiley also publishes its books in a variety of electronic formats and by print-on-demand. Some content that appears in standard print versions of this book may not be available in other formats.

Limit of Liability/Disclaimer of Warranty
The contents of this work are intended to further general scientific research, understanding, and discussion only and are not intended and should not be relied upon as recommending or promoting scientific method, diagnosis, or treatment by physicians for any particular patient. In view of ongoing research, equipment modifications, changes in governmental regulations, and the constant flow of information relating to the use of medicines, equipment, and devices, the reader is urged to review and evaluate the information provided in the package insert or instructions for each medicine, equipment, or device for, among other things, any changes in the instructions or indication of usage and for added warnings and precautions. While the publisher and authors have used their best efforts in preparing this work, they make no representations or warranties with respect to the accuracy or completeness of the contents of this work and specifically disclaim all warranties, including without limitation any implied warranties of merchantability or fitness for a particular purpose. No warranty may be created or extended by sales representatives, written sales materials or promotional statements for this work. The fact that an organization, website, or product is referred to in this work as a citation and/or potential source of further information does not mean that the publisher and authors endorse the information or services the organization, website, or product may provide or recommendations it may make. This work is sold with the understanding that the publisher is not engaged in rendering professional services. The advice and strategies contained herein may not be suitable for your situation. You should consult with a specialist where appropriate. Further, readers should be aware that websites listed in this work may have changed or disappeared between when this work was written and when it is read. Neither the publisher nor authors shall be liable for any loss of profit or any other commercial damages, including but not limited to special, incidental, consequential, or other damages.

Library of Congress Cataloging-in-Publication Data

Names: Cobb, Andrew, author.
Title: Herbicides and plant physiology / Professor Dr Andrew H. Cobb,
    formerly Emeritus Professor of Plant Science, Dean of Academic Affairs
    and Director of Research at Harper Adams University, Shropshire, UK.
Description: Third edition. | Hoboken, NJ : Wiley-Blackwell, [2021] |
    Includes bibliographical references and index.
Identifiers: LCCN 2021031562 (print) | LCCN 2021031563 (ebook) | ISBN
    9781119157694 (paperback) | ISBN 9781119157717 (adobe pdf) | ISBN
    9781119157700 (epub)
Subjects: LCSH: Plants–Effect of herbicides on. | Plant physiology. |
    Herbicides–Physiological effect. | Weeds–Control.
Classification: LCC SB951.4 .C63 2021 (print) | LCC SB951.4 (ebook) | DDC
    632/.954–dc23
LC record available at https://lccn.loc.gov/2021031562
LC ebook record available at https://lccn.loc.gov/2021031563

Cover Design: Wiley
Cover Image: courtesy of Dr J P H Reade

Set in 10/12 pt TimesLTStd by Straive, Pondicherry, India

10 9 8 7 6 5 4 3 2 1

Contents

Preface

He who has bread may have many troubles; he who lacks it has only one.

(Byzantine proverb)

A peasant must stand for a long time on a hillside with his mouth open before a roast duck flies in.

(Chinese proverb)

The origin of the word herbicide is a combination of the Latin words *herba* (noun, herbaceous plant) and *caedere* (verb, to kill; R.L. Zimdhal, *Weed Science* **17**, 137–139, (1969).

It is now 30 years since the first edition of this book was published and so it is timely to reflect on herbicide use and the environmental consequences of their use. While we now have fewer products in our agrochemical armoury, farmers and growers have become dependent on fewer active ingredients, and glyphosate dominates the global market. This third edition aims to update the reader on how herbicides contribute to modern agriculture, how they are discovered and developed, and how they interact with plant growth and development and the environment. Since the publication of the second edition in 2010 there have been many advances in our understanding of plant physiology, especially regarding how plants function in tune with their ever-changing environment and how post-translational modifications provide regulatory control of most plant processes. Modern agriculture, however, still faces many challenges.

1 *The global food challenge.* To some, the continuing expansion of the human population will inevitably outstrip the growth of our food supply, resulting in global starvation. To others, an expansion of arable land, a growing global food trade and the increases in crop yield predict a more optimistic future. Yet is it inevitable that a growing food supply will continue to meet demand?

2 *The problem.* More than half the global population will suffer some form of malnutrition by 2030 unless urgent action is taken to increase access to food of high nutritional quality. The Food and Agriculture Organization of the United Nations (FAO *et al.*, 2017) estimated that 815 million persons were hungry in 2016 (11% of the global population), an increase of 35 million since 2015. While 155 million children have stunted growth owing to poor nutrition, 2 billion persons suffer from hunger, while 1.9 billion adults and 41 million children are either overweight or obese. In addition, the human population is expected to grow by about 80 million per annum to an estimated 10 billion in 2050 (Oerke and Dehne, 2004). Furthermore, the global impact of the current covid-19

pandemic could result in at least a further 200 million undernourished persons. How can we expect to produce 70% more food to feed them all?

3   *Food security.* Food security is an increasing global problem in the face of climate change, combined with increasing populations and volatile food prices. With the global population growing at 230,000 persons each day and 60% of us now living in cities, the pressure on farmers to increase crop yields is ever present. At the same time, in the UK as an example, the land available per head of population has decreased from 0.8 to 0.2 hectares in the last 50 years. Every year 12 million hectares is degraded globally owing to drought, de-forestation and desertification, an area roughly the size of Nicaragua, the largest country in Central America. Furthermore, global freshwater supply is becoming increasingly limited and unreliable to an estimated 700,000 persons, notwithstanding the fluctuations in weather as a result of climate change. We are experiencing greater extremes of weather, such as flooding or drought, and so we need to use the available fresh- and artesian-water more wisely. This is especially so whether we grow crops or raise animals. For example, it is estimated that 70 litres of water are needed to produce one apple, whereas 15,000 litres are needed for one kg of beefsteak! It is interesting to note that the carbon footprint of beef and lamb is three times that of pork, five times that of chicken, over 30 times that of bread wheat and 50 times that of potatoes. Urbanisation and increasing incomes generate a higher demand for animal protein, yet beef production requires four times more land than dairy, per unit of protein consumed. In addition, beef is seven times more resource intensive than pork and poultry, and 20 times more so than pulses. Not forgetting that animal production results in increased greenhouse gas emissions. It is a further uncomfortable fact that about a third of all food produced never reaches the table. This value is higher for fruit and vegetables, and such losses are even higher in the developing world, owing to the lack of effective storage and/or transport (IFPRI, 2016).

As available land for farming is in ever shorter supply and extremes in climate become more evident, many scientists predict an increased degradation of soils and a need for increased attention to land management. In recent years in the UK, for example, farmers have seen above average rainfall with increasing soil erosion, degradation and run-off. Palmer and Smith (2013) noted that 75% of fields planted with maize or potatoes in the south-west of England were severely damaged by soil degradation, with one in five sites experiencing serious rill and gully erosion. Some 60% of fields growing winter cereal crops, such as wheat and barley also displayed high to severe soil degradation. Techniques to avoid soil compaction, such as topsoil lifting or sub-soiling, are options to loosen soil layers, but the use of increasingly large and heavy machinery increases the risk. These authors concluded that soils with good agricultural properties are over-exploited in crop production and, as a result, can become highly degraded. Conversely, chalk and limestone soils degrade less.

A further definitive study, by Challinor *et al.* (2016), has predicted that gradually rising temperatures in Africa, and more droughts and heatwaves caused by climate change, will have a profound impact on maize yields. Higher temperatures reduce the length of time between planting and harvesting, which results in less time to accumulate biomass and yield. They also predict similar shortening of time to yield for maize crops across the tropics and suggest that maize breeding systems must adapt to increasing temperatures to ensure positive yields in the decades ahead.

A further interesting area of research would be to investigate how the major weeds of maize crops may also adapt to increasing growth temperatures, with especial attention to weeds exhibiting C_4 photosynthesis.

4 *Greater intensification?* In a comprehensive and thought-provoking study, Fischer *et al.* (2014) concluded that greater crop yields are possible through a greater intensification of agriculture, especially in Sub-Saharan Africa. This assumes more agricultural research, development and training in the developing world and a more efficient use of inputs, such as plant protection products. What is certain is that global governments will need to invest substantially in agriculture to achieve the yield increases necessary to feed the world. Greater meat consumption and an expanding human population implies that crop productivity needs to double by 2050.

5 *Sustainability.* One widely used definition of sustainability is 'meeting the needs of the present without compromising the ability of future generations to meet their own needs' (DEFRA, UK). Another considers that 'a sustainable agriculture is ecologically sound, economically viable and socially just and humane' (Alliance for Sustainability, www.afors.org). How can humankind tackle climate change, reduce population growth, cut out waste, educate consumers in the developed world to eat less, conserve freshwater and eat more plant products and fewer animals? Political awareness, more education and an alert and informed media may provide the answer. Indeed, Ehrlich and Harte (2015) 'urge policymakers around the world to move the issue of food security to the top of the political agenda'. And they conclude that 'anything less is a recipe for disaster'.

6 *Can organic agriculture feed the world?* Regarding crop production, organic practices infer that the use of inorganic crop nutrition products is not allowed, genetically modified crop cultivars are not permitted and the use of chemical plant protection products for the control of pests, weeds and diseases is forbidden. Such practices are becoming widely accepted in the developed world, as they are seen as a more 'natural' means of crop production. Those that espouse organic farming practices often say that it is better for the environment, since it requires fewer inputs, but it is generally agreed that organic farming is less productive per hectare than conventional, intensive agriculture. More land would be needed for equivalent yields, and conversion to organic practice would release more organic carbon into the environment. Furthermore, it is doubtful that legume cover crops could replace the nitrogen fertiliser needed to give higher yields (Connor, 2008). Perhaps in the future, nitrogen fixation from the atmosphere will be possible using genetically engineered crop plants?

The EU Farm to Fork Strategy, launched on 20 May 2020, has 'aspirational targets' for 2030 for a 50% cut in pesticide use and a commitment to dedicating 25% of agricultural land to organic farming. One wonders if the consequences of reduced yields and higher prices, to name but two, have been thought through. Furthermore, the aspiration to replace pesticides with 'biocontrol agents' appears idealistic. Although regularly promoted, an agreed definition of biocontrol remains elusive and, at the present time, commercial agents are expensive, are unproven in the field and no viable weed biocontrol methodology exists. At a time of imminent global recession owing to the covid-19 pandemic, a global scarcity of food is predicted, made worse by plagues of locusts in Africa, Arabia, Iran and Pakistan adding to pressures on food security. Can food producers realistically promote organic agriculture and cut pesticide use in such uncertain times? Instead, can we use our existing practices more effectively?

7 *Plants for the future: genetic diversity.* In 2016, the Royal Botanic Gardens, Kew, UK, released a report on the state of the world's plants. It noted that an estimated 31,000 plant species have a documented use for medicines, food and materials. There are an estimated 391,000 vascular plants known to science of which 369,000 species are flowering plants. A further 2000 new vascular species are described each year. Many are wild relatives of known crops that can be a source of genetic variation to improve our crops in the future, such as tolerance to drought or possess a unique metabolism. Some 21% of the world's plants, however, are currently threatened with extinction, especially in declining rainforests. Humankind must preserve this genetic biodiversity in seed banks at all costs for future generations (www.stateoftheworldsplants.org). The human population derives 50% of its calorific intake from only three species, namely rice, wheat and maize. The rest is derived from only 20 species. Having so few staple crops means that we lack diversity in our diets and have an over-reliance on the chosen few. Can conventional plant breeding produce the advances in yield needed to feed our global population? Perhaps it can, with the application and more widespread adoption of gene editing techniques developed in the last decade.

8 *Can we survive without plant protection products?* Currently, global agriculture is heavily reliant on plant protection products, such as herbicides, plant growth regulators, fungicides and insecticides, to maximise crop yields. Without an equivalent process, yields would be reduced by at least 20–40%, so an increase in food prices would inevitably follow, with public unrest and food volatility. The reader is encouraged to note Oerke and Dehner (2004) and Pesticides in Perspective (n.d.) for further details.

   I note with concern that the EU is planning to withdraw as many as 75 active ingredients from the crop protection armoury. In addition to yield losses, this will erode farmers' margins and reduce farm productivity across the EU, and over a million jobs are at risk of being lost. There are no current viable alternatives to the use of agrochemicals. Their judicious use should be promoted and these agents preserved if we are to feed the world and ensure future food security. In order for plant protection products to be used effectively, it is imperative to have an understanding of the biology of the target organisms and how an active ingredient works in both the plant and the environment. Thus, an understanding of weed biology, soil science and plant physiology underpins herbicide choice, use and effectiveness.

9 *Is science and technology the answer?* Scientists and technologists consider that appropriate scientific and technological developments might come to the rescue of humankind. Why such optimism? The answer lies in recent research findings reported in the plant sciences literature, some examples of which are noted below. The first reports a rice cultivar that has been engineered to have fewer stomata, which has resulted in an increased tolerance to drought and water availability, giving equivalent or increased rice yields. The importance of this finding is the knowledge that 2500 litres of water are typically required to produce 1 kg of rice. The authors consider that rice plants with fewer stomata should perform better when limitations on water supply threaten food security (Caine *et al.*, 2018).

   A second innovation is the use of gene editing to understand how plants are able to perceive and respond to environmental signals at the cellular level and respond by alterations in gene expression. In this way, plant scientists are able to understand how biotic and abiotic stimuli, such as responses to disease or environmental change, can alter growth, development and crop yield. This advance is largely due to the generation and testing of mutants that can be incorporated into plant breeding programmes. Examples include resistance to drought, resistance to salinity, temperature and water-logging, insect and disease resistance, potatoes free from late blight, enhanced concentrations of omega oils and vitamins, fruit and vegetables that do not turn brown on impact, and low-gluten wheat, to name but a few.

A third example is the RIPE project – Realising Increased Photosynthetic Efficiency – for sustainable increases in crop yield (www.ripe.illinois.edu). This is a collaboration of US, Australian, Chinese, German and UK universities that began in 2012 with an aim of increasing global agricultural production. Several research strategies have been developed with successful investigations that include:

- relaxing mechanisms of photoprotection;
- by-passing photorespiration;
- optimising enzyme activity in the photosynthetic carbon reduction cycle;
- increasing the efficiency of RuBisCo; and
- optimising canopies for photosynthesis.

Also of note the C4 Rice Project (www.c4rice.com) that is jointly funded by the Bill and Melinda Gates Foundation, in which researchers from seven institutions in five countries are working together to develop high-yielding rice cultivars. Their aim is to use gene editing to introduce C_4 photosynthetic machinery into rice, a C_3 crop, which currently accounts for 19% of all calories consumed in the world. If successful, rice plants could be 50% more productive.

10 Finally, the '*Hands Free Hectare*' project in the UK, demonstrated in 2016 that it is possible to drill, tend and harvest a crop of spring barley without operators of machines or agronomists in the field. It proves that there is no technical barrier to automated field agriculture. Weed control is achieved by aerial sensors that ensure that only weed-infested areas of a field are sprayed, rather than the whole field, thereby reducing inputs. It is assumed that unmanned automation will become an increasingly important part of agriculture in the future. Achieving precision spraying with dedicated robots fitted with associated sensors is a current engineering challenge (Ghaffarzadeh, 2017). Flavell (2016) has argued that we need to generate clear plans to increase the confidence of investors and society in the future of the plant sciences. Our collective challenge is therefore to see technological advances in the engineering and plant sciences lead to new concepts, products and innovations that will improve the efficiencies of agriculture in the future.

11 A key conclusion of the 2019 cross-sector review of weed management, commissioned by the UK Agriculture and Horticulture Development Board and the British Beet Research Organisation, was that the approach to Weed Management in the UK needs to be overhauled, and a major investment is required. The review noted that, *inter alia*: (a) essential information on weed management could be lost to the industry without appropriate key sources of references and an archive; (b) coordinated programmes of research and knowledge transfer are necessary to make the best use of depleted national funding; and (c) the plant protection industry needs to be more unified and strategic to maximise the chances of such methods and research results making an economic difference to farms and growers. I hope that the contribution of research institutes, colleges and

universities are to the fore in any future update in the training of the next generations of plant protection personnel.

12 As we have entered a new decade, agrochemical inputs are becoming increasingly under scrutiny and some would argue that agrochemical technology is reaching its limits (Altieri, 2019). Why is this?

- Large-scale crop monocultures occupy about 80% of the 1.5 billion hectares currently used in global agriculture.
- Approximately 2.3 billion kg of pesticides are applied each year to keep weeds, fungal and insect pests at bay.
- However, less than 1% of pesticides reach the target weed or pest, so that most ends up in the soil, water and the air, leading to declines in biodiversity, especially pollinators, and the natural enemies of pests.
- Monoculture agriculture leads to pesticide resistance.

It follows that the removal of pesticides and herbicides will restore biodiversity and a renewed interest in the biological control of pests. Biodiversity can also be enhanced using cover crops, inter-cropping, rotations, agroforestry and the introduction of livestock into crop fields. Surrounding these fields with hedgerows and corridors also generates more complex habitats, as field margins are reservoirs of the natural enemies of crop pests, and provide over-wintering sites for wildlife. In this way, it is thought that replacing monocultures with more complex agricultural systems will contribute to yield advantages via improved biodiversity, enhanced soil quality and resilience to climate change. Such arguments are ecologically persuasive, but more evidence, including detailed cost/benefit analysis, is required before extrapolation to weed control by herbicides. Nonetheless, the observed global increase in weed resistance to herbicides in recent decades is clearly linked to monoculture, and shows no signs of decline.

13 So how can politicians, growers, farmers and the agrochemical industry become more ecologically aware and promote more sustainable practices?

- The industry should recommend and use technologies for a more precise application of agrochemicals that will reduce application volumes and cumulative dosage.
- Greenhouse gas emissions can be reduced and soils preserved by promoting minimal tillage and fewer, but more targeted agrochemical applications.
- More informed farming practices that are sustainable for the use of agrochemicals should be encouraged by continuing professional development and re-education of farmers and growers.
- Biodiversity should be encouraged by returning to more complex agro-ecosystems.

We have the tools and knowledge to defeat hunger and malnutrition, but do we have the political will and commitment to do so?

Despite these reservations it is important to remember that without herbicides and the sustainable intensification of agriculture we would not be able to feed the existing and growing global population. We must remain alert, however, to the environmental consequences of their use. Furthermore, it is vital that independent research in the plant sciences continues to be supported by national bodies in universities and research institutes. New discoveries and current understanding of how plants are adapted to their ever-changing environments will continue to drive agrochemical research and development in the years to come.

The starting point of this book is weed biology. Subsequent chapters consider the modern plant protection products industry, how herbicides are discovered and developed, how they gain entry into the plant and move to their sites of action, and the basis of herbicide selectivity. Detailed and updated accounts follow of how herbicides interact with the major physiological processes in plants, leading to weed control. This begins with the inhibition of photosynthesis, followed by pigment biosynthesis, interactions with the plant growth regulator, auxin, lipid biosynthesis, amino acid biosynthesis, cell division, cellulose biosynthesis, the plant kinome, herbicide resistance, the development of genetically modified herbicide-resistant crops and a consideration of some new targets for the future development of new herbicides.

In the dozen years since the last edition was written, there have been many advances reported in the plant physiology literature. There has been continuing progress in our understanding of the *Arabidopsis* genome and our model plant species, and gene editing techniques are now commonplace. It is fascinating to recall that 10 years ago gene editing techniques had not been published. We now understand more about the mechanisms whereby environmental change and protein synthesis are in tune with both biotic and non-biotic stresses, enabling plant physiology to adapt to an ever-changing plant environment. Consequently, much of this text is new and many recent references have been added. Note, however, that many older references and figures have been retained because they remain relevant in demonstrating how our understanding has developed, and that the work of previous generations of plant scientists is not forgotten. Of course, the errors are still mine and hopefully will be remedied in time.

It is with regret that the co-author of the second edition of this book, Dr John Reade, has been unable to contribute to this volume, owing to other commitments. He continues to teach the next generations of plant scientists at Harper Adams University and supervises research students with his trademark enthusiasm and intelligence.

And finally,

I think it must be rather nice
to live by giving good advice;
to talk of what the garden needs
instead of pulling up the weeds. (Reginald Arkell, 1882-1959)

Andy Cobb
July 2021

References

Altieri, M.A. (2019). Pesticide treadmill. *Chemistry and Industry* **11**, 37.

Caine, R.S., Yin, X., Sloan, J., Harrison, R.L., Mohammed, U., Fulton, T. *et al.* (2019) Rice with reduced stomatal density conserves water and has improved drought tolerance under future climate conditions. *The New Phytologist* **221**, 371–384; doi: org/10.1111/nph.15344

Challinor, A.J., Koehler, A.-K., Ramirez-Villegas, J., Whitfield, S. and Das, B. (2016) Current warming will reduce yields unless maize breeding and seed systems adapt immediately. *Nature Climate Change* **6**, October; doi: org/10.1038/nclimate3061

Connor, D.J. (2008) Organic agriculture cannot feed the world. *Field Crops Research* **106**, 187–190.

Ehrlich, P.R. and Harte, J. (2015) Opinion: to feed the world in 2050 will require a global revolution. *Proceedings of the National Academy of Sciences* **112**, 14743–14744.

FAO, IFAD, UNICEF, WFP and WHO (2017) *The State of Food Security and Nutrition in the World. Building Resilience for Peace and Food Security*. Rome.

Fischer, R.A., Byerlee, D. and Edmeades, G.O. (2014) Crop yields and global food security: Will yield increase continue to feed the world? Australian Centre for International Agricultural Research Monograph No.158. Canberra.

Flavell, R. (2016) Making plant science purposeful and relevant to all. *Journal of Experimental Botany* **67**, 3186–3187.

Ghaffazadeh, K. (2017) Deep learning for agchems. *Chemistry and Industry* **9**, 36.

IFPRI (2016) Global Food Policy Report: How we feed the world is unsustainable.

Oerke, E.-C. and Dehne, H.-W. (2004) Safeguarding production losses in major crops and the role of crop protection. *Crop Protection* **23**, 275–285.

Palmer, R.C. and Smith, R.P. (2013) Soil structural degradation in SW England and its impact on surface-water run-off generation. *Soil Use and Management* **92**, 567–575; doi: 10.1111/sum.12068

Pesticides in Perspective (n.d.) An introduction to crop protection. The Crop Protection Association, UK. www.pesticidesinperspective.org.uk

Chapter 1
An Introduction to Weed Biology

One year's seed is seven year's weed.

A traditional rhyme

1.1   Introduction

The human race has been farming for over 10,000 years. Weeds have been an unwelcome presence alongside crops ever since the first farmers saved and planted seeds in the region that is now present-day Turkey and the Middle East. Indeed, when these early farmers noticed a different plant growing, decided they did not want it and pulled it up, they were carrying out a form a weed control that is still used today: hand roguing.

But what are weeds? Weeds are all things to all people, depending on the viewpoint of the individual. To some they are plants growing where they are not wanted; to others they are plants growing in the wrong place, in the wrong quantity, at the wrong time; and to some they are regarded as plants whose virtues have yet to be fully discovered! The need to control weeds only arises when they interfere with the use of the land, and this is usually in the presence of a crop, such as in agriculture and horticulture. Weed control may also be necessary in other situations including amenity areas, such as parks and lawns, in water courses, or on paths and drives where the presence of plants may be regarded as unsightly. It should not be overlooked, however, that weeds contribute to the biodiversity of ecosystems and should only be removed when financial or practical implications make their presence unacceptable. With this in mind an appropriate definition of a weed is:

Any plant adapted to man-made habitats and causing interference of the use of those habitats. (Lampkin, 1990)

Or

A plant whose virtues remain to be discovered. (Emerson, 1912, see https://theysaidso.com/quote/ralph-waldo-emerson-what-is-a-weed-a-plant-whose-virtues-have-never-been-discovered)

Herbicides and Plant Physiology, Third Edition. Andrew H. Cobb.
© 2022 John Wiley & Sons Ltd. Published 2022 by John Wiley & Sons Ltd.

1.2   Distribution

On a global basis only about 250 species are sufficiently troublesome to be termed weeds, representing approximately 0.1% of the world's flora. Of these, 70% are found in 12 families, 40% alone being members of the Gramineae and Compositae. Interestingly, 12 crops from five families provide 75% of the world's food and the same five families provide many of the worst weeds (Table 1.1). This implies that our major crops and weeds share certain characteristics and perhaps common origins.

Table 1.1   Important plant families which contain both the major crops and the worst weeds of the world.

Number of species classified as the world's worst weeds (%)	Family	Examples of major crops	Examples of major weeds	Common name
44	Gramineae	Barley, maize, millett, oats, rice, sorghum, sugar cane and wheat	*Elytrigia repens* (L.)	Couch
			Alopecurus myosuroides (L.)	Black-grass
			Avena fatua (L.)	Wild oat
			Sorghum halepense (L.) Pers.	Johnson grass
			Echinochloa crusgalli (L.)	Barnyard grass
4	Solanaceae	White potato	*Solanum nigrum* (L.)	Black nightshade
			Datura stramonium (L.)	Jimsonweed
			Hyoscyamus niger (L.)	Henbane
5	Convolvulaceae	Sweet potato	*Convolvulus arvensis* (L.)	Field bindweed
			Cuscuta pentagona (Engelm)	Field dodder
			Ipomoea purpurea (L.) Roth	Tall morning glory
5	Euphorbiaceae	Cassava	*Euphorbia maculata* (L.)	Spotted spurge
			Euphorbia helioscopia (L.)	Sun spurge
			Mercurialis annua (L.)	Annual mercury
6	Leguminosae	Soybean	*Cassia obtusifolia* (L.)	Sicklepod
			Melilotus alba (Desc)	White sweetclover
			Trifolium repens (L.)	White clover

Source: Radosevich, S.R. and Holt, J.S. (1984) *Weed Ecology: Implications for Vegetation Management.* New York: Wiley. Reproduced with permission of John Wiley & Sons.

1.3   The importance of weeds

Most plants grow in communities consisting of many individuals. If the resources available (such as space, water, nutrients and light) become limiting then each species will be forced to compete. Weeds are often naturally adapted to a given environment and so may grow faster than the crop, especially since the crop species has been selected primarily for high yield rather than competitive ability. A unit of land may therefore be regarded as having a finite potential biomass to be shared between crop and weeds, the final proportion being determined by their relative competitive ability.

1.4   Problems caused by weeds

The most obvious problem caused by weeds is the reduction of yield through direct competition for light, space, nutrients and water. Weeds can have many further effects on the use of land, as illustrated in Table 1.2.

Table 1.2   Problems caused by weeds.

Problem	Mechanism
Reduced crop yield	Interference with access to light, water and nutrients
Reduced crop quality	Admixture of contaminating seeds in arable crops
	Contamination of vegetable crops
Delayed harvesting	Conservation of moisture may delay ripening and increase moisture level when harvested
Interference with harvesting	Climbing plants making combining more difficult
	Vigorous, late-growing weeds interfering with harvesting of potatoes and sugar beet
Interference with animal feeding	Plants with spines or thorns inhibiting animal foraging
Poisoning	Poisoning either through ingestion or through contact
Tainted animal products	Imparting an undesirable flavour, e.g. to milk
Plant parasitism	Competing for nutrients and water
Reduced crop health	Acting as an alternative host for crop pests and diseases
	Increasing the amount of vegetation at the base of the crop, increasing moisture and disease
Reduced animal (and human) health	Acting as an intermediate host or a vehicle for ingestion of pests and parasites
	Photosensitivity
	Teratogens
	Carcinogens
Safety hazard	Reducing vision on roadsides
	Causing a risk of fire under electricity lines and on garage forecourts
Reduced wool quality	Hooked seeds reducing the value of fleece
Water flow prevented	Plant mass blocking ditches and irrigation channels
Allelopathy	Releasing substances toxic to the growth of crop plants
Impacted crop establishment	Vegetation preventing the establishment of young trees
	Competing for space with establishing crops

Source: Naylor, R.E.L. and Lutman, P.J. (2002) What is a weed? In: Naylor, R.E.L. (ed.) *Weed Management Handbook*, 9th edn. Oxford: Blackwell Publishing/BCPC. Reproduced with permission of John Wiley & Sons.

1.4.1   *Yield losses*

Crop losses approaching 100% are recorded in the literature (Table 1.3; Lacey, 1985). Such yield losses will, of course have a profound effect on a national economy in terms of both the need to import foodstuffs and the costs of weed control. Despite the many methods of weed management that are now available worldwide, it is estimated that approximately 13% of crop losses are still due to weeds alone (Table 1.4). Indeed, in 1974 the annual cost of weeds to agriculture in the USA was estimated at US$10 billion, with 50% owing to yield reductions and 50% owing to the cost of weed control (Rodgers, 1978).

In the tropics, parasitic weed species from the genera *Cuscuta* (dodders), *Orobranche* (broomrapes) and *Striga* (witchweeds) can have a profound effect on a range of crops. They absorb nutrients directly from the crop plant, which may not set seed at all in the case of cereals such as sorghum.

Weed control techniques are therefore aimed at the reduction in the competitive ability of weeds in a crop and the prevention of weed problems in a future crop. The former is increasingly based on chemical use, and the latter also requires suitable cultural and agronomic practices.

Yield loss may be usefully related to the number of weeds per unit area causing a defined yield loss in a defined crop, that is, as a Weed Threshold (Table 1.5) or as a Crop Equivalent

Table 1.3   Examples of yield losses owing to weeds.

Crop	Yield loss (%)	Country
Cassava	92	Venezuela
Cotton	90	Sudan
Groundnuts	60–90	Sudan
Onions	99	UK
Rice	30–73	Colombia
Sorghum	50–70	Tanzania/Nigeria
Sugar beet	78–93	Texas, USA
Sweet potatoes	78	West Indies
Wheat*	66	UK
Yams	72	Nigeria

Source: Lacey, A.J. (1985) Weed control. In: Haskell, P.T. (ed.) *Pesticide Application: Principles and Practice*. Oxford: Oxford University Press, pp. 456–485. Reproduced with permission of Oxford University Press.
*From Moss (1987).

Table 1.4   Estimated percentage crop losses owing to weeds, 1988–90 (from Oerke *et al.*, 1995).

	Estimated loss owing to weeds (%)
Africa	16.5
North America	11.4
Latin America	13.4
Asia	14.2
Europe	8.3
Former Soviet Union	13.0
Oceania	9.6
Average	**13.1**

Source: Oerke, E.C., Dehne, H.W., Schonbeck, F. and Weber, A. (eds) (1995) *Crop Production and Crop Protection: Estimated Losses in Major Food and Cash Crops*. Amsterdam: Elsevier.

Table 1.5 Relative competitive abilities of a number of common weeds found in winter cereals.

Weed species	5% yield loss (plants m⁻²)	Weed species	5% yield loss (plants m⁻²)
Galium aparine	1.7	*Poa annua*	50.0
Anisantha sterilis	5.0	*Epilobium* spp.	50.0
Avena fatua	5.0	*Polygonum aviculare*	50.0
Lolium multiflorum	8.3	*Sonchus* spp.	50.0
Alopecurus myosuroides	12.5	*Taraxacum officinale*	50.0
Brassica napus	12.5	*Fumaria officinalis*	62.5
Sinapis arvensis	12.5	*Geranium* spp.	62.5
Tripleurospermum inodorum	12.5	*Lamium purpureum*	62.5
Cirsium spp.	16.7	*Ranunculus* spp.	62.5
Convolvulus arvensis	16.7	*Veronica* spp.	62.5
Fallopia convolvulus	16.7	*Aethusa cynapium*	83.3
Papaver spp.	16.7	*Senecio vulgaris*	83.3
Chenopodium album	25.0	*Anagallis arvensis*	100.0
Myosotis arvensis	25.0	*Allium vineale*	250.0
Persicaria maculosa	25.0	*Aphanes arvensis*	250.0
Silene vulgaris	25.0	*Legousia hybrida*	250.0
Stellaria media	25.0	*Viola arvensis*	250.0

Source: Lutman, P.J., Boatman, N.D, Brown V.K. and Marshall, E.J.P. (2003) Weeds: their impact and value in arable ecosystems. In: *The Proceedings of the BCPC International Congress: Crop Science and Technology 2003* **1**, 219–226.

(the amount of resource an individual weed uses expressed as the number of crop plants this resource would support, although in practice it is the biomass of the weed and the crop which is measured). Generally, these figures have only been determined for weed interaction with major crops, but they give a good indication of the ability of a particular species to compete with all crops.

Yield loss may also occur in addition to direct competition for resources. Allelopathy is the production of allelopathic chemicals by one plant species that may inhibit (or, in the case of positive allelopathy, stimulate) the growth of other species. Anecdotal evidence of negative allelopathic effects has been reported for a number of weed species, although supporting research is often lacking. Recent findings have been reviewed by Olofdotter and Mallik (2001) and others (see *Agronomy Journal* vol. 93). Given the ample evidence of allelopathy exhibited by crop species, it is highly likely that many weed species will also display these effects, and that it is only a matter of time before research demonstrating this becomes readily available.

Further examples of yield loss caused by weeds include the effects on non-plant organisms. One example of this is the presence of dandelion (*Taraxacum officinale*) in fruit orchards. Dandelion flowers are preferentially visited by insect pollinators and so pollination of fruit blossom (and therefore fruit yield) is reduced.

1.4.2   Interference with crop management and handling

Some weeds can make the operation of agricultural machinery more difficult, more costly, or even impossible. The presence of weeds within a crop may necessitate the need for extra cultivations to be introduced. This often leads to crop damage, reduced yields and increased pest and disease occurrence, although in sugar beet crops, where inter-row cultivation is often carried out and has previously been associated with yield loss, recent findings suggest that careful implementation can result in no loss of root yield or sucrose content (Dexter *et al.*, 1999; Wilson and Smith, 1999). This is possibly due to the development of tillage equipment that carries out more shallow cultivation and that is more carefully implemented, resulting in less seedling and root damage. Weeds can also affect the processes carried out prior to crop planting. For example, fat hen stems and leaves block the mesh of de-stoners, which are used prior to potato and other root crop planting. Species with rough, wiry stems that spread close to the ground (e.g. knotgrass, *Polygonum aviculare*) or are more erect in growth habit (e.g. fat hen, *Chenopodium album*) present major problems for the mechanical harvesting of many crops and can result in damage to machinery (e.g. pea viners) and subsequent harvesting delays. Other species can be troublesome when the crops are harvested by hand, such as the small nettle (*Urtica urens*) in strawberries and field bindweed (*Convolvulus arvensis*) in blackcurrants. The result is that fruit is not harvested and spoils on the plant.

1.4.3   Reduction in crop quality

Competition between crop and weed species can result in spindly leaf crops and deformed root crops which are less attractive to consumers and processors. A crop may have to be rejected if it contains weed seeds, especially when the crop is grown for seed, such as barley and wheat, and if the weed seeds are similar in size and shape to the crop, e.g. wild oats (*Avena fatua*) in cereal crops. Similar problems are encountered in the contamination of oilseed rape seed with seeds of weed species such as cleavers (*Galium aparine*). Where a proportion of the seed is saved for planting in subsequent seasons, this can cause a large increase in weed infestation. Contamination by poisonous seeds, such as darnel (*Lolium temulentum*) and corncockle (*Agrostemma githago*) in flour-forming cereals is also unacceptable and once led to vastly increased costs of crop cleaning. Such cleaning, however, has meant that these weeds are now probably extinct in agroecosystems in the UK. A further example that still causes major problems is black nightshade fruit (*Solanum nigrum*) in pea crops (Hill, 1977). In this case, the poisonous weed berry is of similar size and shape to the crop and so must be eradicated. Although grazing animals avoid poisonous species in pasture (e.g. common ragwort, *Senecio jacobea*), they may be difficult to avoid in hay and silage, and some species, notably the wild onion (*Allium vineale*), can cause unacceptable flavours in milk and meat.

1.4.4   Weeds as reservoirs for pests and diseases

Weeds, as examples of wild plants, form a part of a community of organisms in a given area. Consequently, they are food sources for some animals and are themselves susceptible to many pests and diseases. Because of their close association with crops, they may serve as important reservoirs or carriers of pests and pathogens, as exemplified in Table 1.6. Even where crop infestation does not occur, the presence of disease in weeds may cause problems, as is the case where grass weeds are infected with ergot (*Claviceps purpurea*), causing contamination of harvested grain with highly toxic ergot fragments.

Weeds may act as 'green bridges' for crop diseases, carrying the disease from one crop to another that is subsequently sown. Volunteer crops are particularly problematic in this

Table 1.6   Some examples of weeds as hosts for crop pests and diseases.

Pathogen or pest		Weed		Crop
1. Fungi				
Claviceps purpurea	(ergot)	Black-grass	*(Alopecurus myosuroides)*	Wheat
Gaeumannomyces graminis	(take-all)	Couch *(Elytrigia repens)*	Cereals	
Plasmodiophora brassicae	(clubroot)	Many crucifers		Brassicas
2. Viruses				
Tobacco ringspot		Dandelion	*(Taraxacum officinale)*	Tobacco
Cucumber mosaic		Chickweed	*(Stellaria media)*	Many crops
3. Nematodes				
Ditylenchus dipsaci	(eelworm)	Chickweed	*(Stellaria media)*	Many crops
		Spurrey	*(Spergula arvensis)*	
4. Insects				
Aphis fabae	(black bean aphid)	Fat hen	*(Chenopodium album)*	Broad and field beans

Source: Hill, T.A. (1977) *The Biology of Weeds*. London: Edward Arnold.

case and can, in severe cases, negate the use of break crops as a cultural control measure for diseases. In addition, weeds can provide over-wintering habitats for crop pests, resulting in quicker crop infestation in the spring. Ground cover provided by weeds can increase problems with slugs and with rodents, as the weeds provide greater cover and therefore reduced predation.

In 1994 and 1995 there were several severe outbreaks of the disease brown rot in potato in several European countries, especially in The Netherlands, which was possibly exported to other countries via infected seed potatoes. This extremely virulent pathogen (*Pseudomonas solanacearum*, syn. *Burkholderia solanacearum*, syn. *Ralstonia solanacearum*) causes a vascular ring rot in the developing tuber and causes a major loss of yield. Although often considered a soil-borne organism, it was not found to persist for long periods in the soil following the harvest of infected crops. However, it was found to survive in the aquatic roots of infected woody nightshade (*Solanum dulcamara*) growing at the edge of irrigation channels. Thus, it may be the case that the pathogen overwinters in this wild host and is leaching into watercourses used to irrigate the crop, thus spreading the disease. This perennial plant is now being eradicated from potato-growing areas. Several other species could also act as alternative hosts to the pathogen, including *Solanum nigrum* and *Tusilago farfara*, but further work is needed to confirm this.

1.5   Biology of weeds

Knowledge of the biology of a weed species is essential to the design of management strategies for that weed. An understanding of the life cycle of a species can be exploited in order to identify vulnerable times when weed management and control might prove more successful.

1.5.1 Growth strategies

According to Grime (1979), the amount of plant material in a given area is determined by two principal external factors, namely stress and disturbance. Stress phenomena include any factors that limit productivity, such as light, nutrient or water availability; and disturbance implies a reduction in biomass by factors such as cultivation, mowing or grazing. The intensity of both stress and disturbance can vary widely, with four possible combinations. However, only three growth strategies have evolved, as shown in Table 1.7. Although plants are unable to survive both highly stressed and disturbed environments, the other strategies have major significance to weed success.

Ruderals are the most successful agricultural weeds. These plants have typically rapid growth rates and devote most of their resources to reproduction. Because they inhabit recently disturbed environments there is little competition with other plants for resources, which therefore can be obtained without difficulty. They are generally short-lived ephemeral annuals that occupy the earliest phases of succession. Conversely, biennial and perennial weeds often employ a more competitive growth strategy in relatively undisturbed conditions. They use their resources perhaps less for seed production and more for support tissues, for example, to provide additional height for the interception of light, or more extensive root systems to obtain more water and minerals. Rapid growth rate may still be evident with high rates of leaf turnover. The third growth strategy, exhibited by the stress tolerators, is to reduce resource allocation to vegetative growth and seed production, so that the survival of relatively mature individuals is ensured in high-stress conditions. Consequently, they have slow growth rates and are commonly found in unproductive environments.

Many arable weeds have characteristics common to both competitors and ruderals, and are referred to as competitive ruderals. Indeed, most of the annuals listed in *The World's Worst Weeds* (Radosevich and Holt, 1984) fit into this category, and are found in productive sites where occasional disturbance is expected. Examples include arable land that is cultivated, and meadows and grassland that are grazed or mowed. Interestingly, most crop plants also adopt a competitive ruderal strategy with their rapid growth rates and relatively large seed production. Competition between crop and weed is then related to their relative abilities to exploit the resources available.

The practice of growing crops in monoculture has exerted a considerable selection pressure in the evolution of weeds. Many characteristics have evolved that contribute to weed success and the main ones are listed in Table 1.8. Fortunately, not all of these features are present in any one weed species, yet each character may give the weed a profound competitive advantage in a given situation. Some of these characteristics are discussed in more detail in the following sections of this chapter.

Table 1.7 Growth strategies of plants.

	Intensity of stress	
Intensity of disturbance	High	Low
High	Death	Ruderals
Low	Stress tolerators	Competitors

Source: Hill, T.A. (1977) *The Biology of Weeds*. London: Edward Arnold.

Table 1.8 The 'successful' weed.

Characteristic	Example species
1. Seed germination requirements fulfilled in many environments	*Senecio vulgaris*
2. Discontinuous germination (through internal dormancy mechanisms) and considerable longevity of seed	*Papaver* spp.
3. Rapid growth through the vegetative phase to flowering	*Cardamine hirsuta*
4. 'Seed' production in a wide variety of environmental conditions	*Poa annua*
5. Continuous seed production for as long as conditions for growth permit	*Urtica urens*
6. Very high 'seed' output in favourable environmental conditions	*Chenopodium album*
7. Self-compatible but not completely self-pollinating	*Alopecurus myosuroides*
8. Possession of traits for short- and long-distance seed dispersal	*Galium aparine*
9. When cross-pollinated, unspecialised pollinator visitors or wind pollinated	*Grass weeds in general*
10. If a clonal species, has vigorous vegetative growth and regeneration from fragments	*Cirsium arvense*
11. If a clonal species, has brittleness of leafy parts ensuring survival of main plant	*Taraxacum officinale*
12. Shows strong inter-specific competition by special mechanisms (e.g. allelopathic chemicals)	*Elytrigia repens*
13. Demonstrates resistance to herbicides through a number of resistance mechanisms	*Alopecurus myosuroides*

Source: Adapted from Baker, H.G. and Stebbins, G.L. (1965) *The Genetics of Colonising Species.* New York: Academic Press.

Invasive species have received far greater research focus in recent years (see, for instance, Shaw and Tanner, 2008 for a review), and DAISIE (Delivering Alien Invasive Species Inventories for Europe) currently reports 10,822 invasive species in Europe (this figure is for all invasive species, not just plants). Alien species present a real threat to biodiversity, and a number of political drivers have been put into place to combat their spread and reduce the occurrence of a number of alien species, including plants. These measures include the Convention of Biodiversity and the EU 2010 Halting Biodiversity Loss, both of which identify invasive weeds as being a key factor in biodiversity loss.

Most non-native plants in the UK were introduced by plant collectors in the last 200 years. They become invasive when they have negative impacts on native species, our economy and even our health. Alien plant species become a problem because they are growing in habitats away from their natural predators and so can spread with ease and do not need to invest valuable energy in biologically expensive chemical defence mechanisms. In addition, the Novel Weapons Hypothesis proposes that in some cases allelopathic chemicals produced by alien species are more effective against native species in invaded areas than they are against species from the alien's natural habitat (Inderjit *et al.*, 2007). This gives further ecological advantage to some invasive species of plant.

The Royal Horticultural Society notes 1402 invasive plants in the UK, of which 108 can have negative effects, and its website offers advice and guidance on how they should be dealt with (www.rhs.org.uk/advice/profile?PID=530). The EU Regulation on Invasive Alien Species lists 36 plants.

Observations of the pathogens and predators that affect these invasive plants in their natural habitats may identify mycoherbicides and other biological controls that may prove useful in their management. Research into such controls for Japanese Knotweed (*Fallopia japonica*), Himalayan Balsam (*Impatiens glandulifera*) and Giant Hogweed (*Heracleum mantegazzianum*) have so far produced, at best, limited success (Tanner, 2008).

Moles *et al.* (2008) have recently proposed a framework for predicting plant species which may present a risk of becoming invasive weeds. This may prove more useful than Table 1.8 in risk assessment relating to alien species, prior to their becoming major weed problems.

1.5.2 Germination time

The success of some weeds is due to close similarity with a crop. If both the weed and the crop have evolved with the same agronomic or environmental conditions they may share identical life cycles and life styles. Thus, if weed seed maturation coincides with the crop harvest, the chances of weed seed spread are increased. This phenomenon is often apparent in the grasses, such as barnyard grass (*Echinochloa crus-galli*) in rice, and wild oat (*Avena* spp.) in cereals. In Europe, spring wild oats (*A. fatua*) germinate mainly between March and May, and the winter wild oat (*Avena ludoviciana*) shows maximum germination in November. Hence, the date of cereal sowing in relation to wild oat emergence is crucial to weed control. In general, weeds may be considered to occupy one of three categories: autumn germinators, spring germinators and those that germinate throughout the year. Figure 1.1 illustrates the germination patterns for a number of common arable weeds. Autumn-sown crops are more at risk from weeds that germinate during the autumn, when the crop is relatively small and uncompetitive. Conversely, weeds that cause problems in spring crops tend to be spring germinators (including the troublesome polygonums). Knowledge of the germination patterns of weeds plays a very important role in the designing of specific weed management strategies that will disrupt conditions conducive to the survival of the weed.

Further major problems are evident in sorghum, radish and sugar beet crops. Hybridisation of cultivated *Sorghum bicolor* (L.) Moench with the weed *S. halepense* (L.) Pers. results in an aggressive perennial weed that produces few seeds, but demonstrates vigorous vegetative growth. Similarly, hybridisation between the radish (*Raphanus sativus* L.) and the weed *R. raphanistrum* (L.) has produced a weedy form of *R. sativus* with dormant seeds and a root system that is more branched and penetrating than the crop. Lastly, hybridisation of sugar beet (*Beta vulgaris* (L.) subsp. *maritima*) has created an annual weed-beet that sets seed, but fails to produce the typically large storage root. In each of these examples of crop mimicry by weeds, chemical weed control is extremely difficult owing to the morphological and physiological similarities between the weed and the crop.

1.5.3 Germination depth

Most arable weeds germinate in the top 5 cm of soil and this is the region that soil-acting (residual) herbicides aim to protect. Where minimum cultivation or direct drilling is carried out, the aim is to avoid disruption of this top region of soil in an attempt to minimise weed-seed germination. A small number weed species can germinate from greater depths and this may be due to these species possessing larger seed. A good example of this is wild oat (*Avena* spp.), which can successfully germinate and establish from depths as low as 25 cm.

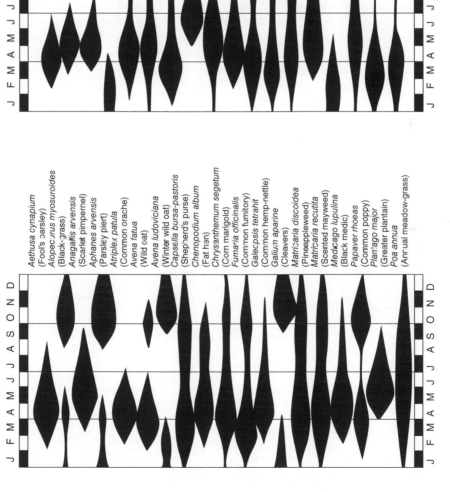

Figure 1.1 Germination periods of some common annual weeds. A greater width of the bar reflects greater germination. Source: Hance, R.J. and Holly, K. (1990) *Weed Control Handbook: Principles*, 8th edn. Oxford: Blackwell. Reproduced with permission of John Wiley & Sons.

1.5.4   Method of pollination

The survival and growth of weed populations are dependent upon successful pollination. Weed species tend to rely on non-specific insect pollinators (e.g. dandelion) or are wind pollinated (e.g. grasses), so their survival is not dependent on the size of population of specific insect pollinators. Annual weeds are predominantly self-pollinated and, when outcrossing does take place, pollination is achieved by wind or insects. This means that a single immigrant plant may lead to a large population of individuals, each as well adapted as the founder and successful in a given site. Occasional outcrossing will alter the genotype, which may aid the occupation of a new or changing niche. Furthermore, many weeds, unlike crops, begin producing seed while the plants are small and young, and continue to do so throughout the growth season. In this way the weed density and spectrum in an arable soil may change quickly.

1.5.5   Seed numbers

Seeds are central to the success of weeds. As with all plants, weed seeds have two functions: the dispersal of the species to colonise new habitats, and the protection of the species against unfavourable environmental conditions via dormancy. Weeds commonly produce vast numbers of seeds, which may ensure a considerable advantage in a competitive environment, especially since the average number of seeds produced by a wheat plant is only in the region of 90–100 (Table 1.9).

1.5.6   Seed dispersal

Many weed species possess methods of both short- and long-distance seed dispersal (Figure 1.2). By recognising these, it is possible to reduce the spread of weed seeds, a vital component of any integrated weed management strategy.

Table 1.9   Seed production by a number of common arable weeds and wheat.

Weed	Common name	Seed production per plant
Veronica persica	Common field speedwell	50–100
Avena fatua	Wild oat	100–450
Galium aparine	Cleavers	300–400
Senecio vulgaris	Groundsel	1100–1200
Capsella bursa-pastoris	Shepherd's purse	3500–4000
Cirsium arvense	Creeping thistle	4000–5000
Taraxacum officinale	Dandelion	5000 (200 per head)
Portulaca oleracea	Purslane	10,000
Stellaria media	Chickweed	15,000
Papaver rhoeas	Poppy	14,000–19,500
Tripleurospermum maritimum spp. inodorum	Scentless mayweed	15,000–19,000
Echinochloa crus-galli	Barnyard grass	2000–40,000
Chamaenerion angustifolium	Rosebay willowherb	80,000
Eleusine indica	Goose grass	50,000–135,000
Digitaria sanguinalis	Large crabgrass	2000–150,000
Chenopodium album	Fat hen	13,000–500,000
Triticum aestivum	Wheat	90–100

Source: Adapted from Radosevich, S.R. and Holt, J.S. (1984) *Weed Ecology: Implications for Vegetation Management*. New York: Wiley; containing information from Hanf (1983).

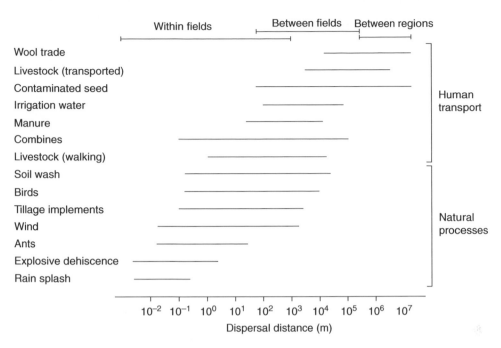

Figure 1.2 Some methods of weed seed dispersal with their estimated range in metres. Source: Liebman, M., Mohler, C.L. and Staver, C.P. (2001) *Ecological Management of Agricultural Weeds*. Cambridge University Press. Reproduced with permission of Cambridge University Press.

1.5.7   *Dormancy and duration of viability*

Although the seed production figures of an individual plant are impressive (Table 1.9), the total seed population in a given area is of greater significance. The soil seed reservoir reflects both past and present seed production, in addition to those imported from elsewhere, and is reduced by germination, senescence and the activity of herbivores (Figure 1.3). Estimates of up to 100,000 viable seeds per square metre of arable soil represent a massive competition potential to both existing and succeeding crops, especially since the seed rate for spring barley, for instance, is only approximately 400 m^{-2}! Under long term grassland, weed seed numbers in soil are in the region of 15,000–20,000 m^{-2}, so conversion of arable land to long-term grassland offers growers a means of reducing soil weed-seed burden.

The length of time that seeds of individual species of weed remain viable in soil varies considerably. The nature of the research involved in collecting such data means that few comprehensive studies have been carried out, but those that have (see Toole and Brown, 1946, for a 39 year study!) show that although seeds of many species are viable for less than a decade, some species can survive for in excess of 80 years (examples include poppy and fat hen). Evidence from soils collected during archaeological excavations reveals seeds of certain species germinating after burial for 100–600 (and maybe even up to 1700!) years (Ødum, 1965).

Dormancy in weed seeds allows for germination to be delayed until conditions are favourable. This dormancy may be innate and contributes to the periodicity of germination, as illustrated in Figure 1.1. In addition, dormancy may be induced or enforced in nondormant seeds if environmental conditions are unfavourable. This ensures that the weed seed germinates when conditions are most conducive to seedling survival.

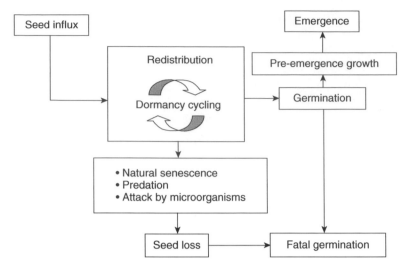

Figure 1.3 Factors affecting the soil seed population. Source: Grundy, A.C. and Jones, N.E. (2002) What is the weed seed bank? In: Naylor, R.E.L. (ed.) *Weed Management Handbook*, 9th edn. Oxford: Blackwell Publishing/BCPC. Reproduced with permission of John Wiley & Sons.

1.5.8 Plasticity of weed growth

The ability of a weed species to make rapid phenotypic adjustment to environmental change (acclimation) may offer a considerable strategic advantage to the weed in an arable context. An example of the consequence of such plasticity is environmental sensing by fat hen (*Chenopodium album*). This important weed can respond to canopy shade by undergoing rapid stem (internode) elongation, although the plant is invariably shorter if growing in full sun. Similarly, many species can undergo sun–shade leaf transitions for maximum light interception (Patterson, 1985).

1.5.9 Photosynthetic pathways

Photosynthesis, the process by which plants are able to convert solar energy into chemical energy, is adapted for plant growth in almost every environment on Earth. For most weeds and crops photosynthetic carbon reduction follows either the C_3 or the C_4 pathway, depending on the choice of primary carboxylating enzyme. In C_3 plants this is ribulose 1,5-bisphosphate carboxylase/oxygenase (RuBisCo) and the first stable product of carbon reduction is the three-carbon acid, 3-phosphoglycerate. Alternatively, in C_4 plants the primary carboxylator is phosphoenolpyruvate carboxylase (PEPC) and the initial detectable products are the four-carbon acids, oxaloacetate, malate and aspartate. These acids are transferred from the leaf mesophyll cells to the adjacent bundle sheath cells where they are decarboxylated and the CO_2 so generated is recaptured by RuBisCo. Since PEPC is a far more efficient carboxylator than RuBisCo, it serves to trap CO_2 from low ambient concentrations (micromolar in air) and to provide an effectively high CO_2 concentration (millimolar) in the vicinity of the less efficient carboxylase, RuBisCo. In this way, C_4 plants can reduce CO_2 at higher rates and are often perceived as being more efficient than C_3 plants.

In addition, because of their more effective reduction of CO_2, they can operate at much lower CO_2 concentrations, such that stomatal apertures may be reduced and so water is conserved.

The C_4 pathway is often regarded as an 'optional extra' to the C_3 system, and offers a clear photosynthetic advantage under conditions of relatively high photon flux density, temperature and limited water availability, that is in tropical and mainly subtropical environments. Conversely, plants solely possessing the C_3 pathway are more advantaged in relatively temperate conditions of lower temperatures and photon flux density, and an assumed less limiting water supply (Figure 1.4).

Returning to the interaction between crop and weed, it is therefore apparent that, depending on climate, light to severe competition may be predicted. For example, a temperate C_3 crop may not compete well with a C_4 weed (e.g. sugar beet, *Beta vulgaris*, and redroot pigweed, *Amaranthus retroflexus*) and a C_4 crop might be predicted to outgrow some C_3 weeds (e.g. maize, *Zea mays*, and fat hen, *Chenopodium album*). Less competition is then predicted between C_3 crop and C_3 weeds in temperate conditions, with respect to photosynthesis alone.

In reality, C_4 weeds are absent in the UK but widespread in continental, especially Mediterranean, Europe. In the cereal belt of North America, however, C_4 weeds pose a considerable problem and it is notable that eight of the world's top 10 worst weeds are C_4 plants indigenous to warmer regions (Table 1.10). It will be of both interest and commercial significance if the C_4 weeds become more abundant in regions currently termed temperate (e.g. northern Europe), with the development of of climate change.

Of the 435,000 plant species on Earth, C_4 photosynthesis is present in fewer than 2%, but accounts for about 25% of plant productivity. It is a carbon dioxide-concentrating mechanism that evolved relatively recently, that makes photosynthesis more efficient. It is noteworthy that many major weeds are C_4 plants (Table 1.10). In the C4 Rice Project (www.c4rice.com), gene editing is being used to introduce C4 genes into crops such as rice. Rice currently accounts for 19% of all calories consumed globally. If successful, GE rice could be 50% more productive than the current C3 rice.

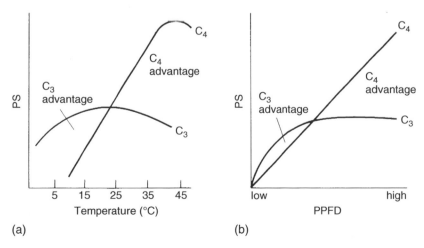

Figure 1.4 Expected rates of photosynthesis (PS) by C_3 and C_4 plants at (a) varying temperature and (b) varying photosynthetic photon flux density (PPFD). Source: Andy Cobb, 1992.

Table 1.10 Photosynthetic pathway of the world's 10 worst weeds.

Latin name	Common name	Photosynthetic pathway	Number of countries where plant is known as a weed
1. *Cyperus rotundus* (L.)	Purple nutsedge	C_4	91
2. *Cynodon dactylon* (L.) Pers.	Bermuda grass	C_4	90
3. *Echinochloa crus-galli* (L.) Beauv.	Barnyard grass	C_4	65
4. *Echinochloa colonum* (L.) Link.	Jungle rice	C_4	67
5. *Eleusine indica* (L.) Gaertn.	Goose grass	C_4	64
6. *Sorghum halepense* (L.) Pers.	Johnson grass	C_4	51
7. *Imperata cylindrica* (L.) *Beauv.*	Cogon grass	C_4	49
8. *Eichornia crassipes* (Mart.) Solms.	Water hyacinth	C_3	50
9. *Portulaca oleracea* (L.)	Purslane	C_4	78
10. *Chenopodium album* (L.)	Fat hen	C_3	58

Source: Holm, L.G., Plucknett, D.L., Pancho, J.V. and Herberger, J.B. (1977) *The World's Worst Weeds. Distribution and Biology*. Hawaii: University Press.

1.5.10 Vegetative reproduction

Not all weeds classified as competitive ruderals are annuals. The exceptions are the herbaceous perennials, which have a high capacity for vegetative growth and include many of the most important weeds in the world. The vegetative production of new individuals can often be a very successful means of weed establishment. This is because the vegetative structures can rely on the parent plant for nutrients, which can confer a competitive advantage, especially at the start of the growth season. There are, however, disadvantages to vegetative reproduction. The principal ones are that since daughter plants are genetically identical to their parents, they may not be well adapted to a changing environment and that widespread dispersal cannot occur by vegetative means alone, unlike with seeds. The vegetative structures themselves include stolons, rhizomes, tubers, bulbs, corms, roots and turions.

Stolons are long, slender stems that grow along the soil surface to produce adventitious roots and shoots; examples include the perennial bermuda grass (*Cynodon dactylon*), the annual crabgrass (*Digitaria sanguinalis*) and creeping buttercup (*Ranunculus repens*). Rhizomes are underground stems from which adventitious roots and shoots arise. Major examples include the perennials Johnson grass (*Sorghum halepense*), couch grass (*Elytrigia repens*), perennial sedges such as purple and yellow nutsedge (*Cyperus rotundus* and *C. esculentus*) and ground elder (*Aegopodium podagraria*). Purple nutsedge (*Cyperus rotundus*) has an extensive underground system of rhizomes and tubers. The rhizomes can penetrate and pass completely through vegetable root crops, and the tubers can remain dormant and carry the plant through very extreme conditions of drought, flooding or lack of aeration. *Cyperus rotundus* is a major weed of tropical and warm temperate regions of sugar cane, rice, cotton, maize and vegetables, groundnuts, soybeans and sorghum. Japanese knotweed

(*Fallopia japonica*), a highly invasive weed that has become a particular problem in many parts of the world, propagates largely by means of rhizomes (both locally and in moved soil; Figueroa, 1989), as colonies rarely result from seed.

Tubers are enlarged terminal portions of rhizomes that possess storage tissues and axillary buds. Examples include the perennial sedges mentioned above, Jerusalem artichoke (*Helianthus tuberosus*) and the common white potato (*Solanum tuberosum*). Another particularly troublesome weed that produces tubers is the horsetail (*Equisetum arvense*). In this case, aerial shoots can be easily controlled, but deep-seated tubers will produce new shoots when conditions permit.

Bulbs are also underground organs that are modified buds surrounded by scale leaves, which contain the stored nutrients for growth, an example being wild onion (*Allium vineale*). Corms are swollen, vertical underground stems covered by leaf bases, for example, bulbous buttercup (*Ranunculus bulbosus*).

Many species produce long, creeping horizontal roots that give rise to new individuals, including perennial sow thistle (*Sonchus arvensis*), field bindweed (*C. arvensis*) and creeping or Canada thistle (*Cirsium arvense*). Some biennials and perennials form swollen, non-creeping taproots capable of regenerating whole plants. Common examples are dandelion (*Taraxacum officinale*) and curled and broad-leaved docks (*Rumex crispus* and *R. obtusifolius*). Several aquatic weeds produce vegetative buds or turions that have specialised nutrient-storing leaves or scales. These separate from the parent plant in unfavourable conditions, or are released after the decay of the parent, to remain dormant until favourable conditions return. Examples include Canadian pondweed (*Elodea canadensis*) and *Ceratophyllum demersum*.

Cultivation and soil disturbance will promote the fragmentation of all these vegetative structures. Propagation will then occur when the vegetative structure is separated from the parent plant. The brittleness of leafy parts ensures that although leaves may be removed manually or by grazing, the means of vegetative reproduction remains in the soil. Only continuous cultivation will prevent the accumulation of stored nutrient reserves and so control these weeds.

1.6   A few examples of problem weeds

Black-grass (*Alopecurus myosuroides*) is of widespread distribution in Europe, temperate Asia, North America and Australia, and has become a major problem weed where winter cereals are planted and reduced cultivation methods employed. In winter cereals about 80% of black-grass seedling emergence occurs from August to November, so crop and weed emergence coincide. The weed shows similar growth rates to the crop over the winter period, but is most aggressive from April to June with substantial grain losses being reported (Figure 1.5).

Black-grass flowers from May to August and is cross-pollinated. Seeds have short dormancy and viability (3% viable after 3 years), so ploughing, crop rotation or spring sowing will remove the problem.

Bracken (*Pteridium aquilinum*) is a widespread, poisonous perennial weed of upland pastures and is found throughout the temperate regions of the world. This fern is thought to occupy between 3500 and 7000 km^2 in the UK alone, and may be spreading at 4% each year. It is difficult to control because of rhizomes that may grow up to 6 m away from the

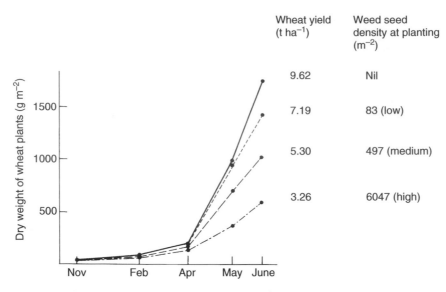

Figure 1.5 Effect of black-grass density on the growth and yield of winter wheat. Source: Moss, S.R. (1987) Competition between blackgrass (*Alopecurus myosuroides*) and winter wheat. British Crop Protection Conference. *Weeds* **2**, 367–374.

parent plant and are capable of rapid growth from underground apices and buds. The rhizome contains a starch reserve which acts as an energy reserve for the developing frond canopy. Translocated herbicides need to be repeatedly used if all rhizome buds are to be killed. Consequently, bracken is capable of rapid canopy establishment and is an aggressive coloniser of new areas. Indeed, some plants are estimated to be over 1000 years old. Other features that contribute to the success of this weed are its mycorrhizal roots, that ensure efficient nutrient uptake, especially in phosphate-deficient soils, and a potential production of 300 million spores per plant, which can remain viable for many years if kept dry.

Bracken creates a profound shading effect, suppressing underlying flora and gradually eliminating grass growth. Bracken also contains various carcinogens and mutagens, and is therefore poisonous to both humans and grazing animals. In addition, bracken may provide a haven for sheep ticks, which can transfer numerous sheep and grouse diseases.

Cleavers (*G. aparine*) is considered by many to be the most aggressive weed of winter cereals. It is of ubiquitous occurrence in hedgerows, and has become most invasive in cereals and oilseed rape. Its climbing and scrambling habit allows it to rapidly outgrow the crop to form a dense weed canopy, eventually causing severe lodging, interference with harvesting procedures, large yield losses and severe crop contamination.

There is nowadays an increasingly widespread occurrence of crop species in succeeding crops when sown in rotation. These 'volunteer' crops include potatoes, cereals, oilseed rape and sugar beet.

Potato 'ground keepers' are usually small tubers that are missed by the harvesters, although some are derived from true seeds. They can last several seasons, and pose a particular problem in subsequent pea and bean crops where they can only be eliminated by hand rogueing. They also pose a considerable threat to the health and certification of subsequent potato crops since they can carry over pests and virus infections.

Volunteer cereals can also carry foliage diseases from one season to the next and to adjacent crops, examples being yellow rust (*Puccinia striiformis*), brown rust (*Puccinia hordei*) and powdery mildew (*Erysiphe graminis*). These cereals may become highly competitive weeds and can smother and kill young oilseed rape seedlings, for example. Even if the seedlings survive, growth is predictably stunted and winter kill more likely. Volunteer oilseed rape plants may also create an additional problem of oil purity. Since modern varieties are grown for low erucic acid and glucosinolate content, the presence of volunteer plants could severely contaminate the crop with unacceptably high levels of these compounds, which could cause the crop to be rejected.

Weed (sugar) beet has also become a serious problem in Europe, such that at least 45% of the UK crop is infested. These bolters also severely reduce crop yield.

In all cases harvesting techniques must be improved to avoid substantial seed return to the soil, and agronomic practices should be altered to reduce rapid germination of volunteers. It is also important that volunteers containing engineered resistance to herbicides are avoided at all costs. These plants would be difficult to control by chemical means, and could have serious consequences to the spread of herbicide-resistant genes in the population at large (Young, 1989).

The water hyacinth (*Eichornia crassipes* Mart. Solms) has been blamed for the loss of 10% of the water in the river Nile, equivalent to 7×10^9 m^3 annually, through increased transpiration, and this loss is in addition to its deleterious effects on irrigation systems, fishing activity, navigation and health (by harbouring vectors of human disease organisms) which, together with its pan-tropical spread, have earned it the name 'the million dollar weed' (Lacey, 1985).

1.7   Positive attributes of weeds

Although this section will outline the positive roles of weeds in agroecosystems, it might more properly be titled 'Positive Attributes of Non-sown Plant Species', to reflect the definition of a weed given at the beginning of this chapter.

Non-sown species of plant, whether native to a piece of land or accidentally introduced, can perform a number of very important roles. These need to be assessed prior to the implementation of any weed management practices as removal might cause more harm than good. Non-sown species have a valuable role in reducing soil erosion by water and wind. This is particularly important when a crop is small and after harvest, when erosion is likely to be more of a problem. An additional benefit is that plants growing in this situation will 'lift' and make available nutrients by absorbing them at depth through their roots and assimilating them into above-ground biomass. When the above-ground biomass dies, then nutrients are returned to the soil surface. If the non-sown species is a legume, then the added benefits of nitrogen fixation can be considered. In this way non-sown species are not only reducing erosion but also reducing nutrient leaching.

Although non-sown species may act as reservoirs and alternative hosts for pests and diseases, they can also act as shelter for beneficial organisms that can contribute to biological control in crops. This shelter may be in the form of hedgerows or artificially created beetle banks, but the role of non-sown species must not be underestimated. In addition, complete removal of all non-crop species gives herbivorous organisms no choice but to eat the crop. Recent research has investigated whether organisms such as slugs might preferentially predate on non-crop species if they are present (Brooks *et al.*, 2003).

Non-sown species also contribute a major food source for birds and insects (Table 1.11), and therefore aid in the support of a biodiverse environment. This positive role must be considered alongside the negative effects of some species. Wild oat and black-grass have a negligible positive effect on biodiversity, while causing high yield losses. In such cases it is likely that the weeds will be controlled. In other cases, where a positive role on biodiversity is identified, removal will depend to an extent on the numbers present and financial implications. Even where a positive role has not been identified then certain species, such as corncockle and darnel, have now become very rare and conservation measures should be adopted where they are found to reduce further decline. With an ever-growing focus on farming in an environmentally sensitive way, it is likely that there will become a greater

Table 1.11   Ranking of the competitive effects of selected weed species and their value for birds and invertebrates.

Species	Competitive index	Value for birds	Value for insects
Alopecurus myosuroides	***		—
Avena fatua	****	—	—
Lolium multiflorum	****		
Poa annua	**	**	***
Aethusa cynapium	**		—
Anagallis arvensis	*		—
Aphanes arvensis	*		
Brassica napus	***	**	—
Chenopodium album	**	***	***
Cirsium spp.	***	*	***
Convolvulus arvensis	***		
Epilobium spp.	**		
Fallopia convolvulus	*	***	
Fumaria officinalis	**	*	—
Galium aparine	****	—	***
Geranium spp.	**		—
Lamium purpureum	**	—	**
Legousia hybrida	*		
Myosotis arvensis	**	—	—
Papaver spp.	***		*
Persicaria maculosa	**	***	**
Polygonum aviculare	**	***	***
Ranunculus spp.	**		
Senecio vulgaris	**	**	***
Sinapis arvensis	***	**	***
Sonchus spp.	**	*	***
Stellaria media	**	***	***
Tripleurospermum inodorum	***		***
Veronica spp.	**		—
Viola arvensis	*	**	—

Source: Lutman, P.J., Boatman, N.D, Brown V.K. and Marshall, E.J.P. (2003) Weeds: their impact and value in arable ecosystems. In: *The Proceedings of the BCPC International Congress: Crop Science and Technology 2003* **1**, 219–226.
The number of asterisks refers to the species' increased importance to birds/invertebrates or increasing competitive impact; '—', no importance; blank, no information).

emphasis on justifying why a non-crop species should be removed rather than justifying why they should remain.

Allelopathy, as mentioned in Section 1.4.1, is usually used to describe the negative effect of one plant on another via the release of natural growth inhibitors. However, incidences of positive allelopathy have been reported, where allelopathic chemicals produced by one species have a positive effect on another species. An example is corncockle (now a rare arable plant) that grows alongside wheat. Corncockle produces agrostemmin which increases the yield and the gluten content of the wheat (Gajic and Nikocevic, 1973).

1.8 The ever-changing weed spectrum

Weed populations are never constant, but are in a dynamic state of flux owing to changes in climatic and environmental conditions, husbandry methods and the use of herbicides. Such change is evident from a consideration of cereal production in the UK. Cereals were traditionally sown in the spring, but MAFF (Ministry of Agriculture, Fisheries and Food) statistics show that by 1985, 75% of the crop was sown in the autumn (Martin, 1987) with major changes in the weed flora. In the 1950s and 1960s the greater area of spring cereals encouraged spring-germinating broadleaf weeds such as knotgrass (*P. aviculare*), redshank (*P. persicaria*), black bindweed (*Bilderdykia convolvulus*), poppy (*Papaver rhoeas*), charlock (*Sinapis arvensis*), chickweed (*Stellaria media*) and fat hen (*Chenopodium album*). The move to winter cereals has meant less competition from these spring weeds because the crop is already established. However, winter cereals have encouraged weeds that germinate and establish in the autumn and early spring, particularly cleavers (*G. aparine*), speedwells (*Veronica* spp.), chickweed (*S. media*) and field pansy (*Viola arvensis*). Similarly, traditional cereal cultivation techniques based on ploughing have given way to direct drilling and minimal cultivations using tines and discs, which are less costly and energy intensive. Indeed, Chancellor and Froud-Williams (1986) have estimated that 40% of arable land in the UK is now cultivated without ploughing. One consequence is a reduction in the density of many annual broad-leaf weeds since they require deep soil disturbance to bring buried seeds to the surface; a lack of seed return owing to herbicide use has also encouraged this decline. On the other hand, minimum cultivation has led to an increased abundance of annual grasses, particularly of species which readily establish near the soil surface and which have relatively short periods of dormancy. The main examples are black-grass (*A. myosuroides*), meadowgrass (*Poa* spp.) and sterile brome (*Anisantha sterilis*). Wild oats (*A. fatua*) have also flourished in minimum cultivation, even though their seed requires moderate burial. These observations have been quantified in a major, large-scale survey of weeds present in winter cereals in the UK (Table 1.12).

The frequency of creeping perennials has also increased with minimal cultivation, for example field bindweed (*C. arvensis*) and Canada thistle (*C. arvense*). Chancellor and Froud-Williams (1986) also point to the occurrence of unusual species in undisturbed arable land, particularly wind-dispersed seeds of the genera *Epilobium*, *Artemisia*, *Conyza* and *Lactuca*, and suggest that these may be the problem weeds of the future. Current increases in minimum and zero cultivation methods for establishment of cereal and oilseed crops, for environmental and financial reasons, will undoubtedly be mirrored by further changes in the weed spectrum of arable land.

Table 1.12   Main broadleaf weeds and grass weeds present in 2359 winter cereal fields.

Broadleaf weeds	Fields infested (%)	Grass weeds	Fields infested (%)
Stellaria media	94	*Poa annua*	79
Veronica persica	72	*Avena* spp.	42
Matricaria spp.	67	*Alopecurus myosuroides*	38
Galium aparine	58	*Elytrigia repens*	21
Lamium purpureum	47	*Lolium* spp.	14
Viola arvensis	45	*Anisantha sterilis*	13
Sinapis arvensis	36	*Poa trivialis*	7
Veronica hederifolia	30	Volunteer cereals	7
Capsella bursa-pastoris	23		
Volunteer oilseed rape	23		
Papaver rhoeas	18		
Fumaria officinalis	17		
Chenopodium album	13		
Aphanes arvensis	12		
Geranium spp.	11		

Source: Whitehead, R. and Wright, H.C. (1989) The incidence of weeds in winter cereals in Great Britain. *Brighton Crop Protection Conference, Weeds* **1**, 107–112.

Herbicide choice and use has also had a profound effect on the weed flora in cereals (Martin, 1987). The use of 2,4-D and MCPA since the late 1940s has caused a decline in many susceptible weeds, such as charlock (*S. arvensis*) and poppy (*Papaver rhoeas*), although more tolerant species, including chickweed (*S. media*), knotgrass (*P. aviculare*) and the speedwells (*Veronica* spp.), have prospered. The introduction of herbicide mixtures in the 1960s with, for example, mecoprop and ioxynil, gave a much wider spectrum of weed control. Other herbicides have since been developed for an autumn-applied, residual action, including, for example, chlorsulfuron and chlorotoluron for improved control of grasses, and new molecules for the control of specific weeds, for example fluroxypyr for cleavers (*G. aparine*). However, the same principles will always apply, namely that the selection pressure caused by sustained herbicide use will allow less-susceptible weed species to become dominant, and their continued use may encourage the selection of herbicide-resistant individuals within a species, as has now occurred to many herbicides (see Chapter 13).

Climate change is predicted to have significant effects on both the geographical distribution of weeds and the severity of weed infestations. Evolutionary rate (for instance, in the development of herbicide resistance) has been demonstrated to vary dependent upon both temperature and moisture availability. This is probably a result of a combination of factors including generation time, population size and relative fitness of herbicide-resistant individuals. All of these factors may be affected by increased average global temperatures and subsequent differences in regional weather patterns (Anon., 2000). In addition, milder winters and warmer summers may allow for the survival and population growth of species that were previously unsuited to a region's climate. Increases in the occurrence of *Phalaris* grasses in the UK in recent years may be a direct result of this (A.H. Cobb and J.P.H. Reade, personal observations).

1.9   Weed control

It is outside the scope of this book to examine the finer details of weed control. Instead, a broad overview is presented. The reader is referred to other, more comprehensive texts for further information, such as Naylor (2002).

According to Lacey (1985), weed control encompasses:

1   the reduction of the competitive ability of an existing population of weeds in a crop;
2   the establishment of a barrier to the development of further significant weeds within that crop; and
3   the prevention of weed problems in future crops, either from the existing weed reservoir or from additions to that weed flora.

The first two objectives are met primarily by chemical means, and the third relies on agronomy and crop husbandry. Cultural practices are forever changing, along with the weed spectrum, and it is now increasingly recognised that an integrated approach utilising both cultural and chemical practices is necessary for optimal weed control.

1.9.1   Traditional methods

It was recognised in medieval times that the rotation of crops with fallow was the best means to conserve soil fertility and to prevent the build-up of pests, diseases and weeds. The later use of 'cleaning' crops (such as turnips and potatoes) allowed weed control by hand during active growth, and was balanced against 'fouling' crops (such as cereals) which could not be similarly weeded. By the mid-nineteenth century fertility was maintained from clover and livestock manure, and weed control by 'cleaning' crops, so that the unprofitable fallow period could be avoided. The advent of chemical fertilisers in the early twentieth century removed the need for clover, and profitability increased by the use of sugar beet as a combined cleaning and 'cash' crop. However, after the Second World War, increased urbanisation and industrialisation have reduced the available workforce, and herbicides have gradually replaced the hoe. Similarly, farm practices have become increasingly mechanised, such that the continuous cultivation of one crop (monoculture) has become widespread, and reduced cultivation techniques are now in vogue.

1.9.2   Chemical methods

Chemical weed control is a twentieth-century technology. Copper sulphate was the first chemical used at the turn of the twentieth century to control charlock (*S. arvensis*) in oats, and soon after came corrosive fertilisers (such as calcium cyanamide) and industrial chemicals (including sodium chlorate and sulphuric acid). Modern synthetic herbicides first appeared in France in 1932 following the patenting of DNOC (4,6-dinitro-*o*-cresol) for the selective control of annual weeds in cereals. Further dinitro-cresols and dinitrophenols soon appeared, but these compounds had variable effectiveness and appeared to kill animals as well as plants. The discovery of the natural plant growth 'hormone' auxin in 1934 led to the further discovery of the synthetic growth regulators 2,4-D and MCPA based on phenoxyacetic acid chemistry. These compounds were the first truly selective herbicides that could reliably kill broad-leaved weeds in cereal crops, and they developed widespread popularity and use after the Second World War (Kirby, 1980). These compounds truly 'replaced the hoe' so that cereals could no longer be regarded as 'fouling' crops, and paved the way to the current practice of cereal monocultures.

Since the 1950s an increasing proportion of world cereal crops has become regularly treated with agrochemicals to achieve the control of an ever-widening variety of weeds. Nowadays, chemical weed control has expanded to probably every crop situation in the world. Modern chemical weed control is not only more economical than traditional methods, but also has important technical advantages as weeds growing closest to the crop, and hence competing most for resources, can be controlled by selective herbicides. Furthermore, less crop-root disturbance is evident than with mechanical hoeing and fewer, if any, weed seeds are brought to the surface in the process. Finally, farmers now have chemical answers for most weed problems at a reasonable price.

1.9.3   An integrated approach

The development of integrated crop management practices means that integrated weed management (IWM) systems have been developed that also embrace environmental and financial factors. The IWM systems need to be effective enough for long-term maintenance of natural resources and agricultural productivity and also to have minimal adverse environmental impact combined with adequate economic returns to the farmer.

Key aspects of IWM systems include the prevention of weed infestation, the identification of weed species that are present, the mapping and monitoring of weed populations, the prioritisation of management, management using a combination of mutually supportive techniques (manual, mechanical, cultural, biological and chemical methods) and evaluation of their success combined with documentation and perseverance.

Through the use of such systems it is hoped that weed management may be carried out in a sustainable manner, giving protection to both financial returns and to agroecosystems.

1.9.3.1   Cultural methods

These involve tillage, where the soil is turned over with a plough and harrow before the crop is planted; flooding, to which rice is tolerant but associated weeds are not; mulching, with fabrics or synthetic membranes, which may be biodegradable, to conserve soil moisture and prevent the germination of weed seeds; and mechanical harrows, removing weeds growing in between rows of crops.

1.9.3.2   Alternative methods

Concerns about the widespread use of glyphosate in public parks and open spaces, including gardens, has led to the marketing of several unconventional alternative methods for weed control that do not rely on chemicals alone. Examples include foams, applied with hot water containing plant oils and sugars; pulses of electricity, which can kill plants by destroying the vascular bundles; microwaves, that can kill weed seeds in the soil; and directed burning with a naked flame. While these treatments may be chemical-free, they can be labour intensive, costly, often require more than one treatment and have their own risks associated with use.

1.9.3.3   Precision weed control

Instead of spraying a whole field, precision technology can now be used to direct sprays to the weedy areas only, reducing blanket spraying and resulting in environmental benefits, such as introducing less herbicide into the environment. Aerial drones can be used to create weed maps, identifying areas to be treated. Image analysis technology is becoming increasingly refined to detect small differences in vegetation owing to weeds, often by spectral information alone and dedicated software that can identify known shapes and growth habits of weeds. Aerial photography of a crop can also be a valuable, non-destructive method to monitor weed spread and control over time. Unmanned ground vehicles with mounted sprayers can then treat the weeds as detected. Indeed, it is perhaps only a matter of time before robotics are more widely used to monitor, detect and ensure precision weed control.

1.9.3.4   Conservation agriculture

Conservation agriculture (CA) is a system that avoids or minimises soil disturbance, combined with soil cover and crop diversification. It is considered to be a sustainable agro-ecological approach to conserving resources in agricultural production. A major challenge to CA is weed management without the use of herbicides.

The review paper by Sims *et al.* (2018) lists and evaluates options for ecological and integrated weed management in CA to prevent weed pressure from building. The choice of methodology will depend on what technical options are locally available and the prevailing economic environment. A summary of methods is presented below:

1 *Quality seed and clean equipment*: use proven quality seed and clean machinery to avoid or lessen weed seed introduction from the external environment.
2 *Reduce the weed seed bank*: to deplete the weed seed bank, seed set should be avoided wherever possible. Reduced tillage and cover crops can favour weed seed predation, especially by insects and birds. 'Beetle banks' are strips of natural vegetation and grasses that provide a haven for seed eaters and can also prevent potential soil erosion if established to follow the contours of the land. Increased germination by minimal soil disturbance followed by mechanical or chemical control may eliminate weed flushes prior to crop establishment. Prevention of the shedding of weed seeds at cereal crop harvest can be achieved by mechanical means. 'Chaff Carts' are trailers attached to the rear of the harvester to collect chaff and weed seeds, which are the collected for disposal, or for animal feed. A 'weed seed mill' can also be trailed behind the combine and contains a mill that pulverises the chaff and weed seeds. The resultant mixture may be returned to the field as nutrients. 'Weed headers' remove weed flowers growing above crop height before the seed heads appear. In this case, horizontal rotors fitted with blades are attached to a tractor. This method has proven effective in low-growing crops, such as sugar beet.
3 *Crop rotation*: this is a very effective cultural approach to weed management. It prevents the proliferation of weed populations that can become dominant in monocultures. Rotations of annual, biennial and perennial crops are effective, as is the inclusion of leguminous crops, which also offers a positive nitrogen input. The inclusion of crops with allelopathic properties to inhibit weed growth has also been reported. In addition, mulching can provide soil cover that prevents light from reaching the soil surface, and so reduces or inhibits weed seed germination.

Sustainable weed seed management therefore consists of a range of alternative options to herbicide use that can prevent or lessen weed seed proliferation, rather than to control them when they interfere with crop growth and development. Finally, farmers and growers are now under increased pressure to reduce agrochemical inputs, yet they struggle with the need to control herbicide-resistant weeds. We may expect new weed control technologies to emerge in the coming decades, such as robots that scan the growing crop, with directed lasers to control weeds.

References

Anon. (2000) *Climate Change and Agriculture in the United Kingdom*. London: MAFF.

Baker, H.G. and Stebbins, G.L. (1965) *The Genetics of Colonising Species*. New York: Academic Press.

Brooks, A., Crook, M.J., Wilcox, A. and Cook, R.T. (2003) A laboratory evaluation of the palatability of legumes to the field slug, *Deroceras reticulatum* Müller. *Pest Management Science* **59**(3), 245–251.

Chancellor, R.J. and Froud-Williams, R.J. (1986) Weed problems of the next decade in Britain. *Crop Protection* **5**, 66–72.

Dexter, A.G., Rothe, I and Luecke, J.L. (1999) Weed control in Roundup Ready™ and Liberty Link™ sugarbeet. In: *30th General Meeting of American Society of Sugar Beet Technologists, Abstracts*, p. 9.

Figueroa, P.F. (1989) Japanese knotweed herbicide screening trial applied as a roadside spray. *Proceedings of the Western Society of Weed Science* **42**, 288–293.

Gajic, D and Nikocevic, G. (1973) Chemical allelopathic effects of *Agrostemma githago* upon wheat. *Fragm. Herb. Jugoslav* XXIII.

Grime, J.P. (1979) *Plant Strategies and Vegetation Processes*. London: Wiley.

Grundy, A.C. and Jones, N.E. (2002) What is the weed seed bank? In: Naylor, R.E.L. (ed.) *Weed Management Handbook*, 9th edn. Oxford: Blackwell/BCPC.

Hance, R.J. and Holly, K. (1990) *Weed Control Handbook: Principles*, 8th edn. Oxford: Blackwell.

Hanf, M. (1983) *The Arable Weeds of Europe with Their Seedlings and Seeds*. Hadleigh, Suffolk: BASF United Kingdom.

Hill, T.A. (1977) *The Biology of Weeds*. London: Edward Arnold.

Holm, L.G., Plucknett, D.L., Pancho, J.V. and Herberger, J.B. (1977) *The World's Worst Weeds. Distribution and Biology*. Hawaii: University Press.

Inderjit, Callaway, R.M. and Vivanco, J.M. (2007) Can plant biochemistry contribute to understanding of invasion ecology? *Trends in Plant Science* **11**(12), 574–580.

Kirby, C. (1980) *The Hormone Weed Killers: a Short History of Their Discovery and Development*. Croydon and Lavenham, UK: BCPC Publications/Lavenham Press.

Lacey, A.J. (1985) Weed control. In: Haskell, P.T. (ed.) *Pesticide Application: Principles and Practice*. Oxford: Oxford University Press, pp. 456–485.

Lampkin, N. (1990) *Organic Farming*. Ipswich, UK: Farming Press.

Liebman, M., Mohler, C.L. and Staver, C.P. (2001) *Ecological Management of Agricultural Weeds*. Cambridge: Cambridge University Press.

Lutman, P.J., Boatman, N.D, Brown, V.K. and Marshall, E.J.P. (2003) Weeds: their impact and value in arable ecosystems. In: *The Proceedings of the BCPC International Congress: Crop Science and Technology 2003* **1**, 219–226.

Martin, T.J. (1987) Broad versus narrow-spectrum herbicides and the future of mixtures. *Pesticide Science* **20**, 289–299.

Moles, A.T., Gruber, M.A.M. and Bonser, S.P. (2008) A new framework for predicting invasive plant species. *Journal of Ecology* **96**(1), 13–17.

Moss, S.R. (1987) Competition between blackgrass (*Alopecurus myosuroides*) and winter wheat. *British Crop Protection Conference, Weeds* **2**, 367–374.

Naylor, R.E.L. (ed.) (2002) *Weed Management Handbook*, 9th edn. Oxford: Blackwell/BCPC.

Naylor, R.E.L. and Lutman, P.J. (2002) What is a weed? In: Naylor, R.E.L. (ed.) *Weed Management Handbook*, 9th edn. Oxford: Blackwell/BCPC.

Ødum, S. (1965) Germination of ancient seeds – floristical observations and experiments with archaeologically dated soil samples. *Dansk Botanisk Arkiv* **24**(2), 1–70.

Oerke, E.C., Dehne, H.W., Schonbeck, F. and Weber, A. (eds) (1995) *Crop Production and Crop Protection: Estimated Losses in Major Food and Cash Crops*. Amsterdam: Elsevier.

Olofdotter, M. and Mallik, A.U. (2001) Allelopathy Symposium: introduction. *Agronomy Journal* **93**, 1–2.

Patterson, D.T. (1985) Comparative ecophysiology of weeds and crops. In: Duke, S.O. (ed.), *Weed Physiology*, Vol. I: *Reproduction and Ecophysiology*, Boca Raton, FL: CRC Press, pp. 101–130.

Radosevich, S.R. and Holt, J.S. (1984) *Weed Ecology: Implications for Vegetation Management*. New York: Wiley.

Rodgers, E.G. (1978) Weeds and their control. In: Roberts, D.A. (ed.) *Fundamentals of Plant Pest Control*. San Francisco, CA: W.H. Freeman, pp. 164–186.

Shaw, D.S. and Tanner, R. (2008) Invasive species: weed like to see the back of them. *Biologist* **55**(4) 208–214.

Sims, B., Corsi, S., Ghebounou, G., Kienzle, J., Taguchi, M. and Friedrich, T. (2018) Sustainable weed management for conservation agriculture: options for smallholder farms. *Agriculture* **8**; doi: 10.3390/agriculture8080118

Tanner, R. (2008) A review on the potential for the biological control of the invasive weed, *Impatiens glandulifera* in Europe, In: Tokarska-Guzik, B., Brock, J.H., Brundu, G., Child, L., Daeler, C.C. and Pyšek, P. (eds) *Plant Invasions: Human Perception, Ecological Impacts and Management*. Leiden: Backhuys, pp. 343–354.

Toole, E.H. and Brown, E. (1946) Final results of the Duvel buried seed experiment. *Journal of Agricultural Research* **72**, 201–210.

Whitehead, R. and Wright, H.C. (1989) The incidence of weeds in winter cereals in Great Britain. *Brighton Crop Protection Conference, Weeds* **1**, 107–112.

Wilson, R.G. and Smith, J.A. (1999) Crop Production with glyphosate tolerant sugarbeet. In: *30th General Meeting of American Society of Sugar Beet Technologists*, Abstracts, p. 50.

Young, S. (1989) Wayward genes play the field. *New Scientist* **123**, 49–53.

Chapter 2
Herbicide Discovery and Development

Farming looks mighty easy when your plow is your pencil and you're a thousand miles from the cornfield.

Dwight D. Eisenhower (1890–1969)

2.1 Introduction

Most crop protection problems can now be solved using chemicals, at a reasonable price to the farmer and grower. The use of herbicides in the developed countries has been particularly successful, with an estimated near-maximum market penetration in major crops (e.g. 85–100% in soybean, maize, and rice crops in the USA; Finney, 1988). The reason for this high usage is that it has proved financially rewarding to the farmer. Indeed, estimates suggest that each US$1 spent in the USA on pesticides results in an additional income of US$4 to the farmer. Similarly, £1 spent on pesticides may generate an additional £5 in the UK. However, reduced subsidies and lower farm incomes have led to lower chemical inputs by farmers, so that the growth rate of the herbicide market has declined in recent years. Also apparent has been an increase in price competition between companies, particularly with mixtures of chemicals whose patents have expired. Farmers, growers and consumers have all benefited from increased competition within the agrochemical industry, since chemical prices have dropped in real terms, while at the same time old or unsound products have been replaced by safer, more environmentally acceptable, lower dosage compounds. Furthermore, it should not be forgotten that expenditure on pesticides has not only proved financially rewarding to producers, but has also increased crop yield and quality, and contributed to self-sufficiency in many crops, particularly in Europe.

While the current armoury of agrochemicals is under threat by the 'green' lobby and the European Union (EU) legislature which has banned the use of many herbicides, the benefits of agrochemical use must not be forgotten. The judicious use of agrochemicals ensures the sustainable yields typical of conventional agriculture with the high quality and safety of agricultural products expected by the consumer. In this context, it is worth noting the study of Oerke *et al.* in 1994. They concluded that when no weed, disease or pest control was practised, 70% losses of potential yields of our major crops (rice, wheat, barley, maize,

Herbicides and Plant Physiology, Third Edition. Andrew H. Cobb.
© 2022 John Wiley & Sons Ltd. Published 2022 by John Wiley & Sons Ltd.

potato, soybean, cotton and coffee) could be predicted. With the use of agrochemicals this loss was estimated at 42%. Even then, crop losses owing to pests, diseases and weeds are estimated at US$60 billion each year. Thus, the development of new crop protection agents and the more effective use of existing ones are important global priorities if we are to feed the world.

2.2 Markets

The agrochemical industry, based on the production of chemicals for crop protection, largely came into being after the Second World War with the commercialisation of the first truly selective broadleaf weed herbicides: 2,4-D (1945) and MCPA (1946). These non-toxic molecules were effective at low doses and were cheap to produce. Furthermore, they became available when maximum food production was essential and farm labour was scarce. Their success stimulated European and North American chemical companies to invest in research that led to the discovery of the wide range of herbicides now available. Early successes created a market value approaching US$3 billion in 1970 and an average of 6.3% real growth per annum was recorded over the following decade. By 1980 it had slowed to 4.5% per annum, averaged 2.2% growth during the 1980s and was predicted to average below 2% for the 1990s. By 1998 the market had become static, with only 0.1% real growth, but still worth US$31 billion.

The global pesticide market was estimated at US$60 billion in 2017, with little or no growth in the last decade. Herbicides now represent up to 70% of this global market, and glyphosate alone makes up about 50% of the herbicide market. Indeed, perhaps glyphosate will become the first US$10 billion agrochemical, since it is now off patent, generics are increasingly widespread and it has a valuable use in herbicide-tolerant crops. This contrasts with the estimated sales of pharmaceuticals of US$1 trillion in 2014. It costs about US$2 billion to bring a new drug from initial discovery to market, and there are high attrition rates, since, on average, only one in every 5000 compounds screened eventually comes to market. Since a successful drug can earn in excess of US$1 million in sales each day, the rewards currently far outweigh the risks and costs of development.

What has caused this slowdown in market growth? Five contributing factors have emerged. First, past successes in chemical crop protection have led to a near-maximum market penetration in all of the major crops, especially in Western Europe and the USA. Thus, there is an increasingly competitive market place for agrochemical companies to operate in. Second, new active ingredients are taking longer to discover, develop and register. Stricter legislative requirements have also led to delays in returns from investments, so that product profitability has declined. Indeed, estimates of at least US$250 million are often given for the cost of bringing a new product to the market (Figure 2.1). Third, the agrochemical arsenal is becoming increasingly mature as fewer examples of new chemistry acting at novel target sites are reported. Fourth, owing to the successes of modern intensive agriculture in recent decades, there has been an overcapacity in farming and a marked decline in commodity prices. Consequently, sales of agrochemicals have declined. Fifth, the growth of the 'green' lobby and the more recent introduction of genetically modified (GM) crops have led to increasing consumer opposition in some nations to agrochemical use.

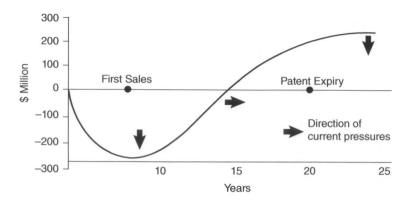

Figure 2.1   Typical cumulative cash flow for a successful new product. Sources: Adapted from Finney, J.R. (1988) World crop protection prospects: demisting the crystal ball. *Brighton Crop Protection Conference, Pests and Diseases* **1**, 3–14; Rüegg, W.T., Quadranti, M. and Zoschke, A. (2007) Herbicide research and development: challenges and opportunities. *Weed Research* **47**, 271–275.

Consequently, the agrochemical industry itself has contracted significantly in recent years. The following chemical companies were involved in discovery research and development in the 1980s, but are no longer active: Celamerck, Chevron, Diamond Shamrock, Dr Maag, Duphar, Mobil, PPG Industries, Shell, Stauffer, 3M, Union Carbide and Velsicol. Mergers in the 1990s saw the agrochemical interests of Schering and Hoechst form AgrEvo, which in 1999 itself merged with Rhône-Poulenc to form Aventis, now a part of Bayer CropScience. In the USA, Dow and Eli Lilly formed DowElanco, now Dow AgroSciences, and the Swiss companies Ciba Geigy and Sandoz merged to become Novartis, which has now merged with Zeneca to form Syngenta, creating the biggest player in the crop protection industry at that time, with combined sales of over US$8 billion and a quarter of the global market.

In 2010 the agrochemical industry was dominated by six major multinational companies: Syngenta, Bayer, Monsanto, Du Pont, BASF and Dow, which invested between 8 and 11% of their sales income in research and development, including the search for new active ingredients. However, Monsanto has since reduced this activity, preferring to focus on the use of glyphosate in GM crops. In recent years there has been further consolidation in the industry. Bayer has now acquired Monsanto, at a cost of US$66 billion, with estimated global sales of US$23 billion. Dow and DuPont have merged to form Corteva and Syngenta has been taken over by ChemChina, with a joint value of US$15 billion. Thus, with BASF, there remain only four major companies with large research and development budgets.

Further consolidation is predicted as other global players consider offloading their agrochemical divisions to concentrate on what is perceived to be the more lucrative pharmaceutical markets.

The acquisition of Monsanto by Bayer has resulted in a range of businesses and assets to pass to BASF. These include the global glufosinate–ammonium business of Bayer; seeds, research and breeding capabilities; some trademarks; seed treatment products; some non-selective herbicides and nematicide research projects. Furthermore, about 4500 Bayer employees will join BASF. The net effect will move BASF up the global sales table by an estimated addition of US$2 billion. It is also reported that Corteva will offload some of its products and research portfolio to the FMC Corporation in the USA. Other companies with

Table 2.1 Estimated annual sales of the four major agrochemical companies, after mergers during the decade of 2010–2020.

Company	Global sales (US$billion)
Bayer/Monsanto	23
Syngenta/ChemChina	15
Corteva (Dow/DuPont)	15
BASF	6

much smaller global sales include NuFarm (Australia), United Phosphorous (India), Platform Agriculture (USA), Albaugh (USA), Sumitomo (Japan) and AMVAC (USA). Table 2.1 summarises estimated sales for the four major agrochemical companies after the wave of mergers.

Further mergers and takeovers are likely in the coming years, as ChemChina is reportedly in discussion with Sinochem. The latter company is the largest seller of seeds, agrochemicals and fertilisers in China. A further interesting example is the decision of DuPont and Syngenta in 2016 to submit a joint patent, entitled 'Substituted cyclic amides and their uses as herbicides'. A new product may be launched in 2023.

The four major players now account for 70% of the global agrochemical market and 60% of the seeds market, investing up to 25% of total budgets in research and development. They are also investing in strategic partnerships with biotechnology start-up companies and academics, to identify new methodologies for candidate agrochemicals and their targets. Innovative companies validate new targets, while the major company optimises chemical structures, efficacy in the field, product development and marketing.

Agrochemical research and development is increasingly global. While the North American and European markets dominated the industry in previous years, expansion into Eastern Europe, Asia, Africa, South America and China is now evident, representing the major climatic zones in the world.

The literature on agrochemicals often overlooks markets outside North America and Europe, while the use of plant protection products is growing elsewhere, especially in India and China. India is currently (2019) the fourth largest producer of agrochemicals, after the USA, Japan and China, and the 13th largest exporter of pesticides in the world. As Nair (2019) points out, the per hectare consumption of pesticides in India is amongst the lowest in the world, at 0.6 kg/ha, compared with 13kg/ha in China, with a predicted growth of 15% for the next 15 years.

To give an indication of the scale of operation of one of the major companies, Syngenta:

- invests US$1.3 billion each year in research and development;
- employs over 5,000 scientists;
- has 116 research and development sites in the world; and
- collaborates with over 400 universities, research institutes and other organisations (Syngenta.com/innovation-agriculture/research-and-development, accessed 20 March 2020).

In a survey of the leading crop protection companies in 2015, Phillips McDougal found that the discovery and development costs of a new crop protection product was approaching US$300 million, twice that in 1995, made up by US$107 million for research,

US$146 million for development and US$30 million for environmental chemistry. The greatest rises were seen in the costs of environmental chemistry and field trials. This additional work has meant that the time between product discovery and development increased from 8.3 years in 1995 to 11.3 years in 2014. Furthermore, the number of products processed leading to a successful commercial launch rose from 52,500 in 1995 to 159,574 in 2014.

It follows that only the biggest companies have the resources and finances to convert such an investment into portfolio products. Indeed, it is estimated that such research and development costs equate to at least US$50 for every minute! These costs continue to rise with the increasing emergence of weed resistance to herbicides, increasing environmental and safety margins and inflationary costs of goods.

It is also worth noting that commodity prices of maize, soybean, cotton, wheat and rye have fallen by over 50% since 2008, resulting in an estimated loss of US$15 billion to the agricultural market. As a consequence, growers are cutting back on their spend on plant protection products and the development of new products for niche markets or small hectarage crops is no longer viable. Problems created by herbicide resistant weeds growing in monoculture crops has led to new approaches and opportunities in herbicide development. These factors, together with the relative lack of herbicides with new modes of action, conspire to make further challenges to the agrochemical industry. It remains to be seen whether the merged industry can continue to thrive with fewer genuinely novel products in an increasingly saturated market, especially in the developed world. The need for genuine innovation and discovery is paramount for the future success and growth of the industry.

A further and final point is worthy of consideration. It is without doubt that take-overs and mergers widen the product portfolio of the new company and increase their global presence and market penetration. On the other hand, there is growing concern from farmers and environmentalists that fewer but larger companies will or can exert a major influence on global food production, by controlling price, while the quality, choice and variety of seed varieties may diminish. It also may follow that competition could be less evident in future, resulting in less innovation, as a handful of global giants concentrate on defending their existing intellectual property. Indeed, in the last two decades, the major agrochemical companies have acquired over 200 companies and their patents.

Only time will tell how many agrochemical companies will survive this new wave of restructuring within the industry. What is clear is that the survivors will be those who have adapted best to change by embracing new technologies and innovative marketing, while at the same time providing competitively priced products that enable profitable farming.

All is not doom and gloom, however, and a significant market for crop protection products awaits the slimmed-down industry. There are expanding agricultural markets in Eastern Europe and Asia to be penetrated and there is much new technology to exploit. For example, the precise delivery of more optimally formulated agents will lead to lower chemical inputs into the environment, and recent developments in agricultural biotechnology are already changing both farming practices in general and agrochemical use in particular. Indeed, the ability to transfer genes for favourable traits to crops by recombinant DNA technology will undoubtedly transform agriculture and crop protection into totally new directions in the next decade.

While the first decade of GM was focused on the gene products controlling weeds (e.g. glyphosate resistance) and insects (via the BT toxin), the next decade is predicted to identify perhaps hundreds of genes that may secure increased yields in our key crops.

Furthermore, it is interesting to note that in 2007 BASF and Monsanto agreed to collaborate to fund the search for stress-tolerant strains of maize, soybean, cotton and oilseed rape using GM technologies. The companies agreed a joint budget of 1.2 billion euros to find unique genes for stress-tolerant traits (Evans, 2010).

2.3 Prospects

There is great potential for parts of the world to increase their production of cereal crops. This is especially the case with wheat in central and eastern Europe and Russia. These areas account for about 30% of the global crop area devoted to wheat. With yields of about 2 tonnes/ha, they generated about 20% of the global crop in the early part of the century. On the other hand, wheat yields in the UK over the same time were about 8 tonnes/ha, mainly owing to the application of modern agricultural practices and the use of agrochemicals and fertilisers. A similar comment can be applied to China, with a population of 1.3 billion and a growing demand for cereals. In this case, China is now a major producer and exporter of agrochemicals.

A view often promulgated in Western agricultural nations is the promotion of more intensive agriculture to optimise land use. Such an approach would enable the use of less land for crops, allowing more species biodiversity. Conversely, an organic, low-yielding farm system would require more land to enter production for current yields to be matched. About two-thirds of arable production, however, is to provide feed grains for animals. It follows that, if this was reduced, farmers could grow more fruit, vegetables, pulses and nuts that would contribute to a healthier human diet. Feeding livestock on grassland alone would serve to improve soil structure and fertility without the need for more land. Indeed, the reader may wish to read D.T. Avery (1995) for a fuller account. The title of his book *Saving the Planet with Pesticides and Plastic: the Environmental Triumph of High-yield Farming* is very persuasive to many in the agrochemical industry in the debate on sustainable food production.

The top 10 herbicides according to sales for each of the main global crops in 2015 are presented in Table 2.2. Note that with one exception, rice, glyphosate is far and away the best-selling herbicide. Why is this? The answer is the almost global adoption of genetically modified crops that are tolerant to glyphosate, the exception being member states of the European Union. In 2018, 191.7 million hectares of genetically modified crops were planted globally by 17 million farmers. Indeed, 70 countries have adopted biotech crops since 1996, up from 1.7 million hectares in the first year of commercial planting. The adoption of biotech crops is the fastest development in crop technology in the history of modern agriculture.

The four major biotech crops in 2018 were soybeans, maize, cotton and oil-seed rape. Some 78% of the global soybean crop was GM (95.9 million hectares), 76% of the cotton crop (24.9 million hectares), 30% of the maize crop (58.9 million hectares) and 29% of the oil-seed rape crop (10.1 million ha). The USA has the highest number of approvals (584) for GM crops in food, feed, processing and cultivation, while maize is the crop with the largest number of approvals, 137 in 35 countries. More diverse crops with various traits became commercially available in 2018, including potatoes that show reduced bruising, reduced acrylamide levels and resistance to late-blight; insect-resistant and drought-tolerant sugarcane; apples with less browning; and oil-seed rape and safflower with

Table 2.2   Leading herbicides in 2015, according to sales for key crops (PhillipsMcDougall, Agribusiness Intelligence).

	Soybean	Cereals	Maize	Rice
1.	Glyphosate (a)	Glyphosate (a)	Glyphosate (a)	Cyhalofop (b)
2.	Flumioxazin (d)	Pinoxaden (b)	Atrazine (c)	Penoxsulam (e)
3.	Clethodim (d)	Flufenacet (g)	Mesotrione (h)	Butachlor (g)
4.	Imazethapyr (e)	Mesosulfuron (e)	Metolachlor (g)	Pretilachlor (g)
5.	Paraquat (i)	Clodinafop (b)	Acetochlor (g)	Pyraclonil (d)
6.	Chlorimuron (e)	2,4-D (f)	Nicosulfuron (e)	Bispyribac-Na(e)
7.	Diclosulam (e)	Florasulam (e)	Tembotrione (h)	Quinclorac (f)
8.	Fomasafen (d)	Fluroxypyr (f)	Isoxaflutole (h)	Bensulfuron (e)
9.	2,4-D (f)	Pyroxsulam (e)	Paraquat (i)	Bentazone (c)
10.	Haloxyfop (b)	Fenoxaprop (b)	Thiencarbazone(e)	Propanil (c)
	US$15.2 billion*	US$8.12 billion	US$16.31 billion	US$4.76 billion

Key: (a) inhibitor of aromatic amino acid biosynthesis; (b) Inhibitor of ACCase; (c) inhibitor of Photosystem II; (d) inhibitor of protoporphyrinogen oxidase (PROTOX); (e) inhibitor of acetolactate synthase (ALS); (f) auxin herbicide; (g)inhibitor of very-long-chain fatty acid biosynthesis; (h) inhibitor of pigment biosynthesis; (i) active at Photosystem I.
* Total sales in each crop sector for 2015.

increased oleic acid content. Amongst other developments, 2019 has seen the introduction of vitamin-A-fortified rice and blight-resistant rice.

Without doubt, there are major reservations in Europe about the developments of GM crops. Consumers are especially concerned about food safety issues, labelling of products if they contain ingredients from GM crops, the perceived problems of transferring genes from one organism to other organisms, perceived threats to biodiversity, the risk of widespread antibiotic resistance and simply the fear of the unknown. The role of the scientific community must not be to simply list the benefits of this brave new botany, but to also present a balanced view of all the above concerns with impartial statements, based on facts and sensitivity to opposing views.

An 'all or nothing' debate on GM products, promulgated by some pressure groups is sterile. Polarised views only serve to confuse the public when they are attempting to take an informed position on this issue. We must all accept that there are major ethical and environmental concerns about GM foods, and that there are varying degrees of acceptability to all parties. Indeed, the true test of consumer acceptance of GM foods is yet to come. Will the consumer pay a premium for a chemical-free and GM-free product, verified by detailed analytical testing, or in 2030 will we look back to the debate and wonder what all the fuss was about?

Agrochemical companies have invested heavily to acquire new seed companies, specifically to produce and market transgenic crops, and strong sales growth has been evident for herbicide-resistant crops (Chapter 14). What is certain is that agricultural biotechnology is here to stay and will transform the crop protection industry in ways that we are only now beginning to comprehend.

Recent surveys in 2019 suggest that attitudes to GM crops are changing in both the UK and the EU. In a sample of 750 farmers in the UK, 77% responded positively and would adopt GM technology if the regulations were to change in the UK. They perceived real benefits for the environment, farmers and consumers (Dyer, 2019). Furthermore, in 2019, in the first survey in the EU since 2010 regarding food safety, a mere 27% of persons asked were concerned about GM organisms, down from 66% in 2010 (GMOinfo.eu.).

2.4   Environmental impact and relative toxicology

The environmental impact and toxicological considerations of new herbicide development are coming increasingly to the fore. Nowadays, before its commercial release, a new product has to satisfy an extensive battery of tests to determine its potential toxicity to a wide range of organisms, effects on key environmental functions and its movement and fate within the environment. It is of interest to compare some current criteria for environmental acceptability with those tests considered necessary in previous decades (Table 2.3).

The registration and approval of new agrochemicals in Europe is both a time-consuming and a complex process. It is governed by the European Commission (EC) Directive 91/414/ EEC, which harmonises the national product approval requirements throughout Europe and consists of two main phases. In the first, the active ingredient must be approved at EU level and in the second, formulations appropriate to national markets have to be registered by European member states. The European public should be reassured that only active ingredients demonstrated to be without risk to human or animal health or to the environment will be allowed to be used and listed in the Directive.

The whole process of product registration now has three stages, namely:

1   generation of a data package for the active ingredient;
2   preparation and submission of a dossier to the member state; and
3   assessment and approval by the member state and the European Food Safety Authority.

Acquiring the data package can take 4–5 years and cost about 10 million euros. Approximately 250 tests need to be conducted on a new active ingredient in the areas of quality, safety and efficacy. *Quality* involves testing the behaviour of the active ingredient and formulated products, including physicochemical properties, storage stability and methods for analysis. *Safety* involves all aspects of toxicity studies including mammalian

Table 2.3   Typical tests required for the environmental acceptability of pesticides.

1950s	1980s
Acute toxicity to bees, fish and birds	Acute toxicity to three species of fish; effects on fish growth and bioaccumulation
	Acute toxicity to algae, aquatic invertebrates (four species), bees and birds (two species)
	Reproduction study in *Daphnia*
	Dietary toxicity to birds
	Effects on soil fauna including earthworms and rates of leaf litter decomposition
	Effects on nitrification and carbon dioxide evolution from soils
	Rates of biodegradation
	Physicochemical properties

Source: Graham-Bryce, I.J. (1989) Environmental impact: putting pesticides into perspective. *Brighton Crop Protection Conference, Weeds* **1**, 3–20.

toxicity (details of metabolism, acute and chronic toxicity, genotoxicity, developmental and reproductive effects), residues (plant and livestock metabolism), crop and livestock trials performed according to Good Agricultural Practice, environmental fate (behaviour in soil, water and air especially regarding metabolites) and ecotoxicity (acute, short-term and reproductive effects in birds, aquatic organisms and non-target organisms, such as bees, beneficial insects and earthworms). *Efficacy* entails many field trials using the proposed formulated product on many crops in defined climatic areas. These studies determine the proposed label recommendations and the good agricultural practice for the product.

The preparation and submission of the dossier can take a year for a new active ingredient and includes study summaries, risk assessments and efficacy of formulations and co-formulations, that is, mixtures with other components. The dossier needs to be prepared to a set format (Guideline Document 1663/VI/94 Rev 8) and in accordance with the Organisation for Economic Co-operation and Development format (Rev 1, 2001), which means that the dossier should be acceptable by the EU, the USA, Canada, Japan and Australia. The full dossier can count to 60,000 pages!

The dossier then undergoes a complete check by the Member State – which may last a further year, although provisional approval product registration may be granted for a period of 3 years. A draft assessment report by the Member State is then reviewed by the European Food Safety Authority, which can take another year. A positive recommendation is then reviewed by the EC Standing Committee on the Food Chain and Animal Health and either endorsed, changed or rejected. This process can take a further 6 months. Next, the active ingredient needs to be authorised in every Member State where the product will be sold, taking into account local climate, cropping patterns and diet. This National Approval is only applicable to the formulated product and can take a further year. Finally, registration lasts for a maximum of 10 years, although active ingredients can be reviewed at any time.

New EU legislation (EU no. 1907/2006) concerning the registration, evaluation and restriction of chemicals (REACH) looks likely to further reform the way agrochemicals are managed in Europe. REACH regulations (2007) require that approximately 30,000 chemicals produced and marketed in the EU in amounts above 1 tonne per annum will require a full technical dossier. Chemicals used above 10 tonnes per annum will also require a chemical safety assessment. Thus 'old' chemistry will become subject to increasingly rigorous and demanding testing prior to use.

This legislation has been very controversial. On the one hand, some pressure groups would prefer a total ban on untested chemicals, while Trades Unions are positive that better Health and Safety procedures will result in protection for their members. Then again, REACH is perceived by some as a threat to the European chemicals industry, ensuring an exodus of business to the less well-regulated nations, such as the USA, China and India. Clearly, the issue of substituting 'old' chemistry with 'new' products having a fully approved dossier will tax the agrochemicals industry for some time to come. It is not known whether the UK will remain as part of the EU REACH regulations for the registration, evaluation, authorisation and restriction of chemicals post-BREXIT. Currently, compliance is overseen by the European Chemicals Agency.

Herbicides are among the most heavily regulated chemical products in the developed world. Their discovery, development and use are closely scrutinised by independent experts to ensure that they are safe to use before they are marketed. Even then, detailed label recommendations ensure that if they are used appropriately, the risks to users and the environment are minimised. On the other hand, about 35 countries have no regulatory processes at all and so some developing nations may become the last users of chemistry long-since withdrawn in the EU and the USA. It has been proposed that more and better regulation is needed to control how agrochemicals are used and how this may affect the user and the environment. Furthermore, long-term impact and effects of their repeated use over several years are seldom studied or published. To some, marketing strategies are often stronger than the product, and the use of more evidence is often needed to prove efficacious use, as is the case with pharmaceuticals (Milner and Bond, 2017).

The widespread and persistent presence of low concentrations of herbicides in freshwater is a growing issue. Beaulieu, Cabana and Hout (2020) have used fast repetition rate fluorometry to demonstrate the effects of Photosystem II inhibitors on phytoplankton in aquatic environments. They observed that atrazine affected phytoplankton populations at concentrations below the existing national guidelines in Australia, Canada, the USA and EU, suggesting that current guidelines are insufficient to prevent adverse effects of PSII inhibitors on algal physiology in aquatic ecosystems.

Increasingly sensitive analytical methods are now available to detect agrochemicals in the environment. For example, liquid chromatography–mass spectrometry (LCMS) is the method most commonly used to detect agrochemicals at low concentrations in river water, streams and ponds. LC separates the molecules present in a sample and MS determines their mass-to-charge ratios, based on their passage through electric and magnetic fields. Retention time and mass-to-charge ratios identify each molecule by reference to data libraries. LCMS is sensitive to nanogram concentrations (10^{-9} g), although sensitivity to femtogram concentrations (10^{-15} g) can be achieved with PIESI (paired ion electrospray ionisation). In this case, separated compounds are ionised on exiting the LC column by applying a voltage. The charged ions so produced are blown to the MS using nitrogen. This makes LC–PIESI–MS more than 1000 times more sensitive than LCMS alone. Xu and Armstrong (2013) have used it to detect 19 pesticides in water at the subpicogram range, including the herbicides 2,4-D, dicamba, cloprop and mecoprop.

Aquatic microphytes may be affected by exposure to unintentionally released herbicides close to field margins and water courses. Entry from leaching or run-off may last from days to hours. To study iofensuron-sodium effects on microphytes in stream systems, Wjeczorek *et al.* (2017) examined a 24 h exposure on *Elodea canadiensis* and *Myriophyllum spicatum* with a recovery period of 42 days. The former species was most sensitive to the herbicide after 7 days and full recovery was observed after 42 days. On the other hand, *M. spicatum* was more sensitive after 14 days. The no-observed-ecologically-adverse-effect-concentrations (NOEAEC) were 10 and 30 μl/L, respectively.

The unicellular green alga *Chlamydomonas rheinhardtii* has been used as a model system for examining herbicide interactions with this freshwater alga using vital dyes (Brickley *et al.*, 2012). A collapse in mitochondrial potential was observed following incubation with 0.72 mM bromoxynil octanoate after only 93 s. Further experiments revealed that sublethal concentrations of formulated herbicide affected respiration, suggesting that this procedure may be of value in monitoring the effects of herbicides in freshwater systems.

On a larger, marine scale, the lagoon of the Great Barrier Reef, off the north-east coast of Australia, is contaminated each year by the run-off of pesticides during the wet season in the antipodean summer. The most common herbicides detected are atrazine, diuron, hexazinone and tebuthiuron. Flores *et al.* (2013) have studied the effects of these herbicides on non-target organisms, principally the seagrasses *Zostera muelleri* and *Halodule uninervis*, which were demonstrated to be more sensitive to the herbicides than corals and tropical microalgae. Chronic damage to Photosystem II was indicated, lowering seagrass productivity. This study is an example of the global need to recognise that pesticide run-off is not limited to agricultural sites and that actions are needed to preserve sensitive marine environments.

Surprisingly, the monitoring of pesticide residues in soil is not required at the level of the EU, in contrast to the monitoring of water, regulated by the EU Water Framework Directive. Consequently, large-scale international studies on soil contamination by pesticide residues are scarce, and even then, limited to a single pesticide, or to only a few compounds, and the sampling strategies, techniques and analytical methods are seldom consistent. Furthermore, thresholds for admissible levels of pesticides do not currently exist. Predicted environmental concentrations (PECs) are calculated based on, for example, crop type, and considering application rates, number of applications, time after application and degradation rates computed in the field or laboratory. Risk assessment relies on a comparison of toxicity exposure ratios and trigger values. Ecotoxicology uses selected indicator organisms. According to Siva *et al.* (2018), however, the PEC values and aforementioned risk assessments require validation using soils from the field and the monitoring of actual pesticide concentrations. In 2015, these authors analysed for the presence and distribution of 76 pesticides in 317 agricultural topsoil samples throughout the EU, including 11 member states and six cropping systems. They found that over 80% of tested soils contained pesticide residues, with glyphosate and metabolite AMPA, DDTs and broad-spectrum fungicides the most frequently found compounds and at the highest concentrations, usually below the computed PEC values. Between 2 and 3 mg of pesticide per kilogram of soil were the maximum values recorded. Mixtures of pesticide residues were usually the norm rather than the exception. This approach will no doubt lead to a more accurate understanding and assessment of pesticide risk in agricultural soils.

The UK Department for Environment, Food and Rural Affairs is planning an indicator for healthy soils and to establish a new national monitoring scheme. Currently, however, just 0.41% of the funds invested in environmental monitoring in the UK goes on soil monitoring.

The selective toxicity of new herbicidal molecules has also become increasingly important in recent years. Successful products are expected to potentially inhibit target weed processes with negligible risk to the crop or other organisms. Hence, most new herbicides have LD_{50} values for acute toxicity to rats in excess of 2000–3000 mg of active ingredient per kilogram of body weight (Table 2.4) and are therefore less toxic than aspirin, caffeine, nicotine or even table salt.

It is a further requirement for pesticide registration that residues are routinely measured both in foodstuffs and in the general environment. This includes when a crop is harvested and after storage, and in animal products such as milk and meat. The results of such residue analyses have lately created some debate, and much confusion, both in the popular press and in the minds of many consumer groups. Indeed, the proliferation of references to 'organically grown' or 'pesticide-free' produce is evidence of an increasingly held belief that all pesticides are harmful to life even at 10^{-15} g! Pesticide registration authorities must

Table 2.4 Toxicity of some herbicides and common chemicals to rats.

	Acute oral LD_{50} to rats (mg kg^{-1})
Herbicides	
Chlorotoluron	>10,000
Asulam	>5000
Imazapyr	>5000
Sulfometuron-methyl	>5000
Glyphosate	4320
Other chemicals	
Table salt	3000
Aspirin	1750
Bleaching powder (hypochlorite)	850
Fluoride toothpastes	52–570
Shampoo (zinc pyrithione)	200
Caffeine	200
Nicotine	50

Source: Graham-Bryce, I.J. (1989) Environmental impact: putting pesticides into perspective. *Brighton Crop Protection Conference, Weeds* **1**, 3–20.

therefore attempt to regularly define safe limits for each product and so convince the consumer of product safety or otherwise. The definition of a reasonable versus an unreasonable risk to the consumer, however, has become a legal and scientific minefield. The maximum permitted amount of residual pesticide is determined from three factors, namely:

1 the smallest dose in parts per million (ppm) to produce detectable harmful effects in laboratory animals (i.e. Acute Reference Dose, ARfD);
2 a safety factor, which should be large enough to compensate between humans and test animals, usually 100; and
3 a food factor, based on the proportion of the particular food in an average diet, (i.e. acceptable daily intake).

In this way, a minimum harmful dose of, say, 10 ppm, a safety factor of 100 and a food factor of 0.2 will result in a maximum permitted value or maximum residue level of 0.5 ppm (i.e. $10 \div 100 \times 2$). Extensive and regular monitoring of foodstuffs suggests that if manufacturers' recommendations for application are followed, then risks to the consumer are minimal or negligible. Constant vigilance is required by both governments and producers to ensure that pesticide use is regularly scrutinised for both real and imaginary hazards, and so allay unnecessary public anxiety.

Media coverage in recent years has suggested that pesticide residues in the human diet constitute an unacceptable risk and, simply, that any detectable residue is too high and potentially carcinogenic. This perceived hazard of pesticide use, however, seldom takes into account the presence of natural toxins in our food. Ames *et al.* (1990) have compared the abundance and toxicity of natural toxins with synthetic pesticides and have concluded that 99.99% of our dietary toxin intake is from natural foods. They have estimated that Americans consume 1.5 g of natural toxins each day in roasted coffee, potatoes, tomatoes, whole wheat, brown rice and maize, which is about 10,000 times more than the amount of pesticide residues consumed. Furthermore, surprisingly few of these natural toxins have been tested for carcinogenicity.

Thus, Ames *et al*. (1990) concluded that the health hazards of synthetic pesticide residues are insignificant when compared with human exposure to natural toxins in our diet.

2.5   Chemophobia

Chemophobia is defined as an irrational public prejudice against chemicals, borne out of ignorance and misinformation. Recent decades have seen a revolution in social media and communication methods. In general, it is the headlines that catch the eye and opinions soon become cited as facts, often presented in sensational and dramatic prose. In a democratic society, activists and pressure groups are able to promulgate conspiracy theories about the scientific issues of the day, that rapidly enter into the public conscience. More than ever before, independent, science- and evidence-based facts need defending from individual and/or community misunderstandings or ignorance, a case in point being the use of chemicals in general and GM organisms in particular. Gaps in knowledge and understanding between so-called experts and lay persons need to be narrowed by improved and regular communication to overcome irrational fears. Our responsibility as scientists must be to get out of our laboratories and to explain what we do and why, and at all times be conscious of public perception.

What is the scale of the problem? A survey published in *Nature Chemistry* (Siegrist and Bearth (2019) reported that 40% of Europeans want to 'live in a world where chemical substances do not exist'. Some 30% reported being scared of chemicals. A further 82% did not know that the table salt they were eating was the same substance whether it originated in the sea or was synthesized in a laboratory. It follows that 'nature is better' and that a single molecule of toxin is just as dangerous as a tonne of it. This stance fails to recognise that everything around us is made of chemicals, including ourselves, not forgetting the products and materials that we use in our everyday lives!

A 2020 survey in the USA by Pew Research found that 51% of adults think that GM foods were worse for health than non-GM foods, while 41% said that GM foods have a neutral effect on health and 74% think that GM foods are at least more likely to improve global food supply at lower prices. Of those who think that GM foods were worse for health than conventionally grown food, 88% think GM will create health problems for the population and create problems for the environment. The survey also notes that women are more likely than men (58 to 42%) to believe that GM foods are worse for health. Chemophobia is more evident when we consider the food that we eat and how it is produced. Modern methods of crop breeding and the use of agrochemicals are regularly in the limelight. Some of us are prepared to spend more on 'organically grown food', yet there is no evidence that it is better for us than conventionally grown food.

Only an increased scientific literacy in our schools and wider society will lead to a better understanding of relative risk and how it might be mitigated. The author looks forward to such discussions with his grandchildren!

What is the evidence about pesticides in food? Media coverage in recent years has suggested that pesticide residues in the human diet constitute an unacceptable risk and, simply, that any detectable residue is too high and potentially carcinogenic. This perceived hazard of pesticide use, however, seldom takes into account the presence of natural toxins in our food. Ames *et al*. (1990) have compared the abundance and toxicity of natural toxins with synthetic pesticides and have concluded that 99.99% of our dietary toxin intake is from natural foods. They have estimated that Americans consume 1.5 g of natural toxins each day in roasted coffee, potatoes, tomatoes, whole wheat, brown rice and maize, which is about 10,000 times more than the amount of pesticide residues consumed. Furthermore, surprisingly few of these natural toxins have been tested for carcinogenicity. Thus, Ames *et al.* (1990) concluded that the health hazards of synthetic pesticide residues are insignificant when compared with human exposure to natural toxins in our diet.

The European Food Safety Authority (www.efsa.europa.eu) reported in 2015 that the risk to consumers from pesticide residues in food was low. Their main findings were that:

- 97% of samples analysed were within legal limits;
- 53% were free of quantifiable residues and 44% contained residues that were within permitted concentrations;
- of samples originating in the EU, 1.7% contained residues that exceeded legal limits, while the corresponding value for samples from developing countries was 5.4%, down from 6.5% in 2014;
- no quantifiable residues were found in 96% of food samples intended for infants and young children;
- 84% of samples of animal products were free from quantifiable residues.

In compiling this data, 84,341 samples were analysed for 774 pesticides. Such annual monitoring should reassure consumers that they can trust farmers and have confidence in the safety of what they produce.

In an entertaining and thought-provoking article, (Hall, 2013) described a dinner menu (Table 2.5) created by the American Council on Science and Health in 2000. The menu was unremarkable, yet contained naturally occurring ingredients that had all been classified as rodent carcinogens.

Of course, it is the dose and how it is administered that determines toxicity! Caffeine in our morning coffee, or the addition of salt to our food can kill, but the dose that is normally ingested is invariably safe. Similar arguments can be used for natural, but potentially toxic secondary metabolites that feature in our everyday diet, such as potatoes and nuts.

Table 2.5   Carcinogens in food.

The menu	The rodent carcinogen
Cream of mushroom soup	Hydrazines
Carrots, celery, tomatoes, green salad	Caffeic acid
Nuts	Furfural and aflatoxins
Roast turkey and roast meats	Heterocyclic amines
Strawberries (aroma and flavour)	Benzene and phenol
Pumpkin pie	Benzopyrenes
Apple pie	Acetaldehyde
Coffee	A cocktail of 14 carcinogens
Black tea	Benzopyrenes
Jasmin tea	Benzyl acetate
Wine	Ethyl alcohol
Freshly baked bread (aroma)	Furfural
Butter (aroma)	Diacetyl

Source: Hall, B. (2013) What's your poison? *The Biologist* **60**, 29–31.

The inability to distinguish between hazard and risk creates endless misconceptions about chemical safety. Hazard is the capacity of a substance to cause harm, whereas risk changes with actual exposure. Thus, the risk from a lighted match over petrol is very high, but if the petrol is stored correctly below ground, the risk is much lower. As a further example, paracetamol is a widely used and hazardous synthetic molecule that can cause fatal damage to the human liver at high concentrations, but the actual risk of the recommended dose to overcome a headache is very low.

2.6   The search for novel active ingredients

The ideal herbicide should:

- be highly selective to plants and non-toxic to other organisms;
- act quickly and effectively at low doses;
- rapidly degrade in the environment; and
- be cheap to produce and purchase.

This is a difficult list of criteria to fulfil and seldom are all of these properties shown in one active ingredient. However, the search to widen our herbicide portfolio continues.

High plant selectivity is achieved by targetting processes unique to plants, and the chloroplast is a unique organelle where these processes are located. Thus, the inhibition of photosynthesis and the biosyntheses of pigments, cofactors, amino acids and lipids is invariably lethal. Other major targets located outside the chloroplast but still unique to plants include the biosyntheses of cell wall materials, microtubules for cell division and the receptors for plant hormones (Cobb, 1992).

The number of target sites that have been exploited, however, is remarkably few, at between 15 and 20. The consequence of a limited number of target sites is that weed resistance to existing herbicides is becoming increasingly prevalent (see Chapter 13). The problem is so serious that scientists from all of the major agrochemical companies, academic institutions and national organisations have formed the Herbicide Resistance Action Committee to standardise herbicide classification according to mode of action and to highlight management strategies for the control of resistant weeds. An important cornerstone to the prevention of herbicide resistance is the use of herbicides with different target sites in mixtures, sequences and rotations. Scientists are becoming increasingly aware, however, of the problems of cross-resistance to herbicides of different chemical groups, and evidence is emerging that resistance owing to enhanced metabolic detoxification of herbicides is becoming increasingly widespread. Thus, the combination of herbicide mixtures that exploit different target sites and are metabolised by different routes is proposed to keep resistance under control. If resistance is allowed to accumulate within our major weeds, then our current armoury of herbicides will be rendered useless in the foreseeable future, with drastic consequences to crop yields and quality.

How can this fate be averted? The scientific challenge is to discover new chemistry with novel modes of action and commercial potential as herbicides. The literature implies limited success in this regard with two target sites, namely acetolactate synthase (ALS) and protoporphyrinogen oxidase (PROTOX) dominating herbicide discovery in recent years. Indeed, the only novel target site to have emerged and been commercially exploited in the last 20 years is the enzyme 4-hydroxyphenylpyruvate dioxygenase (HPPD), EC 1.13.11.276 (see Chapter 6 for further details).

The discovery of new herbicide activity may involve three lines of approach, namely:

1   the rational design of specific inhibitors of key metabolic processes;
2   the use of known herbicides or phytotoxic natural products as lead compounds for further synthesis; or
3   the random screening of new chemicals.

To date there have been no publications to suggest that the first approach is feasible. Although we can identify target processes or enzymes that may have potential for herbicide design and discovery, we have not yet come to terms with the complexities of plant metabolism to exploit them. The literature contains many published attempts to design herbicides as enzyme inhibitors that were potent *in vitro* but commercially unsuccessful. Only an increased understanding of plant physiology and biochemistry will allow us to proceed beyond this theoretical phase and so turn rational design into a distinct possibility.

Recent advances in molecular biology and enzymology are now being employed to aid the discovery of new active ingredients. For instance, genome sequencing of model and target plants has allowed for the identification of many new genes and their products. An aim of rational design is to examine these previously unknown enzymes, understand their activity and attempt to identify potential inhibitors, and the potential herbicidal consequences of their inhibition. Such an approach may also allow investigations of enzyme structure and activity in different species, such as in monocots and dicots, thereby possibly increasing our understanding of selectivity and metabolism, and quantitative structure–activity relationships. In practice, such investigations add to the complexity and costs of herbicide discovery, and it follows that few companies can fund such detailed and demanding studies. On the other hand, the global leaders in the sector continue to invest additional funds to maintain their commercial lead.

The second approach is essentially imitative, and is often referred to as 'analogue synthesis' or 'me-too' chemistry. It does provide new targets and standards for synthetic chemists in particular, although the search for an analogue with new activity is seldom predictable. Lead compounds may have shown biological activity either commercially or have known activity, for example as secondary metabolites, allelochemicals or other 'natural' plant products (i.e. biorational design). Certainly, the academic and patent literature are closely scrutinised for indications of new activity.

The random screening of novel chemicals against target weeds is the approach most likely to lead to the discovery of a new class of herbicide. In this way, the agrochemical and chemical industries, sometimes in collaboration with academic institutions, employ chemists to synthesise novel molecules and biologists to screen for their activity. The outcome of the random screening process is not simply left to chance but nowadays involves a stepwise assessment of the potential of new compounds in primary, secondary and tertiary or field screens.

The primary screen aims to establish lead structures, namely those with sufficient activity against target species at a suitably low dose to warrant further study. Since most agrochemical companies screen thousands of chemicals each year, this process is both costly and crucial (Table 2.6). Thus, if selection criteria are low, many compounds of marginal activity will be selected and screening costs may become astronomical, but if standards are too high, a potential lead may be missed. Since it is practically impossible to screen all compounds in the field, companies attempt to simulate these conditions in glasshouses or

Table 2.6   Costs for the development of an agricultural chemical.

Development phase	Length(years)	Compounds tested per eventual registered product	Total costs (£ millions)
First synthesis and glasshouse screens	1	22 500[a]	45
Re-synthesis and first field experiments	2	150	1
Optimisation of synthesis, large-scale field trials and product safety	2	7.5	7
Further optimisation of synthesis, full field development, product safety and registration	3	1.5	8
Totals	8	1	61

Source: Modified from Giles, D.P. (1989) Principles in the design of screens in the process of agrochemical discovery. *Aspects of Applied Biology* **21**, 39–50.
[a] Note that Berg *et al.* (1999) considered this 'hit rate' to be in the region of 1 in 46,000 for 1995, as a result of increasing competition, environmental issues and toxicological considerations.

controlled environments. In this way, new chemicals are commonly applied to glasshouse-grown weeds and crops, and any growth regulatory or phytotoxic symptoms are scored visually at regular intervals. Interestingly, new compounds are routinely tested in other primary screens, for example molecules synthesised as herbicides will be tested for fungicidal or insecticidal activity. Indeed, quite surprising results have been reported from such screens with the generation of new leads (e.g. Giles, 1989).

Most companies have now introduced high-throughput technologies into their discovery processes (Figure 2.2). High-throughput technologies refer to a range of tools and techniques that enable rapid and parallel experiments, such as herbicide screening, to increase productivity and the development of new leads. The benefits of high-throughput screening can include:

1   faster discovery and optimisation of new lead compounds;
2   greater efficiency and productivity;
3   faster and improved target innovation; and
4   faster optimisation of the formulation process.

Examples in use include *Lemna* plants, green algae, cell suspensions from target weeds and germinating cress (*Lepidum sativum* L.) seeds arranged, for example, in 96 well plates. Robotics and automated assays are used to note changes in control populations. *In vitro* high-throughput screens may reveal new classes of herbicide chemistry. Processes such as combinatorial chemistry have been developed for this purpose (e.g. see Ridley *et al.*, 1998 for a detailed account). The modern screening process therefore starts with a large chemical library and assumes a maximum hit-rate of 0.1%, at best.

It is the aim of the secondary screen to optimise these initial observations by further chemistry to yield compounds with the desired characteristics for commercial potential. To be successful, a logical sequence of further study is necessary to establish whether the new structures are sufficiently active or novel to warrant further development. Thus, the characteristics shown in Figure 2.3 need to be precisely tested, with a clear list of priorities. Furthermore, biologists and chemists need to work closely if such optimisation is to be achieved.

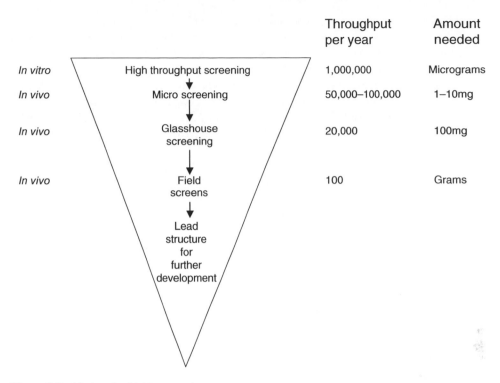

Figure 2.2 Modern herbicide screening.

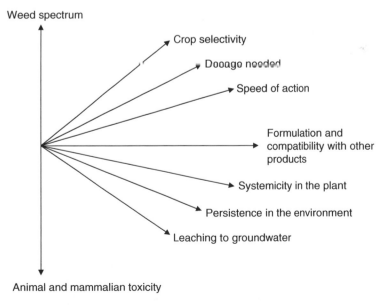

Figure 2.3 Information needed to optimise herbicide development.

Field screens are used to test hypotheses formulated in laboratory and glasshouse testing, and so confirm that molecules are as active in the field as initially predicted. These screens form the basis of many decisions that may yet seal the fate of a new herbicide class. Certainly, promise must be confirmed before very expensive and thorough toxicological and environmental studies are pursued. Compounds showing reliable and reproducible activity in the glasshouse may prove unpredictable in the field when exposed to a wide range of environmental and climatic constraints. Thus, extensive field evaluation is needed to ensure that, for example, the compound is adequately formulated (e.g. for rainfastness, cuticle penetration and compatibility with other pesticides) and used at the correct growth stages to give optimal weed control with minimal crop damage.

To satisfy the aforementioned criteria takes many years and the investment of huge sums of money (Table 2.6). However, as stated by Giles (1989), 'those companies which are able to optimise these processes to reduce their risks but to take advantage of opportunities arising from new types of chemistry . . . have a good chance of discovering products with which to ensure their future; those that do not will have less chance of finding products and a reduced chance of survival'.

2.7   The search for novel target sites

In addition to the challenges of product innovation – reducing the time from discovery to market, global competitiveness and the impact of new environmental legislation – a further issue facing the industry is the discovery of new target sites. The target site being the molecular site where the herbicide binds.

What would constitute an ideal target site? Based on existing and commercially successful target sites, the following properties are relevant:

- specific to plants and is critical for plant survival;
- has no or low impact on non-target organisms, especially mammals;
- can bind a range of chemical inhibitors;
- has a slow rate of metabolic turnover *in vivo*;
- is present in low abundance;
- no alternative pathways exist to by-pass the site;
- inhibition leads to an irreversible and uncontrolled series of biochemical events that overwhelms the cell, its metabolic controls, protective mechanisms and membrane integrity.

There is no proven or simple route to discover the target site of a new herbicide. On the contrary, the investigator requires an appreciation of the vagaries of plant metabolism, an enjoyment of detective work and a large measure of good fortune! A sequential investigation is often followed to establish a target site, as described below. However, this may give the false impression of standardised procedures. Although such an approach is generally sound and valid, it should be noted that the investigator may uncover uncharted areas of plant metabolism, in which the novel herbicide may be regarded as a new and powerful probe. In this way photosynthetic inhibitors have allowed the molecular dissection and characterisation of Photosystem II (Chapter 5), and, as a further example, our understanding of aromatic amino acid biosynthesis was greatly aided from the study of glyphosate action (Chapter 9).

A sequential investigation to discover the target site of a herbicide is as follows:

1 symptom development in target species;
2 structural analysis of sensitive tissues;
3 monitoring of key physiological processes;
4 biochemical studies.

The list of known herbicide target sites is surprisingly short and currently stands at 19 (Table 2.7). However, Berg *et al.* (1999) have listed at least 33 additional targets that have been demonstrated experimentally but not commercialised, and reported 26 other patented herbicide targets (Table 2.8A and B).

This implies a huge investment that may yield novel, commercially important target sites and new chemistry that may generate lead synthesis and ultimately new herbicides.

Table 2.7 Existing targets for herbicide action.

Photosystem II	Tubulin organisation
Photosystem I	Protoporphyrinogen oxidase
Acetolactate synthase	Phytoene desaturase
EPSP synthase	Zeta-carotene desaturase
Glutamine synthase	Dihydropteroate synthase
'Auxin receptor'	HPP Dioxygenase
'Auxin transporter'	'Gibberellin biosynthesis'
ACC carboxylase	Uncouplers
Tubulin assembly plus very long-chain fatty acid biosynthesis inhibitors (Wakabayashi and Böger, 2002)	Cellulose biosynthesis

Source: Berg, D., Tietjen K., Wollweber D. and Hain, R. (1999). From genes to targets: impact of functional genomics on herbicide discovery. *Proceedings of the British Crop Protection Conference, Weeds* **1**, 491–500.

Table 2.8A Thirty-three additional herbicide targets demonstrated *in vitro*.

Adenylosuccinate synthase	Deoxy-xylulosephosphate synthase
Squalene cyclase	Deoxy-xylulosephosphate reductoisomerase
Obtusifoliol demethylase	HMG-CoA reductase
Acetohydroxyacid reductiosomerase	Farnesyldiphosphate synthase
Imidazolglycerol-P dehydratase	Squalene synthase
Isopropylmalate dehydrogenase	'Sphingolipid biosynthesis'
Histidinol dehydrogenase	Panthotenate synthase
Anthranilate synthase	Oxopantoyllactone reductase
Homoserine dehydrogenase	AMP/adenosine deaminase
Ornithine carbamoyl transferase	Dihydrofolate reductase
Aspartate aminotransferase	Aminoacyl-tRNA synthase
Pyruvate dehydrogenase	Protein phosphatase 2A
Glycine decarboxylase	Tyrosine kinase
Acyl-CoA synthase	RNA polymerase III
β-Ketoacyl synthase	
Aminolevulinic acid dehydratase	
Hydroxymethylbilane synthase	
Glutamate-semi-aldehyde amino transferase	

Table 2.8B Twenty-six additional patented herbicide targets.

Homoserine kinase	Sterol Δ14 reductase
Threonine synthase	Adenylosuccinate lyase
Dihydrodipicolinate synthase	Ribose-5P isomerase
Desoxyarabinoheptulosonate-7P synthase	Transketolase
Dihydroxyacid dehydratase	Fructose-6P 1-P transferase
Branched-chain amino transferase	Isocitrate dehydrogenase
Isopropylmalate dehydratase	Galactose dehydrogenase
Anthranilate phosphoribosyl transferase	D1 protease
ATP phosphoribosyl transferase	Rubisco methylase
Cysteine synthase	Cytokinin receptor
Deoxyxylulosephosphate synthase	Glutamate receptors
Isopentenyldiphosphate isomerase	Sodium channel cyclin-dependent protein kinase
	'Cytoskeletal components'

Genome analysis (or functional genomics) is likely to be a powerful tool in future herbicide discovery and the model plant being used for this approach is *Arabidopsis thaliana* L. Heynh (thale cress). The publication of the complete *A. thaliana* genome sequence in December 2000 was a key step in our understanding of the higher plant genome. This cruciferous weed has been used as an increasingly important model system for the study of plant molecular biology. Scientists have been able to exploit its short generation time coupled with a small genome that is amenable to molecular techniques. Hence, of the estimated 23,000 protein-coding genes it is thought that nearly 40% have currently unknown cellular roles, which in theory are all potential herbicide targets. The challenge is to identify which of these genes code for proteins, the inhibition of which would lead to the cessation of plant growth or the identification of a rate-limiting metabolic step or the accumulation of toxic intermediates – or a mixture of all three.

How this may be achieved in practice has been described by Berg *et al.* (1999). In essence, lethal mutants are generated by various means, and the gene or genes involved are identified. This requires a large database and sequence analysis of each of up to 30,000 genes and powerful computing backup. Once the target has been identified its importance requires validation, for example by antisense technology. Thereafter it can be used to design an assay for high-throughput screening to identify new leads for herbicide development.

A comprehensive knowledge of gene function will surely provide innovative targets in the future and may lead to new active ingredients, justifying the major investment required to pursue this approach. Indeed, Berg *et al.* (1999) estimate that up to 3,000 new possible targets may be identified by this method.

2.8   Mode of action studies

The term 'mode of action' may be defined as the complete sequence of events leading to plant injury, and therefore includes all areas of interaction between a herbicide and a crop or test species. A great many features contribute to successful weed control and may be categorised into areas of herbicide uptake, movement and metabolism (Figure 2.4).

The term 'primary or target site' is commonly used to describe the biochemical location at which a herbicide may potently inhibit an important process. This site is expected to

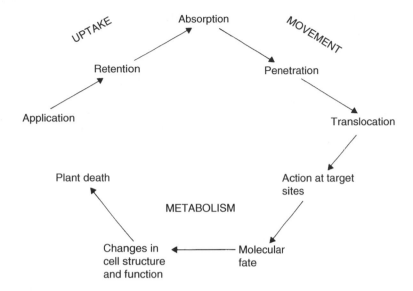

Figure 2.4   Herbicide mode of action.

show the fastest response to the herbicide and should also be the most sensitive site to yield a commercially useful effect. This precise site of action is therefore the location of a molecular interaction which triggers a series of secondary events, leading to the death of a target weed. An example of such a target site is the D1 protein in the thylakoid membranes of the chloroplast. Many herbicides bind to this protein and so inhibit photosynthetic electron flow through the thylakoid. Consequently, active oxygen species are generated which cause plant death by photo-oxidative damage (Chapter 5). The D1 protein is a well-known example of a target site with a specific and strategic membrane location. Other examples of known target sites include natural receptors to plant growth substances (e.g. the auxin-type herbicides, Chapter 7) and specific enzymes in the biosyntheses of lipids and amino acids (Chapters 8 and 9, respectively).

Detailed observations of symptom development provide the starting point for herbicidal detective work. If these observations are recorded daily the speed of response is quickly established and the most sensitive tissues identified. In addition, the development of secondary effects may have useful diagnostic value, as will the addition of 'standard' herbicides to the study. Thus, if symptoms are similar to those produced by existing herbicides, a shared target site may be implied. For example, photosynthetic inhibitors typically induce leaf chlorosis before necrosis, and auxin-type herbicides cause characteristic twisting and bending effects in young, elongating tissues of susceptible plants. In the case of soil-applied herbicides, the emergence of 'white' cotyledons is a good indication of the inhibition of carotenoid biosynthesis by the so-called 'bleaching' herbicides. These observations may be followed by ultrastructural comparisons of sensitive tissues from treated and untreated plants. The progressive development of abnormalities in both extracellular and intracellular organisation is a powerful indicator of which metabolic compartments are involved in herbicide action.

Isotopic tracers and autoradiography may be used at this stage to pinpoint this cellular site with precision.

The clues generated from the visual and microscopic assessment of treated tissues may then be used to monitor key physiological processes in the presence or absence of the herbicide. Such studies point to the most likely areas of inhibition in the discovery of the target site. For example, photosynthesis and respiration may be measured in intact tissues, or isolated chloroplasts or mitochondria using oxygen electrode techniques. In recent years, many groups and companies have developed model systems for such studies, including the use of isolated cells and protoplasts from target species, cell cultures and green algae. In general, various macromolecular syntheses can be examined in these systems, especially by monitoring the incorporation of radiolabelled precursors into stable products, including leucine into protein, thymidine into DNA, uracil into RNA and acetate or mevalonate into lipids. In addition, physiological viability may be assessed using fluorescent stains, and membrane integrity and function examined by measuring potassium flux. In each case, comparison with herbicide standards should be encouraged and both time-courses and dose-responses should be established to determine the precise sensitivity of the process to herbicidal inhibition.

Once a likely target pathway is indicated, a sequence of reactions is examined in more detail. Initially, concentrations of pathway intermediates should be investigated, both in the presence and in the absence of the herbicide. If one compound accumulates in the presence of the inhibitor, then this is good evidence that the target site of an enzyme has been found. For example, glyphosate treatment results in accumulation of shikimate 3-phosphate, implying the inhibition of 5-enoyl-pyruvyl shikimic acid 3-phosphate (EPSP) synthase, and the inhibition of protoporphyrin oxidase by acifluorfen is pinpointed by the accumulation of protoprophyrin IX in treated plants. Such an accumulation of a metabolic intermediate implies that further key molecules are not being formed. This can be confirmed by studying whether the addition of the key metabolite is able to overcome the inhibition. In this way the inhibition of branched-chain amino acid biosynthesis by the sulphonylureas and imidazalinones can be overcome by the addition of leucine, isoleucine and valine. Additional proof that a true target site has been discovered may come from further studies with mutants. Probably the best-documented example of such mutagenesis is the herbicide site of the D1 protein in the chloroplast thylakoid. That a single substitution at position 264 from serine to glycine is sufficient to cause total resistance to atrazine in some weeds is surely powerful evidence of a target site! Further studies may then focus on the chemical and biochemical properties of the target site. For instance, the kinetics of the inhibition and the precise mechanism of inhibition may be investigated. Such studies are not only of academic value, but are also central to our understanding of herbicide selectivity at the molecular level and may establish precise structure–activity relationships for the development of molecules with optimal activity. Additionally, site-directed mutagenesis may then be employed to create crop tolerance.

Grossmann (2005) has described a functional array of bioassays conducted at BASF, Limburgerhof, aimed at diagnosing the mode of action of a new herbicide molecule. These assays are designed to differentiate between the distinct responses of complex structures (plant, tissue, meristematic cell, organelle), developmental stages, types of metabolism and physiological processes. He has coined the term 'physionomics' to describe this physiological profiling as providing the first clues to mode of action. This term follows the use of other '-omics' technologies in studying herbicide discovery and mode of action:

- *Functional genomics*: generating mutants and screening them for functional gene identification.
- *Transcriptomics*: profiling gene expression utilising DNA microarrays and RNA extraction.
- *Proteomics*: protein profiling by gel electrophoresis of extracted proteins.
- *Metabolomics*: metabolic profiling using metabolite extraction and separation by gas chromatography/liquid chromatography–mass spectrometry.
- *Physionomics*: physiological profiling following functional bioassays.

Duke *et al*. (2013) have reviewed -omics approaches. While physical observations of treated plants often hint at a mode of action, they are often proved wrong, and so physiological and biochemical studies are invariably needed to pinpoint areas of plant metabolism likely to be affected.

Omics methods have been refined in recent years, with responses compared with a chemical library of molecules with known target sites. Confirmation of the target site is accomplished by enzyme assays (i.e. biochemistry) or by the generation of mutants (i.e. genetics). The amount of data generated and the complexity of omics approaches requires complex analytical capabilities, such that databases are invariably the confidential preserve of the agrochemical industry. Helpful databases have been published, based on the physicochemical properties of lead molecules. As an example, Gandy *et al*. (2015) report detailed analyses of 324 successful herbicides, allowing for a rapid comparison of new with known herbicides.

New screens also appear in the literature, from time to time, for example Chuprov-Netochin *et al*. (2016) have described a novel, high-throughput screen for potential growth inhibitors and stimulators of pollen growth. The main value of this assay appears to be its speed, since results are available in only 8 h. It can be automated and promises to generate large chemical libraries in a short time.

From a chemistry perspective, Lamberth *et al*. (2013) report that 70% of recent agrochemical active ingredients are based on heterocyclic scaffolds and contain halogen groups. About a third are chiral and over 90% of herbicides have two or fewer hydrogen bond donors. Zakharychev *et al*. (2020) report that 95 agrochemicals, generating US$5 billion per annum, are based on the pyridine scaffold, making it one of the most successful heterocycles in the history of plant protection products in the twenty-first century.

Rapid technological progress in the sequencing of genes has provided the DNA sequences from many organisms as useful databases. Furthermore, antisense approaches offer methods for validating new potential target sites for herbicides.

Mergner *et al*. (2020) have produced a molecular map of the *Arabidopsis* proteome. Using liquid chromatography–tandem mass spectrometry they have mapped the proteome of this model plant and made their findings freely available in an online database (www.proteomicsdb.org/). Of approximately 27,000 genes, more than 18,000 gene products are proteins, expressed at different locations within the plant, in varying quantities. Intriguingly, they also identified more than 43,000 sites that can be phosphorylated, indicating the possibility of many phosphorylation-mediated signalling events. While we cannot extrapolate all these findings to crop and weed species, we can anticipate similarities, which should be of great value to plant scientists, especially in investigations of how plants can adapt to changes in their environment and to establish changes in the proteome in response to biotic attack and herbicide application.

Finally, Ambrust and Palumbi (2015) reported on the findings of a 3 year global expedition to collect and study plankton organisms. They discovered 35,000 species of bacteria, 5000 new viruses and 150,000 single-cell plants and protozoa, most of which are thought to be new to science. Plankton organisms constitute 90% of the mass of all marine life and, as primary producers, form the base of marine food-chains. The authors report 40 million genes of which 80% are new to science, and a major task is now needed to discover their roles and functions. This is one of the largest DNA databases available to the scientific community, representing a potential 'treasure trove' that may lead to new screens and lead synthesis for new agrochemicals and drugs, including antibiotics and new treatments for cancer.

Biochemical studies are often routinely performed in cell-free systems on enzymes of fungal or bacterial origin. Several dangers are implicit in this *in vitro* route, and particular care is needed not to extrapolate results obtained *in vitro* to predict *in vivo* activity in the whole organism. Thus, the perfect herbicide *in vitro* may have negligible practical value if it cannot reach its active site. Indeed, many a promising, lipophilic candidate has not been developed further owing to its being confined to the cuticle and not entering the weed! To this end, increasing use is being made of the physicochemical properties of a molecule that will allow a prediction of its systemicity and stability in both target species and the environment. These include the octan-1-ol–water partition coefficient (K_{ow}), and the dissociation constant (pK_a). No single measure of organic phase/water distribution can be predicted for herbicides because of the wide variety of organic phases within a plant, such as hydrocarbon waxes, triglycerides, proteins, lignins or even carbohydrates. However, Briggs and coworkers (Bromilow *et al.*, 1986) have found that the partition coefficient can give a good prediction of systemicity. For acids and bases the dissociation constant of the chemical and the natural pH of the various plant compartments will also determine the proportion of ionised and non-ionised forms present. Ionisation decreases log K_{ow} and so can have a dramatic effect on the movement of a compound in a plant. Most phloem-mobile, systemic compounds are therefore weak acids with log K_{ow} in the range of −1 to 3, but the immobile soil-applied herbicide diflufenican has a log K_{ow} value of 4.9 (Table 2.9).

Table 2.9   Herbicide systemicity and log K_{ow}.

Mobility	Log K_{ow}		
	−3 to 0 (hydrophilic)	0 to 3 (intermediate)	3 to >6 (lipophilic)
Non-systemic			Trifluralin, diflufenican
Xylem mobile		Triazines, phenylureas	diphenyl ethers
Both xylem and phloem mobile	Glyphosate, aminotriazole, glufosinate	Auxin-type herbicides, sulphonylureas, imidazolinones, sethoxydim	

Source: With permission after Bromilow, R., Chamberlain, K. and Briggs, G.G. (1986). Techniques for studying the uptake and translocation of pesticides in plants. *Aspects of Applied Biology* **11**, 29–44.

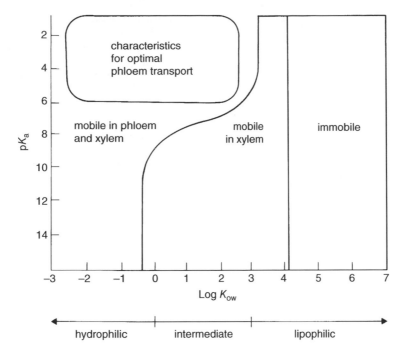

Figure 2.5 Relationship between herbicide mobility, log K_{ow} and pK_a. (After Bromilow, R., Chamberlain, K. and Briggs,G.G. (1986). Techniques for studying the uptake and translocation of pesticidesin plants. *Aspects of Applied Biology* **11**, 29–44).

Metabolism of the herbicide within a plant may also have a profound effect on movement. Metabolism in general attempts to reduce lipid solubility and prevent toxic action. For example, aryl hydroxylation may be expected to cause increased mobility from the parent molecule (pK_a change from neutral to 10), and its subsequent conjugation to glucose will create a less mobile glucoside (pK_a change from 10 to neutral) that becomes subject to further compartmentation or inactivation within the cell (Bromilow *et al.*, 1986). Figure 2.5 presents an overview of how mobility may be related to log K_{ow} and pK_a.

Mode of action studies greatly aid, and in many cases totally explain, the selectivity shown by many herbicides. Indeed, susceptibility, tolerance and resistance are being increasingly defined at the metabolic level. Selectivity may of course be due to differential leaf interception, retention or uptake, but once inside the plant several metabolic criteria are now evident. Generally, selectivity may be achieved at several steps as suggested in Figure 2.6.

Many herbicides are themselves inactive (Figure 2.6A) and need to be metabolically activated before phytotoxicity is observed. Thus, paraquat and diquat are activated by light in the thylakoid to generate toxic active oxygen species (Chapter 5); the butyl esters of MCPA and 2,4-D are converted to active acids in susceptible species by β-oxidation (Chapter 7), and some graminicides, such as the aryloxyphenoxypropionates, require conversion from ester to acid for optimal activity (Chapter 8). Active herbicides may be metabolised before they can reach their target site (Figure 2.6B).

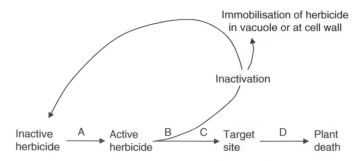

Figure 2.6 Herbicide metabolism in relation to selectivity. Points A–D are stages where selectivity may be achieved.

Many enzymes, often mixed-function oxygenases, have been implicated in herbicide metabolism studies. Maize and sorghum contain a high concentration of glutathione *S*-transferase, so that atrazine is conjugated and detoxified before it reaches its thylakoid site of action; wheat readily hydroxylates chlorsulfuron and the metabolites are inactivated by conjugation with glucose. Similarly, soybean selectivity hydroxylates and glycosylates bentazone; metamitron is selectively deaminated in sugar beet, and diuron is initially demethylated in some species prior to inactivation, again by hydroxylation and glycosylation.

Herbicide metabolites may not always be inactive. On the contrary, evidence is accumulating that glycosyl ester formation may be reversible in many instances. Thus, conjugates stored in the vacuole or bound to cell wall components may be regarded as herbicide reservoirs, which may be called into play at a later stage.

Selectivity may also be achieved at the target site either by differential binding or inactivity (Figure 2.6C). For example, the spectacular selectivity shown by the aryloxyphenoxypropionate and cyclohexanedione graminicides appears to be primarily due to the inability of these compounds to inhibit acetyl coenzyme A carboxylase in dicotyledonous crops (Chapter 8). Monocot–dicot differences are also implied in the mode of action of auxin-type herbicides. In this case, it is speculated that the monocotyledonous auxin receptor is not accessible to the phenoxyalkanoic acids (Chapter 7).

Finally, toxic species may be inactivated before phytotoxicity is observed (Figure 2.6D). An example is the enhanced activity of superoxide dismutase in some grasses that enables relative tolerance to paraquat.

A final example of the importance of mode of action studies lies in an appreciation of the increasing problem of herbicide resistance. The prolonged use of persistent and potent herbicides with a common target site is a certain recipe for the eventual development of resistant weeds. Only by herbicide, and preferably crop, rotation can this problem be overcome. Thus, the use of herbicides with different target sites will lessen the selection pressures that favour resistance.

The most successful herbicides with negligible mammalian toxicity are those which inhibit, often selectively, metabolic processes that are unique to plants. These are principally photosynthesis, the action of plant growth regulators and the biosyntheses of pigments, lipids, and amino acids. These target processes will be considered in separate chapters of this book.

It is, of course, most likely that new modes of action have been detected, but their herbicidal inhibition has never been commercialised. This may be due to:

- challenges with non-target organisms;
- adverse environmental toxicology;
- product volatility and formulation issues;
- production costs and comparative efficacy in comparison with existing products.

In summary, to discover a new herbicide depends on:

- the quality and diversity of the chemistry for screens;
- an awareness of the strengths and weaknesses of each screen;
- an ability to capitalise on unexpected findings;
- luck!

Furthermore, to develop a commercially successful herbicide is essentially a compromise between:

- potency at the target site;
- the generation of irreversible phytotoxic effects;
- selectivity in crops;
- safety in the environment;
- safety in non-target organisms.

2.9   The role of natural chemistry

Natural compounds have played a major role in the discovery of pharmaceuticals and agrochemicals, although only 8% of herbicides have been derived from natural secondary metabolites. These natural molecules can provide leads for the discovery of new herbicides and the identification of new target sites for further exploitation. Since more than 200,000 secondary metabolites are known, and more are being isolated and characterised each year, they can enter existing screening processes. Of course, the challenge then is to convert a promising candidate in a screen into a commercial product by chemical derivatisation and optimisation of activity. For further information the reader is referred to the comprehensive review by Dayan and Duke (2014), from which Table 2.10 has been collated.

2.10   Recent developments

While no new herbicides with new modes of action have entered the market since isoxaflutole in 1994, new compounds have emerged, and some have become widely accepted and successful for selective weed control in major crops (Table 2.11) (Kraehmer *et al.*, 2014).

Kraehmer *et al.* (2014) note that recently introduced ALS inhibitors have been commercially successful and depend on tank-mixtures with safeners for crop selectivity (see Chapter 4, Section 4.3 for further details on crop safeners). An example is thiencarbazone-methyl which can be used for both pre- and post-emergence control of a broad range of broad-leaf weeds and grasses in cereals with the safener mefenpyr-diethyl and cyprosulfamide in maize. Pyroxulam achieves crop selectivity in wheat, rye and triticale when used in

Table 2.10   Target sites of some natural compounds that inhibit plant growth.

Target site	Example of natural compound
Amino acid biosynthesis	
Glutamine synthetase	Phosphinothricin
Tryptophan synthase	5-Methyl tryptophan
Aspartate aminotransferase	Gostatin
Orn carbamoyl transferase	Phaseolotoxin
β-Cystathionase	Riboxitine
Electron transfer	
CF1 ATPase	Tentoxin
Uncoupler of photophosphorylation	Nigericin
Photosystem II electron transport	Sorgoleone
Photosystem I electron divertor	Pyridazocidin
biosynthesis of photosynthetic pigments	
Tyrosine aminotransferase	Cineole analogues
HPPD	Leptospermone
Deoxyxylulose-5-P reductase	Fosmidomycin
Glutamate-1-semialdehyde aminotransferase	Gabaculin
PROTOX	Cyperin
Lipid biosynthesis	
Enoyl-ACP reductase	Cyperin
β-Ketoacyl-ACP synthase	Thiolactomycin
Ceramide synthase	AAL toxins
Membrane function and lipid stability	
H+ ATPase	Juglone
NADH oxidase	Fusicoccin
Membrane destabiliser	Pelargonic acid
Gene expression and regulation	
Adenylosuccinate synthase	Hydantocydin
Isoleucyl tRNA synthase	Pseudomonic acids
Peptide deformylase	Actinonin
Serine/threonine protein phosphatases	Cantharidin
RNA polymerase	Tagetitoxin
Aminopeptidase	Bestatin
Lycine deacylase	Carbonum toxin
AMP deaminase	Carbocyclic coformycin
Calmodulin	Ophiobolin A
Hormonal regulation	
Jasmonate mimic	Coronatine
Auxin signalling	Toyocamycin
ACC synthase	Rhizobitoxine
Gibberrellic acid mimic	GA3
Gibberrellic acid oxidase	Myrigalone
Cell macrostructure	
Microtubule polymerisation	Citral
Cellulose biosynthesis	Thaxtomin
Golgi body assembly	7-Dehydrobrefeldin A
Cell cycle	
DNA polymerase	Aphidicolin
Ribonucleotide reductase	Mimosine
Proteosome interference	Lactacystin

After Dayan and Duke (2014).

Source: Dayan, F.E. and Duke, S.O. (2014) Natural compounds as next generation herbicides. *Plant Physiology* **166**, 1090–1105. Reproduced with permission of Oxford University Press.

Table 2.11 Some new herbicides developed since 2000.

Mode of action	Example	Launch date
Auxins	Aminopyralid	2005
	Aminocyclopyrachlor	2010
	Halauxifen-methyl	2015
	Florpyrauxifen benzyl	2018
Cellulose biosynthesis	Indaziflam	2010
ALS inhibitors	Mesosulfuron-methyl	2001
	Thiencarbazone-methyl	2008
	Pyroxulam	2008
HPPD inhibitors	Topramezone	2006
	Tembotrione	2007
	Pyrasulfotole	2008
	Tefuryltrione	2010
PROTOX inhibitor	Saflufenacil	2010
ACCase inhibitor	Pinoxaden	2006
VLCFA biosynthesis	Pyroxasulfone	2012

Sources: adapted from Kraehmer, H., van Almsick, A., Dietrich, H., Eckes, P., Hacker, E., Hain, R. *et al.* (2014) Herbicides as weed control agents: State of the Art 2: Recent achievements. *Plant Physiology* **166**, 1132–1148; Epp, J.B., Alexander, A.L., Balko, T.W., Bussye, A.M., Brewster, W.K., Bryan, K., Daeuble, *et al.* (2016) The discovery of Arylex Active and Rinskor Active: two novel auxin herbicides. *Bioorganic and Medicinal Chemistry* **24**, 362–371.

combination with the safener cloquintocet-mexyl. These recently introduced herbicides and safeners offer a further commercial advantage in that they offer alternative treatments for the control of glyphosate-resistant weeds.

The HPPD inhibitors are very effective for weed control in maize. Isoxaflutole was the first of its class to be introduced, followed by tembotrione, sulcotrione and mesotrione. Pyrasulfotole is an effective HPPD inhibitor in rice, and combination with the safener mefenpyr-diethyl gives post-emergent control in wheat and barley. The addition of a Photosystem II inhibitor, such as bromoxynil, broadens the spectrum of weed control in cereals, while the addition of fenoxaprop-ethyl gives good additional control of grasses. The HPPD inhibitor tembotrione with safener isoxadifen-ethyl is effective for post-emergence weed control in maize.

Indaziflam, a novel inhibitor of cellulose biosynthesis, has entered the non-selective, pre-emergence market for weed control in perennial crops, such as sugarcane, turf grass and tree plantations.

This century has also seen the re-emergence of auxin-type herbicides, with the introduction of aminopyralid, for weed control in pasture and at industrial sites, aminocyclopyrachlor (broad-leaf weed control in lawns and turf), halauxifen-methyl for broad-leaf weed control in cereals and florpyrauxifen-benzyl ester that controls herbicide-resistant barnyard grass in rice.

Pinoxaden, when applied with the safener cloquintocet-mexyl, is a selective post-emergence graminicide in cereals.

Finally, pyroxasulfone is an inhibitor of very-long-chain fatty acid biosynthesis in both grass and broad-leaf weeds, and is active when applied both before and after crop emergence. It also gives control of ALS-and ACCase-resistant weeds. Fenoxasulfone is a similar inhibitor being developed for selectivity in rice.

A further strategy, introduced in the last decade, has been for the industry to counter the growing threat of herbicide-resistant weeds by marketing herbicide mixtures with different modes of action. Examples include;

- a PROTOX inhibitor with dicamba (BASF);
- glyphosate and dicamba (Bayer);
- pyroxulam, clopyralid and fluroxypyr (Corteva);
- bicyclopyrone, mesotrione, *S*-metolachlor, with the safener benoxacor, for the control of atrazine-resistant weeds in maize (Syngenta).

The top 10 leading herbicides in 2015 for use in the four key crops, as ranked by sales, are presented in Table 2.2.

At the time of writing (spring 2021) there have been glimpses of new herbicides reported on specialist websites and in specialist publications. Time will tell if these new molecules enter the marketplace. A brief summary is presented below.

1 FMC have developed a new, selective herbicide for the control of grass weeds in rice, termed tetraflupyrolimet. It inhibits dihydroorotate dehydrogenase and so prevents the synthesis of nucleic acids. Specifically, this enzyme is located on the outer surface of the inner mitochondrial membrane where it catalyses the ubiquinone-mediated oxidation of dihydro-orotate to orotate in pyridine biosynthesis.

2 Bayer Crop Science have discovered that an older molecule first launched in 1983 as a diphenyl ether, aclonifen, inhibits homogentisate solanesyl transferase, the first committed step in the biosynthesis of plastoquinone-9 in the chloroplast. Owing to structural similarity, it was previously thought to be a diphenyl ether herbicide, but it does not inhibit PROTOX or chlorophyll biosynthesis. Instead, phytoene accumulates but it does not inhibit phytoene desaturase. The new retail name will probably be Bandur.

3 The Japanese company, Mitsui Chemicals Agro, have discovered cyclopyrimorate, for pre-emergence weed control in rice, that also inhibits homogenistate solanesyl transferase. This enzyme is weakly inhibited by cyclopyrimorate, but strongly inhibited by a metabolite, des-morpholinocarbonyl cyclopyrimorate.

4 BASF have two products going through the registration process in Australia in 2020/2021 for the control of annual ryegrass and herbicide-resistant weeds: Luximo, reportedly cinmethylin, while Trexor is a PROTOX inhibitor.

2.11 A lower limit for rates of herbicide application?

One of the great successes of the agrochemical industry in the last six decades has been the vast reduction in the amounts of chemicals applied to obtain weed control, from kilograms to grams of active ingredients per hectare. Indeed, the most recently introduced PROTOX and ALS inhibitors have rates quoted as low as 2–5 g active ingredient (a.i.) ha^{-1} (Wakabayashi and Böger, 2002). This raises the question of whether there may be a lower limit of application that still maintains high efficacy. Clearly, the environmental benefits of lower doses would be enormous, but how low can we go? The following calculations may be of use in this discussion.

Assumptions

1   A weed leaf area of 10 cm^2.
2   The leaf has 300,000 cells cm^{-2}.
3   The leaf therefore has 3×10^6 cells.
4   One herbicide molecule needs to bind to one specific site per cell.
5   3×10^6 molecules need to gain entry into the leaf.
6   If the herbicide has a molecular weight of 250 daltons, then 250 g will contain 6.023 $\times 10^{23}$ (Avogadro's number) of molecules.
7   3×10^6 molecules are contained in $3 \times 10^6 \times 250 \div 6.023 \times 10^{23} = 1.25 \times 10^{-15}$ g per plant!
8   This suggests that a lower theoretical limit of 1.25 fg of herbicide are needed to gain entry into the plant for herbicidal action.

However

9   Perhaps only 10% of the dose gains entry into the plant, and only 10% of that reaches the target site in an active form. Thus, 1.25×10^{-13} g must be applied to each plant, equivalent to approximately 10^{-6} g ha^{-1}.
10 These calculations suggest the possibility of a lowest field rate in the range of micrograms per hectare, some six orders of magnitude lower than rates currently used. Can these theoretical values be achieved in the field? The above model contains some major assumptions and extrapolations that may not be experimentally sound. A few important issues are listed below.
11 Leaves are seldom arranged perpendicular to the spray, so issues of leaf interception, sprayer technique, formulation and retention, will strongly influence the penetration of the herbicide.
12 Environmental variables have a profound impact on herbicide efficacy and are beyond the control of the most efficient operator and agrochemical company! These will influence entry into the plant, the movement of the active ingredient and the metabolic state of the plant to tolerate chemical attack.
13 The assumption that one herbicide molecule per cell may be phytotoxic is likely to be invalid, taking into account the metabolic defences mechanisms that we know are present in each leaf cell and the probability of many sites per cell.

Given the above caveats, it may still be expected that at least another order of magnitude may be achieved in reducing herbicide dosage to less than 1 g per hectare in the decades ahead. This target is surely not beyond the innovative scientists in the modern agrochemical industry.

References

Agrow (2006) Global agrochemical market flat in 2005. Issue 490, p. 15, 24 February.

Ambrust, E.V. and Palumbi, S.R. (2015) Uncovering hidden worlds of ocean biodiversity. *Science* **348**, 865–867.

Ames, B.N., Profet M. and Gold L.S. (1990) Dietary pesticides (99.99% all natural). *Proceedings of the National Academy of Sciences, USA* **87**, 777–781.

Avery, D.T. (1995) *Saving the Planet with Pesticides and Plastic: The Environmental Triumph of High Yield Farming*. Hudson Institute.

Beaulieu, M., Cabana, H. and Hout, Y. (2020) Adverse effects of atrazine, DCMU and metolachlor on phytoplankton cultures and communities at environmentally relevant concentrations using fast repetition rate fluorescence. *Science of the Total Environment* **712**, 136239. doi: org/10.1016/j.scitotenv.2019.136239

Berg, D., Tietjen K., Wollweber D. and Hain, R. (1999) From genes to targets: impact of functional genomics on herbicide discovery. *Proceedings of the British Crop Protection Conference, Weeds* **1**, 491–500.

Brickley, M.R., Lawrie, E., Weise,V., Hawes, C. and Cobb, A.H. (2012) Use of a potentiometric dye to determine the effects of the herbicide bromoxynil octanoate on mitochondrial bioenergetics in *Chlamydomonas reinhardtii*. *Pest Management Science* **68**, 580–586.

Bromilow, R.H., Chamberlain K. and Briggs G.G. (1986) Techniques for studying the uptake and translocation of pesticides in plants. *Aspects of Applied Biology* **11**, 29–44.

Chuprov-Netochin, R., Neskorodov, N., Marusich, E., Mishutkina, Y., Volynchuck, P., Leonov, S. *et al.* (2016) Novel, small molecule modulators of plant growth and development identified by high-throughput screening with plant pollen. *BMC Plant Biology* **16**, 192–204.

Cobb, A.H. (1992) *Herbicides and Plant Physiology*, 1st edn. London: Chapman and Hall.

Dayan, F.E. and Duke, S.O. (2014) Natural compounds as next generation herbicides. *Plant Physiology* **166**, 1090–1105.

Duke, S.O., Bajsa, J. and Pan, Z. (2013) Omics methods for probing the mode of action of natural and synthetic phytotoxins. *Journal of Chemical Ecology* **39**, 333–347.

Dyer, A. (2019) GM ready. *Farmers Guardian*, front page, 2 August.

Epp, J.B., Alexander, A.L., Balko, T.W., Bussye, A.M., Brewster, W.K., Bryan, K., Daeuble, *et al.* (2016) The discovery of Arylex Active and Rinskor Active: two novel auxin herbicides. *Bioorganic and Medicinal Chemistry* **24**, 362–371.

Evans, J. (2010) Food security is all in the genes. *Chemistry and Industry*, **9**, 14–15.

Finney, J.R. (1988) World crop protection prospects: demisting the crystal ball. *Brighton Crop Protection Conference, Pests and Diseases* **1**, 3–14.

Flores, F., Collier, C.J., Mercurio, P. and Negri, A.P. (2013) Phytotoxicity of four photosystem II herbicides to tropical seagrasses. *PLOS ONE* **8**, e75798; doi: 10.1371/journal.pone.0075798.

Gandy, M.N., Corral, M.G., Mylne, J.S. and Stubbs, K., A. (2015) An interactive database to explore herbicide physicochemical properties. *Organic and Biomolecular Chemistry* **13**, 5586–5590; doi: org.10.1039/c5ob00469a.

Giles, D.P. (1989) Principles in the design of screens in the process of agrochemical discovery. *Aspects of Applied Biology* **21**, 39–50.

GMOinfo (n.d.) Most Europeans hardly care about GMOs. www.GMOinfo.eu (accessed19 June 2019).

Graham-Bryce, I.J. (1989) Environmental impact: putting pesticides into perspective. *Brighton Crop Protection Conference, Weeds* **1**, 3–20.

Grossmann, K. (2005) What it takes to get a herbicide's mode of action. Physionomics, a classical approach in a new complexion. *Pest Management Science* **61**, 423–431.

Hall, B. (2013) What's your poison? *The Biologist* **60**, 29–31.

ISAAA (n.d.) Biotech crops continue to help meet the challenges of increased population and climate change. www.isaaa.org/resources/publications/annualreport/2019 (accessed 22 August 2019).

Kraehmer, H., van Almsick, A., Dietrich, H., Eckes, P., Hacker, E., Hain, R. *et al.* (2014) Herbicides as weed control agents: State of the Art 2: Recent achievements. *Plant Physiology* **166**, 1132–1148.

Lamberth, C., Jeanmart, S., Luksch, T. and Plant, A. (2013) Current challenges and trends in the discovery of agrochemicals. *Science* **341**, 742–746.

Mergner, J., Frejno, M., List, M., Papacek, M., Chen, X., Chaudhary, A. *et al.* (2020) Mass-spectrometry-based draft of the *Arabidopsis* proteome. *Nature* **579**, 409–414.

Milner, A.M. and Bond, I.L. (2017) Towards pesticidovigilance. *Science* **357**, 1232–1234.

Nair, A. (2019) Growing business. *Chemistry and Industry* **83**(4), 40.

Oerke E.C., Weber A., Dehne H.-W., Schonbeck F. (1994) Conclusion and perspectives. In: Oerke, E.C., Dehne, H.W., Schonbeck, F. and Weber, A. (eds) *Crop Production and Crop Protection: Estimated Losses in Food and Cash Crops.* Amsterdam: Elsevier, pp. 742–770.

Pew Research (2020) Pewresearch.org/fact-tank/2020/03/18/about-half-of-u-s-adults-are-wary-of-health-effects-of-genetically-modified-foods-but-may-also-see-advantages/ (accessed 25 March 2020).

Phillips McDougall (2016) The cost of new agrochemical product discovery, development and registration in 1995, 2000, 2005–8 and 2010–2014. R&D expenditure in 2014 and expectations for 2019; www.Croplife.org>wp-content>uploads

Ridley, S.M., Elliot, A.C., Yeung, M. and Youle D. (1998) High-throughput screening as a tool for agrochemical discovery: automated synthesis, compound input, assay design and process management. *Pesticide Science* **54**, 327–337.

Rüegg, W.T., Quadranti, M. and Zoschke, A. (2007) Herbicide research and development: challenges and opportunities. *Weed Research* **47**, 271–275.

Siegrist, M. and Bearth, A. (2019) Chemophobia in Europe and reasons for biased risk perceptions. *Nature Chemistry* **11**, 1071–1072.

Silva, V., Mol, H.G.J., Zomer, P., Tienstra, M., Ritsema, C.J. and Geissen, V. (2018) Pesticide residues in European agricultural soils – a hidden reality unfolded. *Science of the Total Environment* **653**, November 2018. doi: org/10.1016/j.scitotenv.2018.10.441

Wakabayashi, K.O. and Böger, P. (2002) Target sites for herbicides: entering the 21st Century. *Pest Management Science* **58**, 1149–1154.

Wjeczorek, M.V. and four others (2017) Response and recovery of macrophytes *Elodea canadiensis* and *Myriophyllum spicatum* following a pulse of exposure to the herbicide iofensuron-sodium in outdoor stream microcosms. *Environmental Toxicology and Chemistry* **36**, 1090–1100.

Xu, C. and Armstrong, D.W. (2013) High performance liquid chromatography with paired ion electrospray ionisation (PIES) tandem mass spectrometry for the highly sensitive determination of acidic pesticides in water. *Analytica Chimica Acta* **792**, 1–9.

Zakharychev, V.V., Kuzenkov, A.V. and Martsynkevich, A.M. (2020) Good pyridine hunting: a biomimic compomd, a modifier and a unique pharmacophore in agrochemicals. *Chemistry of Heterocylic Compounds* **56**, 1491–1516; doi: org/10.1007/s10593-020-02843-w

Chapter 3
Herbicide Uptake and Movement

There are no such things as applied sciences, only the application of science.

Louis Pasteur (1822–1895)

3.1 Introduction

The effectiveness of a herbicide treatment ultimately depends on the amount of active ingredient that reaches the target site. There are, however, many barriers that prevent the herbicide molecule passing from the outside of the leaf through the cuticle and underlying cells, utilising transport systems, to reach the site of action. Indeed, the physical and chemical nature of the cuticle, the contents of the herbicide formulation and the environmental and physiological history of the plants in question will all influence herbicide efficacy. Many herbicides are also applied to the soil and their effectiveness, again, is determined by factors influencing uptake into the root and transport to the target site. This chapter will consider the passage of herbicides from outside the plant to the target tissues.

Herbicides are either foliar- or soil-applied. There are three options for their movement in the treated plant.

1 Movement from source to sink. For example, from the treated leaves to the growing points of the plant, such as the shoot and root meristems. This translocation involves the symplast, defined as the continuum of cell protoplasts, linked by plasmodesmata. Long-distance transport takes place within the phloem. Typical, initial symptoms are observed on new growth, as noted with auxin-type herbicides and glyphosate.
2 Movement in the apoplast. This is the non-cytoplasmic continuum in the plant, such as the xylem and the cell walls. In this case, herbicide translocation follows the movement of water and solutes from root to leaf in the transpiration stream. Typical herbicides are soil-applied, including the triazines and dinitroanilines.
3 Limited movement. These herbicides are often referred to as having a contact action. They do not move greatly within the plant and so are more rapidly phytotoxic. Typical examples include paraquat and the diphenyl-ethers.

Herbicides and Plant Physiology, Third Edition. Andrew H. Cobb.
© 2022 John Wiley & Sons Ltd. Published 2022 by John Wiley & Sons Ltd.

3.2   The cuticle as a barrier to foliar uptake

The outer leaf surface is covered with a waxy cuticle that waterproofs the leaf and provides the first line of defence between the plant and the environment. Its structure and chemical content are both varied and complex, but the successful passage across it is a vital aspect of herbicide efficacy. Generally, the cuticle is 0.1–13 μm thick and contains three components: an insoluble cutin matrix, cuticular waxes and epicuticular waxes (Figures 3.1 and 3.2). It is not a homogeneous layer and varies greatly from species to species.

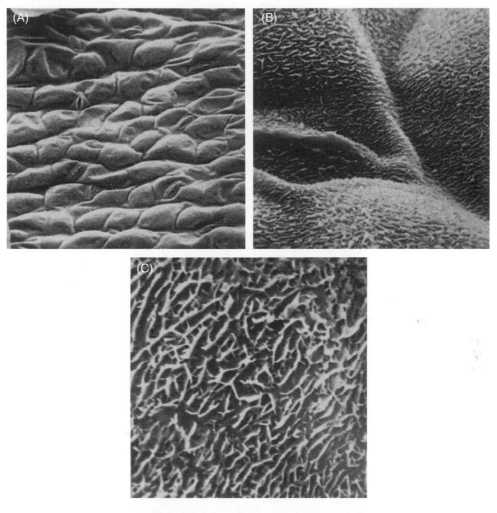

Figure 3.1   The upper leaf surface of fat hen (*Chenopodium album* L.) as shown by scanning electron microscopy at different magnifications: (A) ×540, (B) ×5450 and (C) ×11,000. Source: Taylor, F.E., Davies, L.G., and Cobb, A.H. (1981) An analysis of the epicuticular wax of *Chenopodium album* leaves in relation to environmental change, leaf wettability and the penetration of the herbicide bentazone. *Annals of Applied Biology* **98**(3), 471–478. doi: 10.1111/j.1744-7348.1981.tb00779.x. Reproduced with permission of John Wiley & Sons.

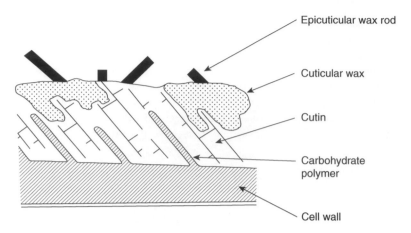

Figure 3.2 A generalised scheme of cuticle structure. Source: Price, C.E. (1982) A review of the factors influencing the penetration of pesticides through plant leaves. In: Cutler, D.F., Alvin, K.L. and Price, C.E. (eds) *The Plant Cuticle*. New York: Academic Press, pp. 237–252. Reproduced with permission of Elsevier.

Table 3.1 Most common epicuticular wax components (after Holloway, 1993).

Class	Formula	Range of n
n-Alkanes	CH_3-$(CH_2)n$-CH_3	C_{17}–C_{35} (often C_{29} or C_{31})
n-Alkyl-monoesters	CH_3-$(CH_2)n$-COO-$(CH_2)n$-CH_3	C_{32}–C_{72}
n-Aldehydes	CH_3-$(CH_2)n$-CHO	C_{16}–C_{34} (often C_{26} or C_{28})
n-1-Alkanols	CH_3-$(CH_2)n$-OH	C_{18}–C_{36} (often C_{26} or C_{28})
n-Alkanoic acids	CH_3-$(CH_2)n$-$COOH$	C_{14}–C_{36} (often C_{26} or C_{28})
Less common components include:		
n-Ketones	$$CH_3\text{-}(CH_2)n\text{-}\overset{\overset{\displaystyle O}{\|}}{C}\text{-}(CH_2)n\text{-}CH_3$$	C_{23}–C_{38} (often C_{29} or C_{31})
n-sec-Alcohols	$$CH_3(CH_2)n\text{-}\overset{\overset{\displaystyle OH}{\|}}{C}H\text{-}(CH_2)n\text{-}CH_3$$	C_{21}–C_{33} (often C_{29} or C_{31})
β-Diketones	$$CH_3\text{-}(CH_2)n\text{-}\overset{\overset{\displaystyle O}{\|}}{C}\text{-}CH_2\text{-}\overset{\overset{\displaystyle O}{\|}}{C}\text{-}(CH_2)n\text{-}CH_3$$	C_{29}–C_{33}

Waxes found on the surface of the cutin matrix are termed the epicuticular waxes and have a very diverse structure and composition. They can easily be removed by brief immersion of the leaf in organic solvents and analysis reveals a complex mixture of very long-chain fatty acids (VLCFAs), hydrocarbons, alcohols, aldehydes, ketones, esters, triterpenes, sterols and flavonoids (Table 3.1 and see Holloway, 1993; Post-Beittenmiller, 1996). Alkanes and ketones predominate in leek and brassica leaf epicuticular waxes, but are seldom observed in barley or maize. Similarly, peanuts are enriched in alkanes compared to maize where primary alcohols are abundant (Table 3.2).

Table 3.2 Variations in epicuticular lipid classes. Values are percentages of total.

Class	Leek	Barley	Maize	Brassica	Peanut	*Chenopodium album*
Fatty acids	6.4	10.3	0	1.9	38.1	0
Aldehydes	18.0	1.7	20.0	3.9	2.4	30.3
Alkanes	31.0	0	1.0	40.3	35.7	6.6
sec-Alcohols	0	0	0	11.9	0	0
Ketones	51.8	0	0	36.1	0	0
Primary-alcohols	0	83.0	63.0	1.9	23.8	44.7
Esters	0	4.7	16.0	3.9	0	17.7

Source: Post-Beittenmiller, D. (1996) Biochemistry and molecular biology of wax production in plants. *Annual Review of Plant Physiology and Plant Molecular Biology* **47**, 405–430.

Table 3.3 Common cutin monomers, normally C16 and C18, of fatty acids.

Monomer type	Abundance (%)
Unsubstituted fatty acids	1–25
ω-Hydroxy fatty acids	1–32
α,ω-Dicarboxylic acids	<5
Epoxy fatty acids	0–34
Polyhydroxy fatty acids	16–92
Polyhydroxy dicarboxylic acids	Trace
Fatty alcohols	0–8
Glycerol	1–14
Phenolics (ferulic acid)	0–1

Source: modified from Pollard, M., Beisson, F., Li, Y. and Ohlrogge, J.B. (2008) Building lipid barriers: biosynthesis of cutin and suberin. *Trends in Plant Science* **13**, 236–246.

When one class of homologue predominates, characteristic crystals of epicuticular wax form, which are very distinctive as rods, granules, crusts or aggregates. These structures may not be uniformly distributed over the whole leaf surface and differences may exist between upper (adaxial) and lower (abaxial) surfaces (e.g. Figure 3.1), and are often less evident on stomatal guard cells. Their presence often gives the leaf a dull or transparent appearance, while leaves with no epicuticular wax projections appear shiny or glossy.

Cutin is a polyester based on a series of hydroxylated fatty acids, commonly with 16 or 18 carbon atoms, the relative proportion of which varies according to species. Although found in all plants, it is one of the least understood of the major plant polymers. In most cutins the dominant monomer is an ω-hydroxy fatty acid, the self-polymerisation of which will produce a linear polyester chain (Table 3.3). The mid-chain oxygen-containing functional groups (such as hydroxyl and epoxy) may be esterified to other ω-hydroxy fatty acids by polyester synthases, creating a branched structure. We still do not know, however, how these components are precisely arranged or how they contribute to cutin function. The outer surface of the cuticle is most lipophilic and becomes more hydrophilic towards the underlying epidermal cells. Various workers have suggested that polar pathways may exist through the cuticle where herbicide movement can take place. Miller (1985) has reported the occurrence of such channels in many plant families, but their significance in herbicide uptake remains to be demonstrated. Further transcuticular pathways may be provided by

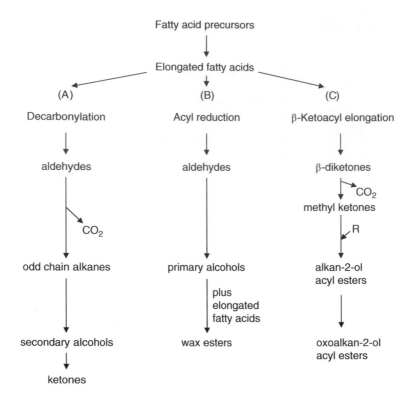

Figure 3.3 An overview of the three primary pathways of epicuticular wax biosynthesis. Source: Post-Beittenmiller, D. (1996) Biochemistry and molecular biology of wax production in plants. *Annual Review of Plant Physiology and Plant Molecular Biology* **47**, 405–430.

carbohydrate polymers extending into the cuticle from the cell wall (Figure 3.2), but their role in herbicide movement is again obscure.

The biosynthesis of the cuticular waxes occurs almost exclusively in the epidermal cell cytoplasm. As detailed in Table 3.1, the majority of epicuticular wax components are derived from VLCFAs (Chapter 8), primarily 20–32 carbons in length. They are synthesised from C_{16}–C_{18} plastidic fatty acid precursors by cytoplasmic membrane-bound elongases, using malonyl-CoA as the two-carbon donor. The wide diversity of wax components arises from the operation of three parallel pathways (Figure 3.3).

The use of forward and reverse genetic approaches in *Arabidopsis* has led to the identification of oxidoreductase and acyltransferase genes involved in cutin biosynthesis (as reviewed by Pollard *et al.*, 2008).

There appear to be three enzyme families involved:

- fatty acid oxidases of the CYP86A subfamily;
- an acyl-activating enzyme of the long-chain acyl-CoA synthase (LACS) family;
- acyltransferases of the glycerol-3-P acyl-CoA *sn*-1 acyltransferase family (GPAT).

These reactions of acyl activation, ω-oxidation of acyl chains and acyl transfer to glycerol could take place in several sequences and pathways, although the ω-oxidised-acylglycerols are considered to be the polyester building block.

These enzymes are thought to be located in the endoplasmic reticulum, although the cellular site of the polyester synthases is not known. How the waxes find their way to the cuticle remains uncertain. Passage along the polar pores has been suggested, or simple diffusion may occur through spaces in the cell wall. Lipid transfer proteins (LTPs) have now been demonstrated in the epidermal cells of several species and located in the cell wall. It is speculated that these LTPs transport lipids through the endoplasmic reticulum and deposit them outside the cell, although this pathway remains to be proven. LTPs are small (9–10 kDa), basic proteins that are widespread and abundant in plants, constituting as much as 40% of the soluble protein pool in maize seedlings. They consist of 91–95 amino acids differing widely in sequence but always containing four disulphide bridges, with a three-dimensional structure that contains an internal hydrophobic cavity. They are able to bind acyl chains and transport them from the endoplasmic reticulum to the cell wall for cutin biosynthesis (Kader, 1997).

Cuticular lipid biosynthesis is very sensitive to environmental conditions and signals such as light intensity, photoperiod, humidity, chilling, soil moisture content and season, which all have an effect on cuticular development and hence herbicide efficacy. In particular, the change from high to low humidity can trigger wax production by more than an order of magnitude, an important factor to consider when extrapolating data on herbicide trials from the glasshouse to the field environment.

Generally, the cuticle will thicken during conditions that are unfavourable to plant growth, including low temperatures, photon flux density and water availability, and so herbicide absorption is maximised when opposite conditions prevail. Users of herbicides are always advised to read the label on the herbicide container for helpful and invariably comprehensive details that recommend the best environmental conditions for use to obtain optimal weed control. Generally, these suggest application when the weeds and crops are actively growing, during the spring and early summer for foliar applications, and when rainfall is not imminent. Extremes of temperature should be avoided as they cause plant stress.

3.3   Physicochemical aspects of foliar uptake

Most chemicals penetrate most plants poorly when applied alone and so require an adjuvant for uptake to occur. The adjuvant increases the amount of uptake into the lipophilic environment of the cuticle. This uptake can be predicted from the octanol–water partition coefficient (K_{ow}), the dissociation constant pK_a and the parameter $\Delta \log P$. This latter term, derived from work on the penetration of the blood–brain barrier, viz.

$$\Delta \log P = \log P_{ow} - \log P_{alk}$$

subtracts the alkane/partition coefficient (generally determined using hexane or cyclohexane) from the octanol–water value and is thought to have considerable relevance to cuticular penetration, since epicuticular wax is essentially a hydrocarbon in character. Thus, Briggs and Bromilow (1994) consider that permeability through the cuticular wax varies inversely with $\Delta \log P$, while permeability along the aqueous, polar cuticular route is inversely related to $\log K_{ow}$. Since $\Delta \log P$ is positive for most compounds, those entering the

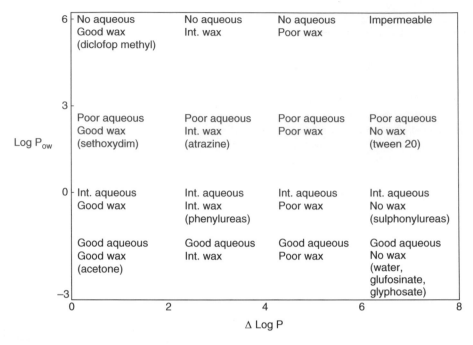

Figure 3.4 Effect of logP_{ow} and ΔlogP on foliar uptake through the cuticle via epicuticular wax or aqueous routes. Int., intermediate; names in parentheses refer to estimates for some active ingredients and adjuvants. Source: Briggs, G.G. and Bromilow, R.H. (1994) Influence of physicochemical properties on uptake and loss of pesticides and adjuvants from the leaf surface. In: Holloway, P.J., Rees, R.T. and Stock, D. (eds) *Interactions Between Adjuvants, Agrochemicals and Target Organisms*. Berlin: Springer, pp. 1–27. Reproduced with permission of Springer Nature.

wax then move readily into the cuticle, as they are more strongly absorbed by the octanol-like cuticle. On the other hand, uptake via the aqueous route occurs for compounds with high water solubility. This relationship is illustrated in Figure 3.4.

Surfactants are amphipathic molecules (i.e. they possess a hydrophilic head and a hydrophobic tail) and are often added to aqueous formulations for many purposes, including cuticle retention on leaf surfaces, absorption and penetration to the target site. Four classes of these surface-active agents have been identified:

1 *Anionic*. Here, surface-active properties are provided by a negatively charged ion. For example, a hydrophobic group is balanced by a negatively charged hydrophilic group, such as a carboxyl (–COO$^-$).

2 *Cationic*. In this case, the surface-active properties are provided by a positively charged ion. Thus, a hydrophobic group is balanced by a positively charged hydrophilic group, such as a quaternary ammonium.

3 *Non-ionic*. No electrical charge is evident. Thus, a hydrophobic group consists of alkyl-phenols, alcohols or fatty acids and is balanced by non-ionisable hydrophilic groups, such as ethylene oxide (–CH$_2$CH$_2$O–).

4 *Amphoteric*. These molecules have hydrophilic groups with the potential to become cationic in an acid medium or anionic in alkaline conditions.

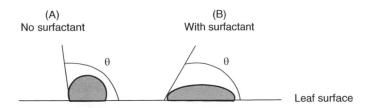

(A)
No surfactant

(B)
With surfactant

θ   θ

Leaf surface

Figure 3.5 Schematic representation of droplets on a leaf surface in the absence (A) and presence (B) of a surfactant. θ, Contact angle.

The ratio between the hydrophilic and lipophilic groups is termed the hydrophilic–lipophilic balance (HLB). Compounds of low HLB are relatively water-soluble and so surfactants can be selected for specific purposes. They can also be further categorised as spray modifiers or activators. Spray modifiers reduce the surface tension of the spray droplets and so improve the wetting and spreading properties of the formulation, resulting in a greater degree of retention on the leaf. Activators are added to improve foliar absorption, an example being a range of polyoxyethylenes with HLB values of 10–13.

How surfactants aid cuticular penetration still remains uncertain and large differences are observed between plant species. The most obvious effect of surfactants is to reduce leaf surface tensions and contact angles, thus increasing the spread of the chemical to cover a greater proportion of the leaf surface. A high contact angle (Figure 3.5A) between the droplet and the surface also means that the droplet is easily dislodged and runs off the leaf. The lowest values for contact angle and surface tension are reached at the critical micelle concentration of the surfactant, usually about 0.1–0.5%, on a volume to volume proportion of the formulation. At such conditions contact angles may be lowered, for example, from 110 to 60° (Figure 3.5B) and surface tensions from 70 to 35 mN m^{-2}.

3.4   Herbicide formulation

Herbicides will not kill weeds as active ingredients unless they are formulated to reach the target site. In the broadest sense, formulated means combined with a liquid or solid carrier so that they can be applied uniformly, transported and perform effectively. Knowles (2008) summarises the main objectives of a formulation as 'To provide the user with a convenient, safe product which will not deteriorate over a period of time, and to obtain the maximum activity inherent in the active ingredient'. He lists the main factors governing the choice of formulation for an active ingredient as:

- physico-chemical properties;
- biological activity and mode of action;
- safety in use;
- formulation costs; and
- market preference.

In addition, formulation ingredients should be inert and produce a stable formulation with at least a 2 year shelf-life during storage under varying climatic conditions (Knowles, 2008).

The formulation, then, is a physical mixture of the active ingredient and formulants, and the formulation chemist has a very broad brief, namely to ensure that the product retains activity with time, is easy and safe to use, and is as cheap as possible. Thus, the chemical may need to be stored below freezing-point in the winter and yet be exposed to 30°C in the summer, and so freezing and combustion with loss of activity and safety are unacceptable. Similarly, the control of foaming when the formulation is mixed is necessary, since excessive foam leads to lack of accuracy of measurement and possible environmental contamination. Thus, the formulation chemist should be included early on in the development of a new product to ensure optimum delivery and biological efficacy.

A single active ingredient is often sold in many formulations, the choice depending on the use. Label recommendations are invariably helpful and detailed. Choice depends on the application equipment available; the risk of spray drift and run-off, both influenced by the weather, proximity of water courses and other crops; the likelihood of wind or rain; safety of the personnel involved and nearby organisms, especially pollinators; growth patterns of the crop and weeds; timing within the growing season; and, perhaps most importantly, the cost.

Granules are most often used in soil treatments where their weight will enable them to pass through the foliage to the ground. They should not be applied to frozen soil nor to steep slopes. On the other hand, they can be relatively expensive, need moisture to activate the herbicide, and can be attractive to non-target organisms, such as birds. Other formulations with their main advantages and limitations are presented in Table 3.4.

Formulations are either liquid or solid, and many variations are possible. Aqueous concentrates contain a soluble active ingredient in a carrier, usually water. The concentration is dependent on the solubility of the active ingredient. An emulsifiable concentrate (EC) consists of a herbicide dissolved in an organic solvent with an emulsifier added to create an oil/water emulsion when water is added. The EC is commonly 60–65% (by weight) of herbicide dissolved in 30–35% solvent with 3–7% emulsifier, and typically forms an opaque or milky emulsion when added to water. The solvent itself can also be phytotoxic and aid herbicide activity. Dusts are finely powdered dry materials that provide good surface coverage. In microcapsules, the herbicide is enclosed in small, porous polymer particles that release the herbicide slowly over the growing season. Suspension concentrates are concentrated aqueous dispersions of herbicides and are virtually insoluble in water. They contain no organic solvent but require other agents to ensure suspension. Ultra-low-volume formulations, also referred to as controlled droplet application, entail the use of specialist sprayers depositing very low volumes on leaf surfaces, often by electrostatic attraction. Water-dispersible granules are solid formulations as fine particles containing 2–10% active ingredient. They are applied directly and are valued for soil-applied herbicides. Wettable powders are finely divided solids that are easily suspended in water. The active ingredient is added to an inert material, such as a clay, and a wetting agent. The relative strengths and weaknesses of the main herbicide formulations are listed in Table 3.4.

Concern over the use of solvents and other ingredients has created pressure for the chemical industry to change or re-examine existing formulations. Formulators are now challenged with investing in solvent-free formulations, which will force a shift away from some traditional solvent-based emulsifiable concentrates to water-dispersible granules and aqueous concentrates (e.g. Morgan, 1993). For example, Syngenta changed their formulation of the graminicide fluazifop-butyl from a solvent-based emulsifiable concentrate to a

Table 3.4   Herbicide formulations.

Name	Ingredients	Advantages	Disadvantages
Aqueous concentrate (SL)	• Active ingredient • Wetter • Surfactant • Water	• Cheap and easy to produce • Low volatility • Low phytotoxicity • Easy to mix	• Expensive to pack and transport • Frost sensitive • May corrode metal • Cannot contain high active ingredient concentrations
Dust (DP)	• Active ingredient • Carrier	• Cheap and easy to produce • Easy to apply • Safe to handle	• Risk of user contamination and drift • Bulky to store and transport • Flowability affected by damp
Emulsifiable concentrate (EC)	• Active ingredient • Solvent • Emulsifier • Activator	• Easy to produce • Easy to handle and mix • Good when active ingredient is insoluble in water • High efficacy	• Expensive to pack and transport • Fost sensitive • Risk of thickening • May corrode containers • Can be phytotoxic • Volatile
Microcapsule (CS)	• Active ingredient • Solvent • Emulsifier • Thickener • Anti-foaming agent • Preservative	• Cheap to produce • Low dusting • Easy to handle • Low solvent	• Expensive production equipment • Frost sensitive • May thicken at high temperatures • Expensive to package
Suspension concentrate (SC)	• Active ingredient • Diluent • Wetting agent • Dispersant • Anti-freezing agent • Anti-foaming agent • Preservative • Water	• No solvent • High active ingredient concentration • Easy to mix and store • Compatible with aqueous concentrates	• May settle out in storage • Frost sensitive • Can cause phytotoxicity
Ultra-low volume (ULV/CAD)	• Active ingredient • Oil • Viscosity modifier	• Applied in very small quantities • Ready to use • Cheap to store, pack and transport	• Application can be labour intensive • Risk of patchy application • High toxicity • Needs specialised equipment
Water-dispersable granule (WG)	• Active ingredient • Carrier • Wetting agent	• Low dusting • Cheap to pack • Easy to handle • Frost-tolerant • No solvent	• Expensive production equipment
Wettable powder (WP)	• Active ingredient • Carrier • Wetting agent	• Cheap and easy to produce and pack • Easy to handle • Frost-tolerant • No solvent	• Produces dust • Difficult to measure and mix

Source: Briggs, G.G. and Bromilow, R.H. (1994) Influence of physicochemical properties on uptake and loss of pesticides and adjuvants from the leaf surface. In: Holloway, P.J., Rees, R.T. and Stock, D. (eds) *Interactions Between Adjuvants, Agrochemicals and Target Organisms*. Berlin: Springer, pp. 1–27. Reproduced with permission of Springer Nature.

water-based, emulsion-in-water formulation. The advantage of the new emulsion-in-water formulation is that the herbicide is now dissolved in much less solvent and surrounded by emulsifying agents in water, and this leads to improved handling, transport and storage characteristics.

The possible contamination of drinking water is also a highly emotive issue in some European countries, such as Denmark, where groundwater provides a large proportion of the drinking water. In this case, considerable effort is needed to reduce herbicide leaching, and formulations such as slow-release microcapsules may show promise. Solid formulations are also possibly safer than liquid ones and so may be expected to gain an increased market share.

The availability and cost of formulation technology is a further consideration. Micro-encapsulation, for example, requires relatively expensive technology that may not be easily accessible outside of Europe and the USA. The trend towards safer, low solvent and solid formulations, however, should be applauded, and perhaps further encouraged by stricter controls being imposed at the product registration stage. Micro-encapsulation is an evolving formulation technology. For example, trifluralin is an effective pre-plant herbicide for the control of annual grasses and broad-leaf weeds in many crops. It is, however, sensitive to photodegradation and volatilisation, and so needs to be rapidly incorporated into the soil. Daneshvari *et al.* (2021) report controlled-release formulations of trifluralin by micro-encapsulation in a polyurethane network. Resultant formulations were shown to significantly reduce volatilisation and photodegradation.

Further concern has recently been expressed about possible adverse consequences arising from the release into the environment of molecules with oestrogenic properties. Implicated agrochemicals include the alkylphenyl polyethoxylates, in particular the nonylphenols, affecting humans (lower sperm quality and count, testicular and male breast cancer), fish (including gender changes) and reptiles (developmental abnormalities). The risks need to be clearly defined and many controlled experiments are needed, so that both the industry and public at large are aware of the real risks involved. Clearly, the agrochemical industry is very concerned with this issue and the replacement of the nonylphenols with safer products is awaited.

The convention used to name the non-ionisable alcohol ethoxylates is to give the alcohol chain followed by the number of $-CH_2CH_2O-$ units. Ethoxyethanol, $CH_3CH_2OCH_2CH_2OH$, is therefore $C_2 E_1$.

Volatility is expressed as the equivalent hydrocarbon (EH), which is the number of carbon atoms in the benzene or alkane groups with the same expected boiling point and vapour pressure. Thus, chlorine has a carbon equivalent of 2 and hence the EH of chlorobenzene is 8. The boiling point of chlorobenzene is 132°C, similar to that of the C8 hydrocarbons *p*-xylene (135°C), octane (125°C), dimethylcyclohexane isomers (120–125°C) and allylcyclopentane (125°C). Oxygen in an ether link and silicon are both equivalent to one carbon. So, for the ethoxylated trisiloxanes contained in Silwet L-77 (mean ethylene oxide content 8, oligomer range 2–14; Figure 3.6).

It is predicted that a compound EH20 will only have a half-life of 30 min on a leaf, while EH25 will undergo considerable vapour loss over 1 day and EH30 is equivalent to the loss of a few g ha^{-1} per day, which may be acceptable in the field.

These values explain why many common solvents are all highly volatile with a very short-lived influence on penetration (e.g. xylene and chlorobenzene (EH8), acetone and isopropylamine (EH5), heptylacetate, acetophenone and dimethyl sulphoxide (EH11)).

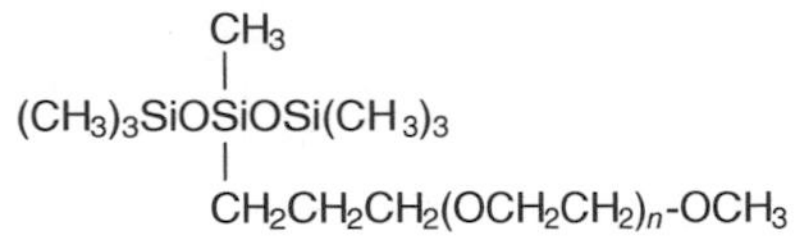

$$CH_3$$
$$|$$
$$(CH_3)_3SiOSiOSi(CH_3)_3$$
$$|$$
$$CH_2CH_2CH_2(OCH_2CH_2)_n\text{-}OCH_3$$

Figure 3.6   Structure of ethoxylated trisiloxanes in Silwet L-77. When $n = 2$, the equivalent hydrocarbon is C15 + 3 (Si) + 5 (ether O), i.e. EH = 23.

Table 3.5   Estimated physical properties of ethoxylated alcohols and allyl phenols.

Adjuvant	EH	log P_{ow}	Water solubility (mol 1^{-1})	logP_{alk}
C8 alcohol				
E2	19	2.7	2×10^{-3}	1.1
E15	28	2.4	4×10^{-3}	−0.1
E10	>30	1.9	1×10^{-2}	−2.1
E15	>30	1.4	4×10^{-2}	−4.1
E20	>30	0.9	1×10^{-1}	−6.1
C10 alcohol				
E5	30	3.4	4×10^{-4}	1.1
E10	>30	2.9	1×10^{-3}	−0.9
E15	>30	2.4	4×10^{-3}	−2.9
E20	>30	1.9	1×10^{-2}	−4.9
C12 alcohol				
E5	>30	4.4	4×10^{-5}	2.3
E10	>30	3.9	1×10^{-4}	0.3
E15	>30	3.4	4×10^{-4}	−1.7
E20	>30	2.9	1×10^{-3}	−3.7
C18 alcohol				
E5	>30	7.4	4×10^{-8}	5.9
E10	>30	6.9	1×10^{-7}	3.9
E15	>30	6.4	4×10^{-7}	1.9
E20	>30	5.9	1×10^{-6}	−0.1
Octylphenol				
E5	>30	4.1	7×10^{-5}	1.6
E10	>30	3.6	2×10^{-4}	0.1
E15	>30	3.1	8×10^{-4}	−2.4
E20	>30	2.6	2×10^{-3}	−4.4
Nonylphenol				
E5	>30	4.4	4×10^{-5}	2.5
E10	>30	3.9	4×10^{-4}	0.5
E15	>30	3.4	4×10^{-4}	−1.5
E20	>30	2.9	1×10^{-3}	−3.5

Source: Briggs, G.G. and Bromilow, R.H. (1994) Influence of physicochemical properties on uptake and loss of pesticides and adjuvants from the leaf surface. In: Holloway, P.J., Rees, R.T. and Stock, D. (eds) *Interactions Between Adjuvants, Agrochemicals and Target Organisms*. Berlin: Springer, pp. 1–27. Reproduced with permission of Springer Nature.

Fatty acids, esters and alcohols are less volatile and possess higher lipophilicity (e.g. butyl oleate has an EH = 24 and log P_{ow} = 9.5). However, small chain fatty acids (e.g. C_8–C_{10}) appear phytotoxic in their own right.

Perhaps the most common surfactant adjuvants are the ethoxylated alcohols and phenols (Table 3.5). Other than the short-chain alcohols with E2–E15, volatility does not present a problem. The log P_{ow} values suggest that the C_8–C_{10} alcohol ethoxylates are potentially mobile in plants following penetration, but the C_{12} and higher alcohol ethoxylates and the alkylphenol ethoxylates with E-values <15 would not be expected to be mobile.

Increasing ethoxylation shows only a small decrease in $logP_{ow}$, equivalent to a small increase in water solubility. However, $logP_{alk}$ decreases sharply with ethoxylation by about 0.4 per E unit, indicating decreased lipophilicity, and so $\Delta logP$ increases. Consequently, penetration rate should be lower for the long-chain higher ethoxylates, which are too polar to penetrate via the epicuticular wax and at the same time too water-insoluble to penetrate by an aqueous route.

An adjuvant is anything added to a herbicide solution, to modify or enhance its activity. It is worth noting that a foliar application of an agrochemical is a very inefficient process. Perhaps only 15% will reach a target leaf and only 0.1% will reach the target site. Consequently, the use of adjuvants can lead to a more efficient, cost-effective and environmentally benign product.

Activator adjuvants include surfactants, wetting agents, penetrants and oils added to the herbicide solution in the spray tank. Spray modifier adjuvants include stickers, spreaders, thickening agents, film formers and foams, usually present as part of the formulation. Some of these are briefly considered below.

Surfactants reduce the surface tension of spray droplets on leaves and so improve the coverage of the spray. Agricultural surfactants do not form precipitates, as is often the case with soaps and detergents, and do not foam in the spray tank. Non-ionic surfactants are good dispersal agents and have low toxicity to plants and animals. They are stable and have no overall charge. The organo-silicone surfactants are perhaps the most effective at lowering surface tension so that stomatal penetration can take place.

Oils, such as crop oils, crop oil concentrates and methylated seed oils, can also improve spray coverage. They can keep the leaf surface moist for longer than water alone, enhancing herbicide uptake. An example is the control of *Chenopodium album* by the herbicide bentazone in the presence of Actipron.

The most commonly used spray modifies are stickers, drift inhibitors and anti-foaming agents. Stickers cause the herbicide to adhere to the foliage, so reducing run-off from the leaves and stems. Drift-inhibitors or thickeners cause the spray solution to be more cohesive and reduce the number of small droplets that may drift away from the crop and the field. Anti-foaming agents reduce froth production during tank filling and agitation, so that the spray tank can be filled more easily. They are usually silicone-based and used in small amounts. The reader is referred to Knowles (2008) for more examples of formulation types and recent trends in their development and use.

Also abundant on leaf surfaces are trichomes (leaf hairs) and stomata (pores for gaseous exchange). The involvement of the former in foliar penetration remains unknown. Trichomes may form a dense, impenetrable mat on the leaves of some species and so may represent a further physical barrier to herbicide uptake, since they can prevent or slow down the passage of the herbicide formulation to the cuticle. On the other hand, stomata have frequently been implicated in foliar uptake and the general view held in the 1970s was that stomata could provide a direct route for herbicide penetration. In order to do so, a surface tension of less than 30 mN m^{-2} would be necessary, which is seldom the case with most herbicide formulations. An exception to this is the organo-silicones, which includes the ethoxylated trisiloxane, Silwet L-77. These molecules cause excellent surface spreading and so can enhance or maximise the deposition of the active ingredient. Their concentration is crucial, however, since high doses may lead to an excessive run-off. These 'super-spreaders' do allow sufficient reductions in surface tension to permit stomatal infiltration. This response can only take place immediately following application, while the spray deposit remains in a liquid form, after which cuticular uptake is the sole pathway (Stephens *et al.*, 1992). This may be a useful property when rainfall is expected within a few hours after application, imparting rainfastness to the formulation, rather than rain washing off the herbicide from the leaf.

A further example of developments with uptake and formulation is with glyphosate, used for total vegetation control. This important herbicide is traditionally formulated as an isopropylamine salt with a cationic surfactant and exhibits relatively slow uptake into perennial grass weeds. A new trimesium (i.e. trimethyl sulfonium) salt with a novel polyalkylglucoside adjuvant has been claimed to show higher activity owing to enhanced uptake, and to be rainfast within an hour after application (Figure 3.13).

Since most adjuvants evolved from empirical screening, it is important for the user and the public to be informed how the additive works and why it is included in the formulation. Unfortunately, this information is seldom, if ever, available and there is a current lack of regulatory harmonisation, especially in Europe, where different laws operate in the different Member States. Forthcoming legislation will mean more stringent registration requirements for adjuvants in the future. Holloway (1994), in anticipation of these changes, has proposed a wide-ranging list of possible criteria for the safe and efficacious use of existing and new products. He proposes several important questions: does the adjuvant affect spray atomisation, droplet evaporation during flight, droplet drift, deposition retention or spreading, overall target coverage, rate of droplet evaporation, physical form and moisture content of the spray deposit, rate and amount of uptake, translocation and metabolism? He also suggests the need to know: on what plant species the adjuvant is most beneficial; at what concentration it is most effective; whether there are any problems with intrinsic phytotoxicity; whether the product is compatible with other formulants; how safe the adjuvant is to use; what the toxicity is to non-target organisms; how rapidly the product is biodegraded in soil and plants; and what the potential cost benefits are. This information would ensure higher standards for adjuvant efficacy and provide a more rational and scientific footing for future adjuvant use.

The hypothesis that surfactants alone might be phytotoxic to plant tissues was tested at the molecular level by Madhou *et al.* (2006), who studied gene expression in *Arabidopsis thaliana* in response to a foliar application of the etheramine surfactant NUL1026 at 0.2% (w/v). They found that the expression of 196 genes was significantly altered 1 h after treatment. A number of genes were upregulated, coding for enzymes involved in both detoxification and signalling pathways. This is consistent with a stress response of overlapping gene expression to wounding, pathogen, abiotic stress and hormone treatment. It may therefore be that surfactant use prepares the plant defences for external attack. Follow-up studies are awaited to further understand plant response to xenobiotic attack.

3.5   Uptake by roots from soil

The uptake of herbicides by plant roots and their movement in the xylem or phloem is becoming increasingly well understood on the basis of their physicochemical properties. Uptake can take place from the soil via either the air or water phase. Knowing that diffusion constants are about 10,000 times greater in air than in water, and the ratio of concentration in air and water, it is possible to predict which phase is most important in root uptake. Hence, herbicides such as trifluralin, EPTC, triallate and bifenox are likely to move as vapour in moist soils, while movement will be by diffusion in the aqueous phase by monuron and simazine. Lipophilic herbicides with a high vapour pressure will be strongly absorbed by roots from a moist soil. Volatility should also be considered and soil incorporation is often necessary (e.g. with trifluralin) to prevent rapid loss by volatilisation.

Uptake via the soil aqueous route has been studied in detail by Briggs *et al.* at Rothamsted Experimental Station, UK, using barley plants grown in nutrient solution and radiolabelled test compounds (e.g. Briggs *et al.*, 1982, 1987; Briggs and Bromilow, 1994). Distribution of a non-ionised compound between the roots and the bathing solution was defined by the root concentration factor (RCF), where

$$RCF = \frac{\text{Concentration in roots}}{\text{Concentration in nutrient solution}}$$

It was found that the RCF was directly related to the log of the octanol–water coefficient ($\log P_{ow}$), thus uptake increased with increasing lipophilicity.

The uptake of acidic compounds by roots is very different from that of the non-ionised herbicides above, but is dependent on the pH of the soil solution. Thus, the uptake of 2,4-D into barley roots over a 24 h period from nutrient solution was 36 times greater at pH 4.0 than at pH 7.0 (Briggs *et al.*, 1987). This may be explained as an ion-trap effect, whereby weak acids are accumulated in compartments of higher pH by virtue of the greater permeation rates across membranes of the undissociated form compared with the anion (Figure 3.7).

Once inside the root hair, the herbicide has to be transported to the vascular system for long-distance transport to the target site. Two routes are possible via the apoplastic or symplastic pathways. The former entails movement down a concentration gradient along the cell walls, while the latter entails the cytoplasmic continuity of the root cortical cells via plasmodesmata. Water and solutes cannot enter the xylem by an entirely apoplastic route, but must move through the symplast of the endodermis. This is because the tangential walls of the endodermal cells are thickened by the deposition of suberin to form the water-impermeable Casparian strip (Figure 3.8). Efficiency of transport from the root cortex to

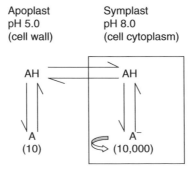

Apoplast
pH 5.0
(cell wall)

Symplast
pH 8.0
(cell cytoplasm)

Figure 3.7   Accumulation of a weak acid (pK_a 4.0) within root cells by the ion-trap effect.

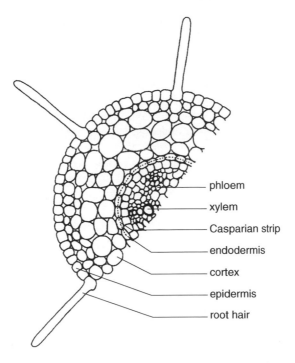

Figure 3.8   Cross-section of a dicotyledonous root, showing the position of the endodermis and the Casparian strip. Source: Bromilow, R.H. and Chamberlain, K. (1991) Pathways and mechanisms of transport of herbicides in plants. In: Kirkwood, R.C. (ed.) *Target Sites for Herbicide Action*. New York: Plenum, pp. 245–284. Reproduced with permission of Springer Nature.

the xylem is low, at less than 10%, which further illustrates the difficulty that weak acids have in crossing biological membranes when they are largely ionised at a physiological pH.

3.6   Herbicide translocation from roots to shoots

Movement in plants can be predicted from pK_a and $logP_{ow}$ values and is therefore largely controlled by the physicochemical properties of the herbicide (Figure 3.4). As a general rule, lipophilic compounds (i.e. $logP_{ow}$ >4) are non-systemic, while compounds of intermediate lipophilicity ($logP_{ow}$ −0.5 to +3.5) move in the xylem, and weak acids are also mobile in the phloem to the physiological sinks, or growth points, in the plant. Most compounds enter the phloem quite freely, in contrast to sucrose, which requires an active transport system, and leave it equally readily. Movement is therefore according to sink strength. Movement in the xylem is at least 50–100 times faster than in the phloem and is therefore dependent on the transpiration stream at any particular time. These rates can reach huge values, especially in trees where, for example, it has been calculated that an oak tree can transpire a ton of water a day, or a maple tree growing in the open was measured to transpire 48 gallons of water per hour! Even in young seedlings (two to three leaf stage) of black-grass, rates of 0.25 g m^{-2} leaf h^{-1} have been recorded from plants growing in greenhouse conditions (Sharples *et al.*, 1997).

Thus, water can move through the xylem at velocities of greater than 10 m h^{-1}, while movement through the phloem may be 50–100 times less.

The above patterns of systemicity have been confirmed for many herbicides by autoradiography using radiolabelled molecules, and further details for individual herbicides can be found in Bromilow and Chamberlain (1991).

Metabolism of a herbicide within the plant will also influence its movement, since metabolism generally reduces lipophilicity. Whereas aryl hydroxylation may be expected to increase the mobility of the parent molecule, its subsequent conjugation to glucose, for example, may create a less mobile glucoside that becomes compartmentalised or inactivated within the cell vacuole. An example of the consequences of metabolism on mobility is illustrated in Table 3.6.

3.7   A case study: the formulation of acids

Most commercial formulations of the phenoxyalkanoic acids or 'auxin-type' herbicides (Chapter 7) contain the active ingredient (e.g. Figure 3.9) in the salt or ester form (Figure 3.10). Of the salts, the amine forms are commonly used, although the sodium, potassium and ammonium salts are also found. Some examples of the cationic amine salts

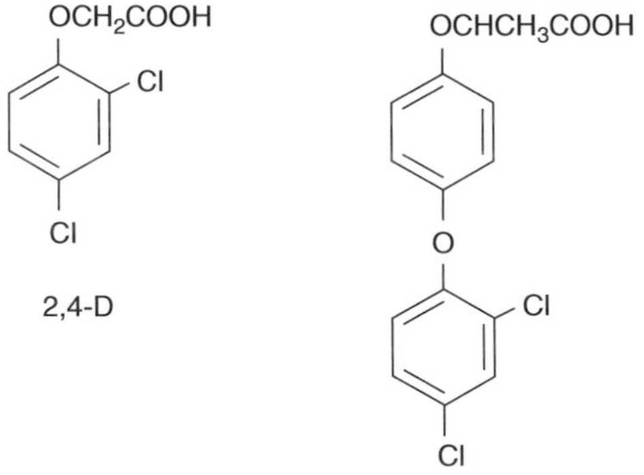

Figure 3.9 Structures of a typical phenoxyalkanoic acid (2,4-D) and an aryloxyphenoxypropanoate (diclofop-acid).

Table 3.6 Metabolism of 2,4-D butyl ester and systemicity of its metabolites (from Bromilow *et al.*, 1986 with permission).

Compound	logK_{ow}	pK_a	Mobility
2,4-D butyl ester	4.4	Non-ionised	Immobile
2,4-D acid	2.9	3.0	Xylem/phloem
2,4-D glucose	0.6	Non-ionised	Xylem
Ring-OH 2,4-D	2.2	3.0	Xylem/phloem
Ring *O*-glucosyl 2,4-D	0.0	3.0	Xylem/phloem
Diglucosyl conjugate of –COOH and ring-OH	−2.3	Non-ionised	Immobile

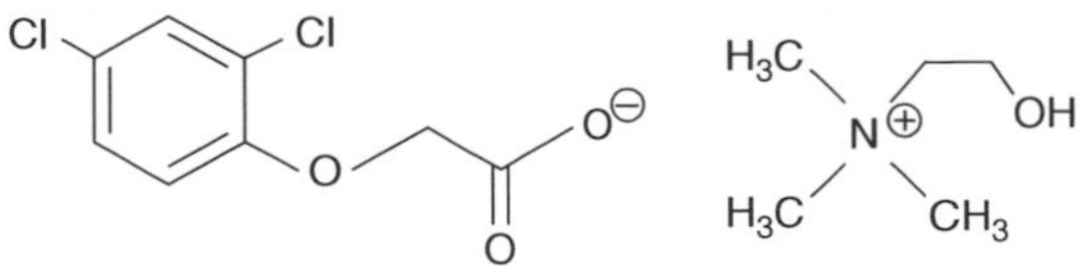

Figure 3.10 The quaternary salt of 2,4-D

Table 3.7 Amine cations of phenoxyalkanoic acid amine salts.

Organic group (R)	Cation structure and name		
CH_{3-} CH_3CH_{2-} $HOCH_2CH_{2-}$ $\overset{\displaystyle OH}{\underset{\displaystyle \|}{}}$ CH_3CHCH_2-	Methylamine Ethylamine Ethanolamine Isopropanolamine	Dimethylamine Diethylamine Diethanolamine Diisopropanolamine	Trimethylamine Triethylamine Triethanolamine Tri-isopropanolamine

Source: Loos, M.A. (1975) Phenoxyalkanoic acids. In: Kearney, P.C. and Kaufman, D.D. (eds) *Herbicides; Chemistry, Degradation and Mode of Action*. New York: Marcel Dekker, pp. 1–128.

are shown in Table 3.7. These salts are highly soluble in water and formulated as aqueous concentrates. This ensures that they can be applied at relatively high rates of active ingredient but at relatively low volume.

Phenoxyalkanoic acid esters may also be applied as emulsifiable concentrates and as such exhibit greater herbicidal efficacy than the parent acids. This is because the ester is more lipophilic and more rapidly absorbed by the target weed. The esters of low-molecular-weight alcohols (such as methyl, ethyl and propyl), however, are volatile and may cause unwanted phytotoxicity to non-target plants, including crops. This problem has been overcome with the use of low-volatility esters (Figure 3.11).

A further example has been the introduction of the butoxy-methyl ethyl ester of fluroxypyr (Figure 3.12; Snel *et al.*, 1995), which can be formulated as an emulsion-in-water or a wettable powder without the aromatic solvents required by its EC predecessor, the methylheptyl ester.

The aryloxyphenoxypropanoates, such as diclofop (Figure 3.9), are also formulated as esters, traditionally as emulsifiable concentrates that partition rapidly into the cuticle. In this case, the esters of the lower-molecular-weight alcohols are relatively stable, as evidenced by the commercialised examples diclofop-methyl, quizalofop-ethyl and fluazifop-butyl. The ester itself is not the active moiety but requires conversion to the potent acid as a prerequisite for herbicidal activity.

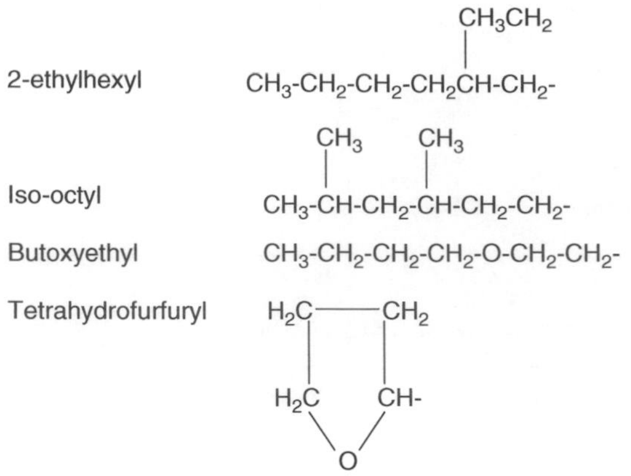

2-ethylhexyl

CH₃-CH₂-CH₂-CH₂CH-CH₂-
with CH₃CH₂ branch

Iso-octyl

CH₃-CH-CH₂-CH-CH₂-CH₂-
with CH₃, CH₃ branches

Butoxyethyl

CH₃-CH₂-CH₂-CH₂-O-CH₂-CH₂-

Tetrahydrofurfuryl

H₂C——CH₂
H₂C   CH-
   O

Figure 3.11   Examples of some low volatility esters used in herbicide formulation.

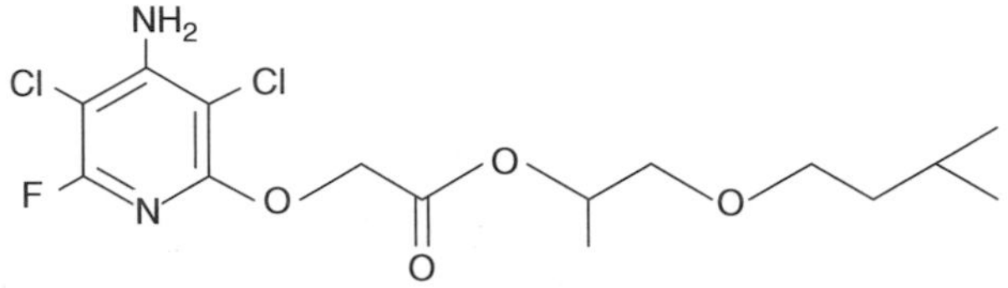

Figure 3.12   Butoxy-methyl ethyl ester of fluroxpyr. Source: Snel, M., Banks, G., Mulqueen, P.J., Davies, J. and Paterson, E.A. (1995) Fluroxypyr butoxy-1-methylethyl ester; new formulation opportunities. *Brighton Crop Protection Conference, Weeds* **1**, 27–34.

De-esterification or hydrolysis *in vivo* was first demonstrated by the pioneering work of Crafts (1960) and an enzyme responsible was first partially purified and characterised from wild oat by Hill *et al.* in 1978. These authors found that this enzyme was non-specific and could de-esterify several ester types. Hence, the esters may be regarded as 'pro-herbicides', requiring enzymatic activation in the plant tissues.

The phenoxyacetic acids (including 2,4-D, MCPA and mecoprop) have $\log K_{ow}$ values in the range of 2.9–3.6 and pK_a values close to 3.0. Thus, they move effectively in the phloem but are strongly held in adjacent tissues by ion-trapping. On the other hand, the aryloxyphenoxypropanoates are more lipophilic ($\log K_{ow}$ 3.4–4.4) and have pK_a values in the region of 3.5. As may be predicted from such lipophilic molecules, their phloem mobility is very poor. However, the potency of these herbicides at the grass meristem is such that even though less than 1% of applied dose is mobile, it is sufficient to control grass weeds (see Chapter 8, and Carr *et al.*, 1986).

In the formulation of the auxin-type herbicide 2,4-D, the dimethyl-amine salt and the 2-ethylhexyl ester account for at least 90% of global use. Generally, esters penetrate foliage and salts are absorbed by roots. The choline salt is less prone to volatilisation than other salts, and is now used in combination with glyphosate. A further development is the quaternary ammonium salt (Figure 3.10) which is less irritating to the eyes than more commonly used salts. It is also less volatile than other formulations and is less prone to spray drift.

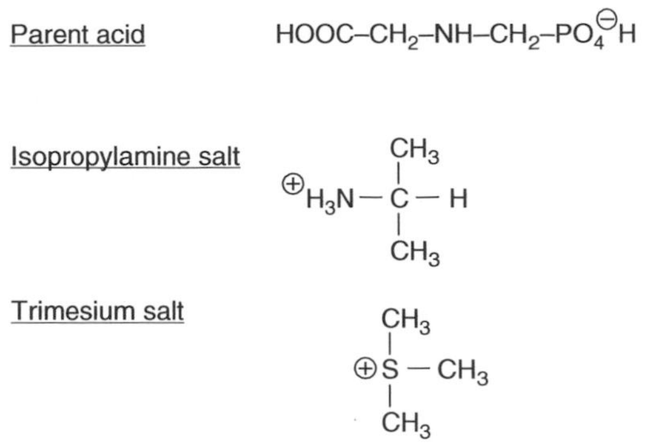

Parent acid $HOOC-CH_2-NH-CH_2-PO_4^{\ominus}H$

Isopropylamine salt

$$\overset{\oplus}{H_3N}-\overset{\overset{\displaystyle CH_3}{|}}{\underset{\underset{\displaystyle CH_3}{|}}{C}}-H$$

Trimesium salt

$$\overset{\oplus}{S}-CH_3$$

with CH_3 groups attached

Figure 3.13   The formulation of glyphosate.

3.8   The formulation of glyphosate

Glyphosate is the world's most widely used agrochemical with annual sales in excess of US$7 billion. It was first launched by Monsanto in 1974 and, following patent expiry in 2001, over 200 different formulations are sold in the UK alone. Glyphosate alone shows almost no penetration into leaves, so salts and surfactants are added for the molecule to be active. It was traditionally formulated as an isopropylamine salt with a cationic surfactant, and exhibits relatively slow uptake into perennial grasses. It was re-formulated in 2014 as the trimesium salt (i.e. trimethyl sulfonium, Figure 3.13) with a novel polyalkylglucoside adjuvant at 720 g kg^{-1} in a granular product, to give better weed control, owing to enhanced uptake, rain-fastness within an hour and less spray-drift.

Travlos *et al.* (2017) provide an overview of the effects of various glyphosate formulations and adjuvants on weed control. Glyphosate is a week acid and the acidic carboxyl hydrogen can be replaced with a salt or reacts with an alcohol to form an ester. The many commercialised formulations include isopropylamine, monoammonium, diammonium, potassium, trimethyl sulfonium and sesquisodium salts. Monovalent salts have high water solubility, but since only the parent acid acts at the target enzyme, the acid equivalent should be noted. Glyphosate is often applied with ammonium sulphate as an adjuvant, to reduce water hardness and increase activity. As 1% of the spray mixture (weight/volume), it can also result in higher absorption and translocation of the herbicide. Glyphosate is absorbed by all aerial plant surfaces. It translocates in both the symplasm and the phloem, following the movement of photosynthates, and accumulates in the meristems, i.e. the actively growing parts of the plant.

3.9   Further developments

A new class of adjuvant producing increased efficacy and an improved environmental profile is the amido propyl amines. *N,N*-Dimethyl-1,3-propanediamine is condensed to fatty acid molecules (Figure 3.14). Surface properties are dependent on the pH environment. At

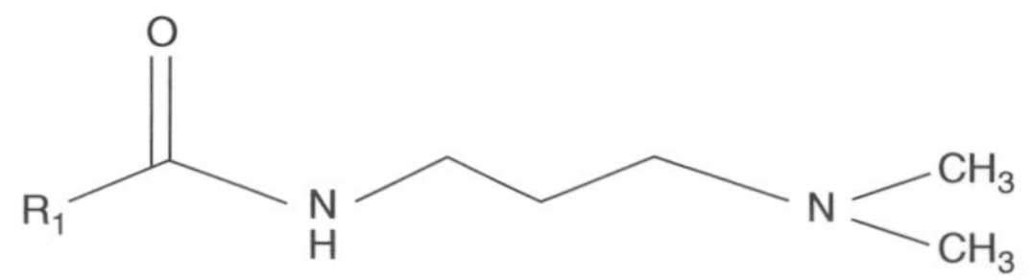

Figure 3.14 Structure of amido propylamines, where R1 can be C8/C10, COCO or SOY fatty acids.

high pH, these molecules are non-ionic, but at pH values below 8 the amine will protonate and the molecule will become more cationic. This may lead to more absorption to the negatively charged leaf surface.

Van der Pol *et al.* (2005) have developed lactate esters (*n*-propyl lactate, *n*-butyl lactate and 2-ethylhexyl lactate) that enhance the efficacy of the dimethylamine salt of 2,4-D and the 2-ethylhexyl ester of 2,4-D on fat hen (*C. album*). The lactate esters may act as solvents or penetrate the leaf cuticle itself and enhance the mobility of the herbicide through the cuticle. No phytotoxic effects were observed on tomato seedlings, suggesting promising potential for these 'green' solvents.

In recent years increasing attention has been paid to the problems associated with spray drift and how they might be overcome. Polymers of polyacrylamide, hydroxypropyl guar and ethyl hydroxyethyl cellulose (EHC) have been incorporated into the tank mix to reduce spray drift. Hazen (2005) found increased glyphosate retention in *Echinochloa crus-galli* when EHC was added to the formulation leading to enhanced bioefficacy.

Daneshvari *et al.* (2021) report that micro-encapsulation of trifluralin may help to reduce its sensitivity to photodegradation and volatilisation to the atmosphere. They have encapsulated trifluralin in polyurethane, resulting in a slower controlled release, which allows more time for this pre-emergence herbicide to be incorporated into the soil.

References

Briggs, G.G. and Bromilow, R.H. (1994) Influence of physicochemical properties on uptake and loss of pesticides and adjuvants from the leaf surface. In: Holloway, P.J., Rees, R.T. and Stock, D. (eds) *Interactions Between Adjuvants, Agrochemicals and Target Organisms*. Berlin: Springer, pp. 1–27.

Briggs, G.G., Bromilow, R.H. and Evans, A.A. (1982) Relationship between lipophilicity and root uptake and translocation of non-ionised chemicals by barley. *Pesticide Science* **13**, 495–504.

Briggs, G.G., Rigitano, R.L.O. and Bromilow, R.H. (1987) Physicochemical factors affecting the uptake by roots and translocation to shoots of weak acids in barley. *Pesticide Science* **19**, 101–112.

Bromilow, R.H. and Chamberlain, K. (1991) Pathways and mechanisms of transport of herbicides in plants. In: Kirkwood, R.C. (ed.) *Target Sites for Herbicide Action*. New York: Plenum Press, pp. 245–284.

Bromilow, R.H., Chamberlain, K. and Briggs, G.G. (1986) Techniques for studying the uptake and translocation of pesticides in plants. *Aspects of Applied Biology* **11**, 29–44.

Carr, J.E., Davies, L.G., Cobb, A.H. and Pallett, K.E. (1986) Uptake, translocation and metabolism of fluazifop-butyl in *Setaria viridis*. *Annals of Applied Biology* **108**, 115–123.

Crafts, A.S. (1960) Evidence for hydrolysis of esters of 2,4-D during absorption by plants. *Weeds* **8**, 19–25.

Daneshvari, G., Yousefi, A.R., Mohammadi, S., Banibairami, S., Shariati, P., Radhar, A. and Kyzas, G.Z (2021) Controlled-release formulations of trifluralin herbicide by interfacial polymerisation as a tool for environmental hazards. *Biointerface Research in Applied Chemistry* **11**, 13866–13877; doi: org/10.33263/BRIAC116.13866113877

Hazen, J.L. (2005) Retention and bioefficacy with ethyl hydroxyethyl cellulose (EHC) as a tank mix adjuvant to reduce spray drift. In: Proceedings of the British Crop Protection Council Congress. *Crop Protection and Technology* **2**, 891–896.

Hill, B.D., Stobbe, E.H. and Jones, B.L. (1978) Hydrolysis of the herbicide benzoylprop-ethyl by wild oat esterase. *Weed Research* **18**, 149–154.

Holloway, P.J. (1993) Structure and chemistry of plant cuticles. *Pesticide Science* **37**, 203–232.

Holloway, P.J. (1994) Evaluation of adjuvant modes of action. In: Holloway, P.J., Rees, R.T. and Stock, D. (eds) *Interactions between Adjuvants, Agrochemicals and Target Organisms*. Berlin: Springer, pp. 143–148.

Kader, J.C. (1997). Lipid-transfer proteins: a puzzling family of plant proteins. *Trends in Plant Science* **2**, 66–70.

Knowles, A. (2008) Recent developments of safer formulations of agrochemicals. *Environmentalist* **28**, 35–44.

Loos, M.A. (1975) Phenoxyalkanoic acids. In: Kearney, P.C. and Kaufman, D.D. (eds) *Herbicides; Chemistry, Degradation and Mode of Action*. New York: Marcel Dekker, pp. 1–128.

Madhou, P., Raghavan, C., Wells, A. and Stevenson, T.W. (2006) Genome-wide microarray analysis of the effect of a surfactant application in *Arabidopsis*. *Weed Research* **46**, 275–283.

Miller, R.H. (1985) The prevalence of pores and canals in leaf cuticular membranes. *Annals of Botany* **55**, 459–471.

Morgan, L.J. (1993) Formulants and additives and their impact on product performance. *Brighton Crop Protection Conference, Weeds* **3**, 1311–1318.

Pollard, M., Beisson, F., Li, Y. and Ohlrogge, J.B. (2008) Building lipid barriers: biosynthesis of cutin and suberin. *Trends in Plant Science* **13**, 236–246.

Post-Beittenmiller, D. (1996) Biochemistry and molecular biology of wax production in plants. *Annual Review of Plant Physiology and Plant Molecular Biology* **47**, 405–430.

Price, C.E. (1982) A review of the factors influencing the penetration of pesticides through plant leaves. In: Cutler, D.F., Alvin, K.L. and Price, C.E. (eds) *The Plant Cuticle*. New York: Academic Press, pp. 237–252.

Sharples, C.A., Hull, M.R. and Cobb, A.H. (1997) Growth and photosynthetic characteristics of two biotypes of the weed black-grass (*Alopecurus myosuroides* Huds.) resistant and susceptible to the herbicide chlorotoluron. *Annals of Botany* **79**, 455–461.

Snel, M., Banks, G., Mulqueen, P.J., Davies, J. and Paterson, E.A. 1995. Fluroxypyr butoxy-1-methylethyl ester; new formulation opportunities. *Brighton Crop Protection Conference, Weeds*, **1** 27–34.

Stephens, P.J.G., Gaskin, R.E. and Zabkiewicz, J.A. (1992) Pathways and mechanisms of foliar uptake as influenced by surfactants. In: Foy, C.L. (ed.) *Adjuvants for Agrochemicals*. Boca Raton, FL: CRC Press, pp. 385–398.

Taylor, F.E., Davies, L.G. and Cobb, A.H. (1981) An analysis of the epicuticular wax of *Chenopodium album* leaves in relation to environmental change, leaf wettability and the penetration of the herbicide bentazone. *Annals of Applied Biology* **98**, 471–478.

Travlos, I., Cheimona, N. and Bilalis, D. (2017) Glyphosate efficacy of different salt formulations and adjuvant additives on various weeds. *Agronomy* **7**, 60; doi: 10.3390/agronomy7030060

Van der Pol, J.F., van der Linden, J.T. and de Ruiter, H. (2005) Phytotoxicity and adjuvancy of lactate esters in 2,4-D based agrochemical formulations. Proceedings of the British Crop Protection Council Congress, *Crop Science and Technology* **1**, 447–452.

Chapter 4
Herbicide Selectivity and Metabolism

All things are poisons, nothing is not poisonous, and that it is the dose which makes a thing non-poisonous.

Theophrastus von Hohenheim Paracelsus (1493–1541)

4.1   Introduction

The main reason for the considerable success of modern herbicides is their selective phytotoxic action. Thus, only the weed is killed and its competitive effects are overcome by the crop. However, selectivity is a very relative term and totally dependent on dose, such that an application rate may alone determine crop tolerance and weed susceptibility. Since the concentration that reaches a target site is crucial to phytotoxicity, it follows that any factor that alters the concentration of active herbicide *in vivo* will then contribute to selectivity. How selectivity is achieved is therefore a complex subject, usually based on exploiting differences between the crop and the weed. These differences are many and varied; some of them are briefly considered below.

At the level of the whole plant, differences in morphology can be a major contributor to herbicide selectivity, especially in the case of a broadleaf weed in a cereal crop. Foliar arrangement ensures that more surface area of the weed leaf is exposed for greater spray interception than the more upright, narrower leaf of the crop. The extent of spray retention may therefore ensure selectivity when the same dose may be equally phytotoxic to both crop and weed. Furthermore, the growth areas (meristems) of a broadleaf weed tend to be more accessible to the spray compared with the 'protective sheath' of coleoptile or leaves that surround the cereal meristem.

In the case of soil-applied herbicides, the relative positions and growth rates of the weed and the crop are the most important factors. Depth protection is valuable in the pre-emergent control of many weeds in large-seeded, slow growing crops. Sugar beet, for example, may be sown deeper in the soil to gain protection from surface-incorporated herbicides. Indeed, many weeds that are shallow-rooted may rapidly establish before the sugar beet seedlings, and may therefore be controlled by a fast-acting contact treatment before the crop emerges. In this way the timing of herbicide treatment is an important contributor to selectivity.

Herbicides and Plant Physiology, Third Edition. Andrew H. Cobb.
© 2022 John Wiley & Sons Ltd. Published 2022 by John Wiley & Sons Ltd.

Differences in absorption and uptake may be exploited by choice of herbicide formulation, as may be rates of both short- and long-distance transport within the plant. Translocation patterns will also alter sites of herbicide accumulation. Selectivity is commonly achieved by the presence of detoxifying enzymes and their relative rates of activity. Furthermore, it is now established that differences in herbicide sensitivity at a target site may provide the basis of selectivity in some cases. Usually, however, selectivity results from a complex interaction of several factors, although differential metabolism is nowadays regarded as the major contributor. Plants show large differences in their ability to metabolise herbicides, and a significant difference in metabolism may often be closely correlated with tolerance or susceptibility.

Plants have evolved a wide range of metabolic systems to detoxify the chemicals which they encounter in the environment. This is well illustrated in the differential metabolism of selective herbicides between tolerant crops and susceptible weeds. Indeed, it is now clear that in many cases, the basis of herbicide selectivity is the ability of the crop to metabolically detoxify the herbicides, whereas the weeds are less able to do this or may even activate the herbicide. It should also be noted, although beyond the scope of this text, that plant metabolism is also able to modify the action of systemic fungicides, insecticides and industrial pollutants that are encountered.

Plants rely on their biochemical mechanisms of defence against foreign chemicals (xeno-biotics). As a general rule, the enzymes involved have a broad substrate specificity. Thus, the foreign compounds are often structurally related to the intermediates of plant secondary metabolism, especially flavonoids. These enzymes are normally present and functional throughout the life of the plant (i.e. they are constitutive). In some cases, however, their activity is induced by the xenobiotics.

For detoxification to be successful, the phytotoxic chemical must be quickly metabolised to a less or non-toxic product. The rate of this process can determine whether a plant survives or succumbs to chemical attack. Often both crop and weed may have the same biochemical pathways, but metabolic rates are higher in the crop. Also, differences in herbicide metabolism and compartmentation are observed in crops and weeds.

4.2   General principles

In order to penetrate the waxy surfaces of leaves, pesticides are invariably highly lipophilic molecules. As a general rule, oxidative metabolic attack serves to enhance both the reactivity and the polarity of the molecule. Herbicides modified in this way become less phytotoxic, although there are some notable examples in which bioactivation can take place. The products of these primary reactions are then conjugated, often with sugars or amino acids, rendering them highly water-soluble, biologically inert and readily stored in the plant cell vacuole away from any potential sites of action. These storage forms may become susceptible to hydrolysis and, in some cases, may constitute a pool of potentially active herbicide. Immobilisation by binding to lignin or other insoluble cellular constituents may become the final fate of a herbicide residue or, in some instances, water-soluble metabolites may be excreted from the roots.

It is now clear that herbicides can be metabolised in many ways and this can have a direct bearing on crop tolerance or susceptibility. An understanding of how herbicides are metabolised is therefore central to enhancing herbicide design and selectivity.

4.2.1   *Phases of herbicide metabolism*

Herbicide metabolism has been well documented in the last three decades and much progress has been made in our understanding of the enzyme systems involved and where they act in the plant cell. The four phases commonly observed are:

Bioactivation	
Phase I	Metabolic attack
Phase II	Conjugation
Phase III	Sequestration

   Some herbicides have been shown to undergo bioactivation within or at the cell wall of plant cells, where a pro-herbicide is converted to a phytotoxic agent by the action of plant enzymes. Before this activation they may be less or non-phytotoxic, so the plant can be instrumental in manufacturing the substance that will eventually kill it. Bioactivation can involve removal of chemical groups that have aided in herbicide uptake and this can have the added benefit of trapping the herbicide within the cell. Herbicide detoxification is achieved by key enzymes that carry out two major functions. First, they alter the chemical structure of the herbicide to render it biologically inert. Second, these reactions serve to increase both the reactability and the polarity of the herbicide, so that it can be removed from the cytoplasm and either stored in the vacuole or bound to the cell wall. Metabolically, this is achieved by introducing or uncovering polar groups (phase I metabolism). In some instances polar groups may already be in place in the original herbicide structure, in which case phase II metabolism can be carried out without the need for phase I reactions. In other cases, conjugate formation may be reversible and this may serve to store phytotoxic molecules.

4.2.1.1   *Bioactivation*

Bioactivation is the process whereby 'pro-herbicides', often herbicidally inactive molecules, are enzymatically converted to phytotoxic compounds in the plant cell. Many herbicides are therefore formulated as inactive hydrophobic esters to enable them to penetrate the waxy leaf cuticle. Ester hydrolysis then reveals the biologically active acid or alcohol groups. Examples include the conversion of bromoxynil octanoate to bromoxynil, and diclofop-methyl to diclofop acid (Figure 4.1). Both the aryloxyphenoxypropionate graminicides, such as fenoxaprop-ethyl, and the phenoxy-carboxylic acids, such as 2,4-DB, are rapidly de-esterified in both crops and weeds. De-esterification may, in some cases, inactivate herbicides, as in the metabolism of sulphonylurea esters, such as chlorimuron-ethyl, in which the de-esterified product is herbicidally inactive. Another example is the conversion of EPTC to the herbicidally active sulphoxide derivative. Imazamethabenz-methyl may also be regarded as a pro-herbicide. In susceptible weeds, hydrolysis results in the potent inhibition of branched chain amino-acid biosynthesis, whereas hydroxylation of the intact ester occurs in resistant maize and wheat. Bioactivation of DPX-L8747 by *N*-dealkylation in susceptible species leads to an active herbicide, whereas in resistant crops hydroxylation following the formation of a glutathione conjugation of the intact pro-herbicide leads to non-toxic metabolites. These examples demonstrate how bioactivation may be a mechanism of selectivity between crop and weed. Opening of the oxadiozolidine ring of methazole to form 1-(3,4-dichlorophenyl) urea and *N*-demethylation of pyridazinone to form a potent phytoene desaturase inhibitor further demonstrate the wide range of chemical reactions that can lead

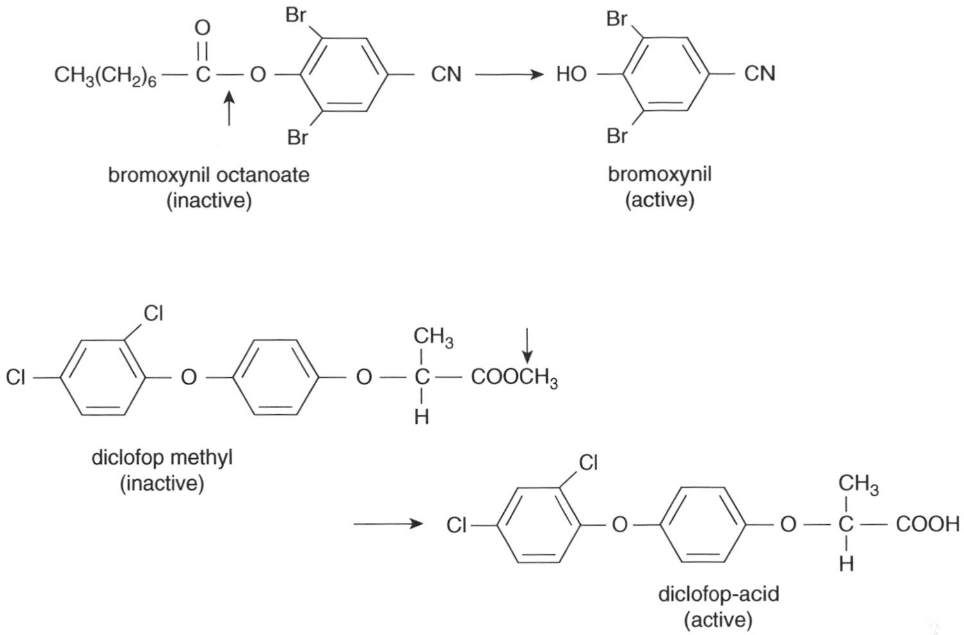

Figure 4.1 Bioactivation of inactive pro-herbicides to active molecules.

to bioactivation. In the case of the bioherbicide bialaphos, a tripeptide obtained from *Streptomyces*, metabolic cleavage results in the release of glufosinate, which is an important herbicide in its own right. Interestingly, resistance to the herbicide triallate in a biotype of *Avena fatua* has been demonstrated to be due to a reduced ability to convert triallate to the phytotoxic product triallate sulphoxide. This is the only reported instance of resistance being due to an inability of a weed to bioactivate a herbicide.

Cummins *et al.* (2001) found a large number of diverse proteins in wheat capable of hydrolysing herbicide esters and these activities differed from those in competing grass weeds. Indeed, crude extracts from black-grass plants were more active in ester hydrolysis than wheat. They purified a 45 kDa esterase from wheat that could bioactivate bromoxynil octanoate, but showed no activity towards diclofop-methyl. Conversely, esterase activity towards diclofop-methyl was higher in the weed than the crop. The observations of Cummings *et al.* support the rapid bioactivation by ester hydrolysis of graminicides in grass weeds, contributing to their bioavailability and selectivity. Gershater *et al.* (2007) investigated carboxylesterase activity in *Arabidopsis thaliana* and observed multiple carboxylesterases with activities towards herbicide esters that become activated to active acids by hydrolysis. They are all serine hydrolases. One, AtCXE12, was shown to hydrolyse the pro-herbicide 2,4-D-methyl to 2,4-D-acid. The location of this esterase activity is thought to be the cell wall.

4.2.1.2 Metabolic attack

This phase of metabolism aims to introduce or reveal chemically active groups, such as –OH or –COOH, which can undergo further reactions. The most common way in which plants attack herbicides is by hydroxylation of aromatic rings or of alkyl groups by a family of enzymes known as the cytochrome P450 mono- or mixed-function oxidases (P450s).

Cytochrome P450 was first observed in 1958 in an absorption spectrum from rat liver microsomes which had a peak of absorbance at 450 nm when the reduced enzyme was bound to the inhibitor carbon monoxide, and 'P' signifies protein. The inhibition by carbon monoxide is overcome by light. We now know that there are more than 18,500 examples of P450 molecules from hundreds of species. Indeed, there are over 300 cytochrome P450 monooxygenase genes in *A. thaliana* alone, of which half are unique to plants, contributing about 1% of the plant genome.

The P450s are a very large family of enzymes now thought to be the largest family of enzymatic proteins in higher plants. They all have a haem porphyrin ring containing iron at a catalytic centre. These enzymes are responsible for the oxygenation of hydrophobic molecules, including herbicides, to produce a more reactive and hydrophilic product. The reaction utilises electrons from NADPH to activate oxygen by an associated enzyme, cytochrome P450 reductase. One atom from molecular oxygen is incorporated into the substrate (R), while the other is reduced to form water.

$$\text{R-H} + O_2 \xrightarrow{} \text{R-OH} + H_2O$$

$$\text{NADPH} + H^+ \qquad \text{NADP}^+$$

The enzymes are located on the cytoplasmic side of the endoplasmic reticulum and are anchored by their *N*-terminus (Figure 4.2). They are found in all plant cells but in very low abundances. This, coupled with their lability *in vitro*, has meant that they are difficult to study biochemically. All P450s have a highly conserved region of 10 amino acids surrounding the haem group and it is this region that is responsible for the binding of O_2, its activation and the transfer of protons to form water. The rest of the P450 amino acid sequences are highly variable and this probably explains the wide variety of reactions and substrate specificity shown by this enzyme superfamily.

The following features are all used in setting criteria for the involvement of P450 enzymes in herbicide metabolism:

- requirement of O_2;
- requirement for NADPH;
- association of enzyme activity with the microsomal fraction produced by centrifugation at 100,000g, enriched in the endoplasmic reticulum;
- inhibition by CO, which is reversible by light;

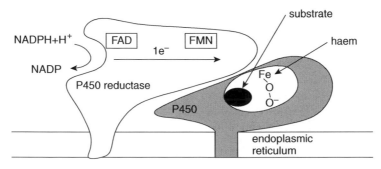

Figure 4.2 A diagrammatic representation of plant P450 and P450 reductase enzymes on the endoplasmic reticulum. Source: Werck-Reichhart, D., Hehn, A. and Didierjean, L. (2000) Cytochrome P450 for engineering herbicide tolerance. *Trends in Plant Science* **5**, 116–123. Reproduced with permission of Elsevier.

- inhibition by anti-reductase antibodies; and
- inhibition of *in vitro* activity by known P450 inhibitors, including aminobenzotriazole, paclobutrazole, piperonyl butoxide and tetcyclasis.

The P450 proteins are between 45 and 62 kDa in size. While their amino acid sequences may vary considerably, their three-dimensional structure is highly conserved, especially in the haem-binding region. The haem binds to the protein at a cysteine residue and the flanking sequence (Figure 4.3) is a characteristic of all P450s.

The conserved oxygen-binding sequence is about 150 residues upstream from the haem and consists of Ala or Gly–Gly–X–Asp or Glu–Thr–Thr or Ser. In both haem- and oxygen-binding sequences, X denotes any other amino acid.

When these conserved sequences were used to study P450s in plants, a surprisingly large number and diversity of P450s were found. Indeed, more than 500 plant P450 genes are now known in over 50 families, indicating that the P450s are the largest group of plant proteins. The precise roles of the proteins encoded by these genes are, however, largely unknown.

A nomenclature has been designed for P450 genes based on the identity of the amino acid sequences of the proteins that they encode (Figure 4.4). The genes have been numbered in chronological order depending on their date of submission to the P450 nomenclature committee (http://drnelson.utmem.edu/CytochromeP450.html).

Typical families are numbered from CYP71 to CYP99. For example, CYP71 C6v1 in wheat is inhibited by glyphosate and CYP76B1 catalyses the dealkylation of the phenylurea herbicides (Wen-Sheng *et al.*, 2005).

The discovery of new P450 genes continues in plants. In contrast, only about 50 P450 genes in 17 families have been described in humans. Why are there so many P450s in plants? The answer seems to be that they play a very wide role in plant secondary metabolism. They have been shown to be involved in the biosynthesis and metabolism of a wide variety of compounds, including terpenes, flavonoids, sterols, hormones, lignins, suberin, alkaloids and phytoalexins. They are also induced by pathogen attack and xenobiotics and by light-induced stress, unfavourable osmotic conditions, wounding and infection. It is currently believed that herbicide molecules also fit the active sites of these P450s involved in biosynthesis, suggesting a broad diversity of substrate selectivity.

Regarding their roles in herbicide metabolism, much remains to be done to establish substrate specificity and both the molecular and the metabolic regulation of these important

$$Fe$$
$$|$$
$$Gly-Cys-X-Arg-X-Gly-X-X-Phe$$

Figure 4.3   The haem-binding sequence that is characteristic of all P450s.

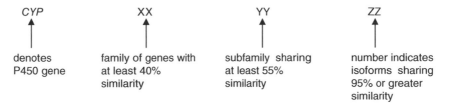

CYP	XX	YY	ZZ
denotes P450 gene	family of genes with at least 40% similarity	subfamily sharing at least 55% similarity	number indicates isoforms sharing 95% or greater similarity

Figure 4.4   Nomenclature for naming P450 genes.

enzymes. Such understanding will be invaluable in predicting and elucidating herbicide selectivity, as well as in the discovery and design of new selective herbicides.

P450s catalyse many reactions in plant metabolism, including the biosyntheses of pigments, defence-related molecules, protectants from ultra-violet light, lignin, fatty acids, terpenes, phytosterols, flavonoids and phytohormones, with further functions being discovered each year (Bak *et al.*, 2011). The diversity of these enzymes is such that that they play important roles in all facets of plant growth and development. Furthermore, Nelson (2013) eloquently describes the central role of these enzymes in the evolution of plants, over 3 billion years. Thus, without them, plants would not have the waterproof cutins and suberins in epicuticular waxes that have enabled the success of terrestrial plant life. The tallest plants on Earth would have been mosses, lacking lignin, which, as coal, later powered the industrial revolution and still contributes significantly to global energy consumption. We should not forget flower colour and the generation of secondary metabolites that form the basis of many modern pharmaceuticals and herbicide leads. Indeed, these enzymes are some of the most versatile enzymes on Earth, and are involved in the detoxification of xenobiotics.

The main reactions catalysed by P450s are shown in Figure 4.5. By inserting an oxygen atom into hydrophobic molecules, such as most herbicides, it makes them more reactive and relatively more water soluble (Table 4.2). In herbicide metabolism these are hydroxylation and dealkylation, which progress via a hydroxylation step. Examples of herbicides metabolised by P450s in plant systems include sulphonylureas (including primisulfuron, nicosulfuron, prosulfuron, triasulfuron and chlorimuron), substituted ureas (chlorotoluron, linuron), chloroacetanilides (metolachlor, acetochlor), triazolopyrimidines (flumetsulam), aryloxyphenoxypropionates (diclofop), benzothiadiazoles (bentazon) and imidazolinones (imazethapyr). Selectivity to herbicides can be due to the ability of the crop to metabolise herbicides via P450s, an ability that may not be possessed by susceptible weeds. In some cases, however, this metabolism is not enough to prevent crop damage, because of either low rates of P450 metabolism or the phytotoxicity of products produced by these reactions. Crop damage may only be prevented if reactions from phase II (conjugation) are successful in carrying out further detoxification.

Some phase I reactions may be catalysed by peroxidases (E.C. 1.11.1.7) that are commonly found in leaves at high concentrations, being able to catalyse oxidations using hydrogen peroxide. They are currently thought to be involved in proline hydroxylation, indole acetic acid oxidation and lignification, and have been implicated in the metabolism of aniline compounds produced in the degradation of phenylcarbamate, phenylurea and acylaniline herbicides.

4.2.1.3   Conjugation

In this phase of herbicide metabolism, the molecule becomes conjugated to natural cell metabolites, such as amino acids, sugars, organic acids or the tripeptide glutathione (γ-glutamyl-cysteinyl-glycine). This has the effect of both further reducing phytotoxicity and increasing the solubility of the herbicide or its metabolite, which may also serve to target the conjugate to the vacuole.

The most widely studied conjugation reaction in relation to herbicide detoxification is that of glutathione conjugation, carried out by the enzyme family glutathione *S*-transferases (GSTs), E.C. 2.5.1.18, first identified in plants in 1970. Glutathione is abundant in plants, often exceeding 1 mM concentration in the leaf cell cytoplasm, where it functions as a

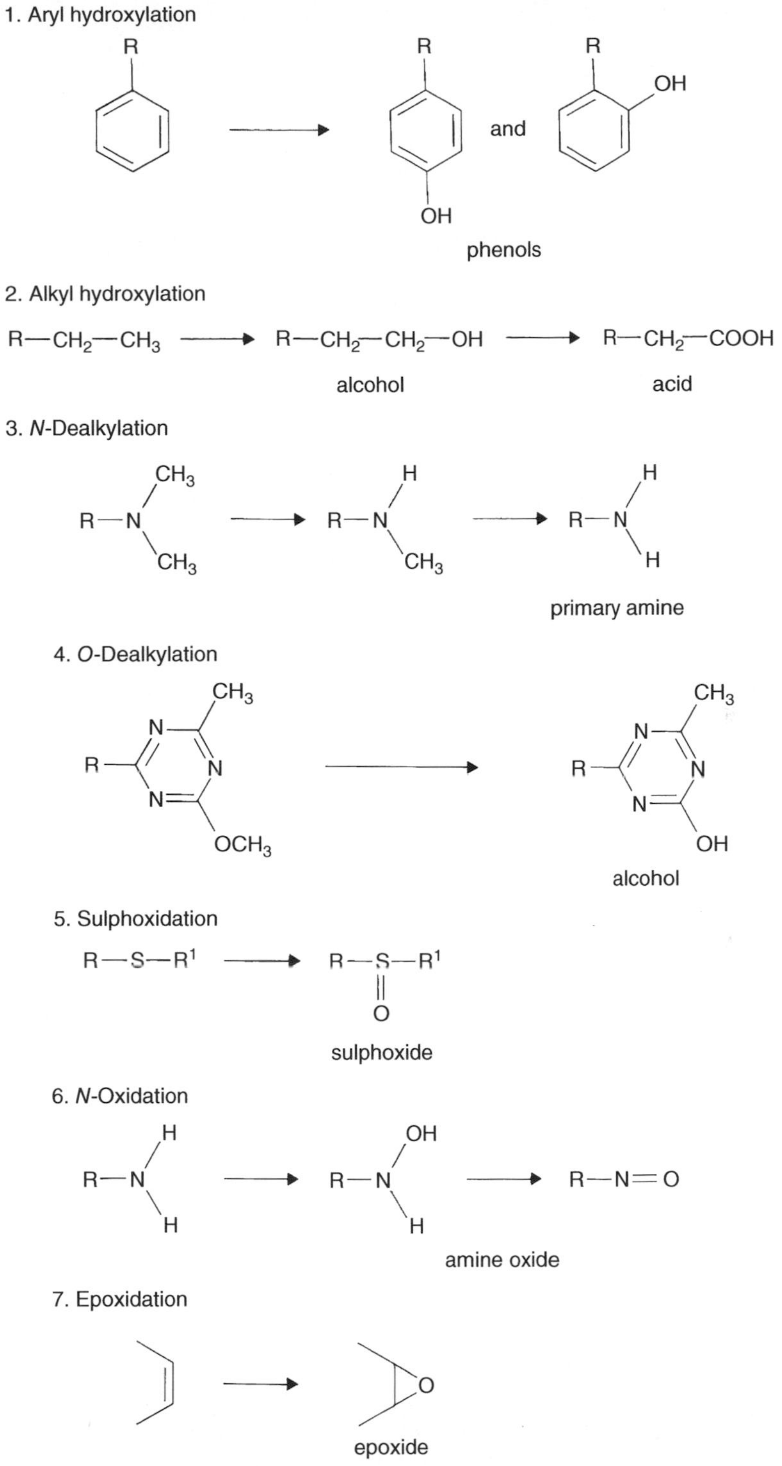

1. Aryl hydroxylation

and

phenols

2. Alkyl hydroxylation

$$R—CH_2—CH_3 \longrightarrow R—CH_2—CH_2—OH \longrightarrow R—CH_2—COOH$$

alcohol   acid

3. *N*-Dealkylation

primary amine

4. *O*-Dealkylation

alcohol

5. Sulphoxidation

sulphoxide

6. *N*-Oxidation

amine oxide

7. Epoxidation

epoxide

Figure 4.5   The main reactions carried out by cytochrome P450 monooxygenases (P450s).

In *Zea mays*:	*Zm* GST F1
	F2
	F3
	U1
	U2
	U3
In *Arabidopsis thaliana*:	*At* GST T1
	F1
	U1
In *Alopecurus myosuroides*:	*Am* GSTU1
	F1

scavenger of free radicals, protecting photosynthetic cells from oxidative damage. As GSTs are a large group of similar enzymes found in all eukaryotes, differences in the spectrum of GSTs present play an important role in the selectivity of herbicides. GSTs have a range of endogenous functions involving their abilities to detoxify and act as redox buffers.

GSTs catalyse nucleophilic conjugation of the reduced form of the tripeptide glutathione to a wide variety of hydrophobic, electrophilic and often cytotoxic substrates. The conjugated products are transported to either the vacuole or the apoplast The GSTs are abundant, soluble enzymes of about 50 kDa, each composed of two subunits of equal size, containing an active site located in the *N*-terminus that binds glutathione and is highly conserved in all GSTs. Herbicides and other xenobiotics are bound at the hydrophobic *C*-terminal half of the subunit. This site varies considerably and accounts for the differing specificity of GSTs towards herbicides.

There are 55 GSTs in *Arabidopsis* and 79 in rice. Plant GSTs are divided into six classes, named phi (F), tau (U), zeta (Z), theta (T), lambda (L) and dehydro-ascorbate reductases (DHAR), classified on the basis of their amino acid sequence and similarities in gene sequence. The latter two classes are probably glutathione reductases, the zeta class functions in tyrosine catabolism, while the theta class may be involved in the metabolism of oxidised lipids. Classes phi and tau are plant specific, are often induced by stress and are associated with herbicide detoxification. The classes are characterised by the species name, for example:

While the importance of GSTs is well understood, their location within plant tissues is less well known. Li *et al.* (2017) have investigated GSTs in maize roots and shoots following treatment with metolachlor. They observed that GST genes were more expressed in roots than shoots in both control and treated plants, leading to faster detoxification in roots.

Why are there so many GSTs in plants and what are their physiological roles? Although often cited in the literature, the involvement of GSTs in the conjugation of secondary metabolites remains far from clear. A function in shuttling anthocyanin pigments to the tonoplast for vacuole uptake appears likely. Plants exposed to environmental stress or infection show elevated expression and activity of GSTs, suggesting a role in maintaining cellular homeostasis following oxidative stress in tissues infected by pathogens. They are also involved in the metabolism of oxylipins (see Section 8.7), again as a result of biotic and abiotic stress. Herbicide-resistant black-grass has also been shown to have higher activity of GSTs (Reade *et al.*, 2004). In addition, GSTs are also involved in plant growth and development. They are able to bind plant hormones and can be induced by ethylene, auxin, methyl-jasmonate, salicylic acid and abscisic acid. Jiang *et al.* (2010) suggested that GSTs are also involved in light signalling in *A. thaliana*, implying a coordinated regulation of plant development with phytochrome A and phytohormones

Edwards and Dixon (2000) consider that GSTs are well placed to use physiologically high cytoplasmic concentrations of glutathione (0.2–1.0 mM) to conjugate electrophilic herbicide residues effectively. As the conjugates can inhibit GST activity, they are actively transported out of the cytoplasm into the vacuole by the ATP-binding cassette (ABC)-transporters. When the ABC-transporter gene from *Arabidopsis* was expressed in yeast, Lu *et al.* (1997) demonstrated the uptake of *S*-metolachlor-glutathione, confirming the involvement of these transporters across the tonoplast membrane.

Once in the vacuole, a specific vacuolar carboxypeptidase hydrolyses the glycine residue and the dipeptide conjugate is re-exported to the cytoplasm. The glutamate group is removed by a glutamyl transpeptidase and the remaining cysteinyl derivatives are further transformed by *N*-malonylation and further oxidation to a complex range of more polar products.

GST activity was first demonstrated in plant tissue against atrazine in maize extracts in 1970. Since this observation, GST activity against a wide variety of herbicides has been reported (Table 4.1). As with other conjugate types, glutathione conjugation can be carried out against the parent herbicide if an appropriate conjugating group is present, or can follow phase I metabolism. An example of the latter is the conjugation of glutathione with thiocarbamates, only after they have undergone conversion to their corresponding sulphoxides. Crops are often reported to possess higher GST activities against herbicides than susceptible weeds and this might offer some degree of selectivity between crop and weed. Activity against chloroacetamides (maize, wheat, sorghum, rice), oxyacetamides (maize), atrazine (maize, sorghum), fenoxaprop, fluorodifen, flupyrsulfuron-methyl, dimethenamid (all wheat) and the sulphoxide metabolite of EPTC (sorghum) have all been reported. In soybean, homoglutathione is found in place of glutathione. Conjugations utilising this against several chloroacetanilides, the diphenyl ethers acifluorfen and fomesafen and the sulphonylurea chlorimuron-ethyl are all reported in this crop. In addition to selectivity, GSTs have also been implicated in playing a role in herbicide resistance in a variety of weeds. In black-grass, biotypes resistant to chlorotoluron and fenoxapropethyl demonstrated approximately double the GST activity of susceptible biotypes. This suggests that GSTs, as well as P450s, may play a role in enhanced metabolism resistance in this species.

Table 4.1 Examples of some herbicides metabolised by glutathione *S*-transferases in various plant systems.

Chemical family	Examples
Chloroacetamides	Alachlor, acetochlor, metolachlor, pretilachlor
Triazines	Atrazine
Aryloxyphenoxypropionates	Fenoxaprop
Thiocarbamates	EPTC
Diphenyl ethers	Acifluorfen, fomesafen
Sulphonylureas	Clorimuron-ethyl, triflusulfuron-methyl

In velvetleaf (*Abutilon theophrasti*), resistance to atrazine has also been demonstrated to be due to higher conjugation of this herbicide to glutathione.

Another commonly encountered phase II reaction in plants is conjugation with glucose, catalysed by the glucosyltransferases (EC 2.4.1.71) utilising uridine diphosphate glucose as the glucose donor. There are 120 UDP-glucosyltransferase genes in *Arabidopsis* of which 107 encode functional enzymes, organised into 14 groups, based on sequence similarity and evolutionary relatedness. They all transfer a sugar residue from an activated nucleotide sugar to a specific acceptor molecule, forming a glycosidic bond. Once conjugated, the glucose may undergo a further phase II reaction by 6-*O*-conjugation with malonic acid, catalysed by the malonyl-CoA-dependent malonyltransferases. These phase II metabolites then undergo ATP-dependent transport into the vacuole.

Interestingly, Pflugmacher and Sandermann (1998) found *O*-, *N*- and *S*-glucosyltransferase activity to be very widely distributed throughout the plant kingdom, not solely confined to higher plants, but even in marine macroalgae. Indeed, they hypothesised that this activity in 'lower' plants may make an important contribution to the detoxification of xenobiotics in the global environment.

Finally, acidic herbicide molecules such as the synthetic auxin phenoxyacetic acids can be conjugated to the amino acids, glutamine, valine, leucine, phenylalanine and tryptophan, although the enzymology of these reactions remains obscure. As an example, crop plants are able to rapidly detoxify the photosynthetic inhibitor bentazone by rapid aryl hydroxylation followed by conjugation to glucose. Susceptible weeds appear unable to metabolise the parent herbicide and phytotoxicity is observed.

A 200-fold margin of selectivity between rice and *Cyperus serotinus* has been attributed to this metabolic route. Similarly, in soybean, where an 8-hydroxy derivative has been detected, Leah *et al*. (1992) isolated and purified two glucosyltransferases from tolerant soybean that were capable of glycosylating 6-hydroxybentazone (Figure 4.6). This soluble enzyme had a relative molecular mass of 44.6 kDa with binding constants for kaempferol and 6-hydroxybentazone of 0.09 and 2.45 mM, respectively. They also found a membrane-bound enzyme, whose primary substrate was *p*-hydroxyphenylpyruvic acid, with a relative molecular mass of 53 kDa and binding constants of 0.11 and 1.96 mM for *p*-hydroxyphenylpyruvic acid and 6-hydroxy-bentazone, respectively. These findings, and those subsequently shown by others, imply an overlapping specificity of aryl hydroxylated herbicides and the synthesis and storage of secondary metabolites.

4.2.1.4   Sequestration

Compartmentalisation of a herbicide metabolite appears to take place in much the same way as products of plant secondary metabolism are moved for storage. The place for storage is either the vacuole or in association with the cell wall. Identification of a membrane-bound glutathione-dependent ABC pump in the vacuolar membrane suggests that phase II conjugation to glutathione or malonate might serve to facilitate the movement of metabolites and could be considered as a way of 'tagging' molecules for movement into the vacuole.

The processing and vacuolar import of herbicide conjugates in plants is a two-step process, involving first glucosylation and then derivatisation of the sugar with malonic acid. The significance of this reaction is not entirely understood, although malonylation appears to act as a tag, directing the conjugates for vacuolar import.

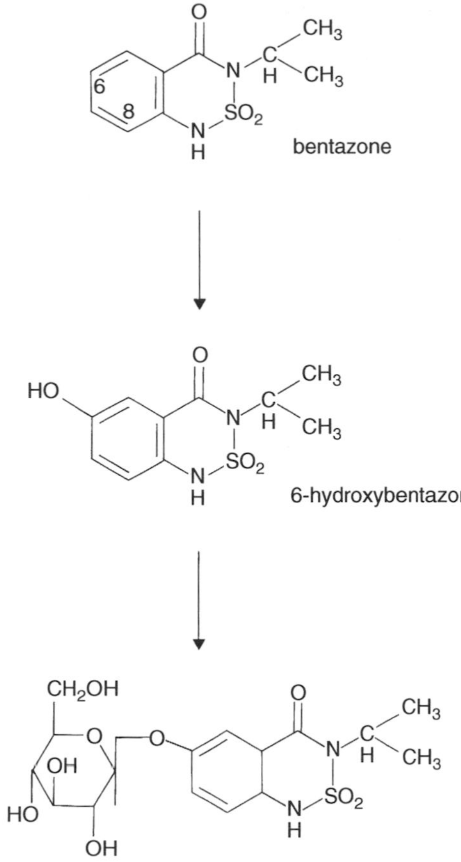

Figure 4.6 Metabolism of bentazone in tolerant plant species.

The malonylation reaction is carried out by malonyltransferases, which can conjugate compounds containing amino (*N*-malonylation) or hydroxyl (*O*-malonylation) residues. Malonylation also appears to prevent digestion by glucosidases and may also facilitate conjugate transport across the plasma membrane. For example, glucosidic conjugates of pentachlorophenol formed in soybean and wheat undergo malonylation, and 3,4-dichloroaniline undergoes *N*-malonylation in soybean and wheat. In each case, the transport of the newly formed conjugates was shown to be routed towards the vacuole. Although malonylation of glucosides is important in directing their importation into the vacuole, it is clear that the glucosides themselves can undergo vacuolar deposition.

Once conjugates are situated in the vacuole, sequential removal of peptides from glutathione is carried out by peptidases. This results in the metabolite being conjugated to glutamylcysteine and possibly just to cysteine. It is postulated that this allows for recycling of amino acids back to the cytoplasm and in addition may prevent the conjugated metabolite from being exported back there, as it is no longer a full glutathione conjugate. This pumping mechanism may have the additional benefit of stopping the build-up of

Table 4.2   Generalised effect of each phase of metabolism on herbicide activity and bio-availability.

	Phase 1	Phase 2	Phase 3	Phase 4
Metabolism	P450s	Conjugation	Compartmentation	Cell wall
Activity	Reduced	Non-toxic	Non-toxic	Non-toxic
Solubility	Less	More	More	Insoluble
Mobility	Less	Limited	Immobile	Immobile

glutathione conjugates from inhibiting cytoplasmic GST activity, as some conjugates have been demonstrated to be powerful competitive inhibitors of GSTs. Once the metabolite conjugate has entered the vacuole it may be further metabolised, stored there or excreted across the plasma membrane to the extracellular matrix. Transport of glucosylated herbicides into the vacuole has also been reported. This is ATP-requiring, so is also active transport. It appears that the membrane pump carrying this out is distinct from the glutathione system.

Serine carboxypeptidase and a lactoyl-glutathione lyase (a carbon–sulphur lyase) may also be involved in phase 4 of the detoxification process. Serine carboxypeptidase contributes to the degradation of glutathione conjugates to their respective cysteine derivatives, which may be further metabolised by carbon–sulphur lyases. The genes for these enzymes are also induced following safener treatment (Beheringer et al., 2011).

Most herbicide metabolites eventually become associated with the insoluble compartments of the cell, bound to lignin or polysaccharides. In these forms they can be released by enzymic hydrolysis, and both hydroxylated and unaltered forms have been detected. It is thought that this process can occur by covalent linkage.

While conjugates usually represent the end product of herbicide metabolism, they should not be regarded as totally inert. Some evidence exists for the hydrolysis of the glucoside to regenerate an active molecule, and so the conjugate may be regarded as a reservoir of potential activity. Such regeneration depends on the nature of the glycoside linkages involved and their proximity and susceptibility to the action of β-glucosidase.

An overview of the different phases of herbicide metabolism in a plant cell is presented in Table 4.2. and Figure 4.7.

4.3   Herbicide safeners and synergists

Herbicide safeners, also termed antidotes in some texts, are chemicals that, when applied before or with herbicides, increase the tolerance of a cereal crop to a herbicide. This activity has been known since the 1970s and the safening effect is not seen with the weeds. A list of herbicide safeners available as commercial products is given in Table 4.3.

As some safeners may show structural homology with herbicides, it was previously thought that they competed with the herbicide molecule for the target site. We now know, however, that their protective effect results from a general enhancement of detoxification processes, including the induction of:

- P450 oxygenases
- glutathione S-transferases

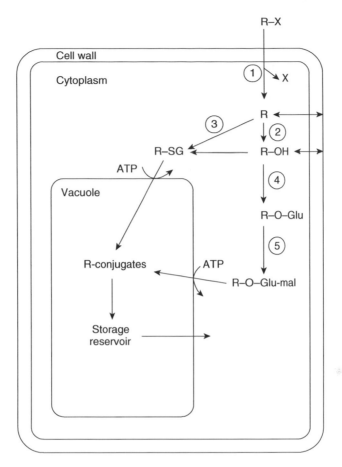

Figure 4.7 A schematic representation of the different phases of herbicide metabolism in a plant cell (R denotes the active herbicide molecule). Key: 1, hydrolase; 2, oxido-reductase; 3, glutathione *S*-transferase; 4, glucosyl transferase; 5, malonyl transferase.

- glucosyltransferases
- vacuolar transport
- glutathione synthesis
- glutathione peroxidase
- sulphate assimilation

Since a crop has to show some degree of tolerance to the herbicide for it to be safened, it would appear that the process is a 'top-up' mechanism for detoxification routes that are already operative. Despite considerable interest in safeners, their activities have mainly been demonstrated in monocotyledonous crops, notably maize, wheat, sorghum and rice.

It is likely that the identification of genes encoding safener-inducible enzymes will allow their transfer to give genetically modified crops with an enhanced ability to detoxify herbicides. This approach has already shown potential (Davies and Caseley, 1999), although further development will require a greater public acceptance of GM crops in future.

Table 4.3 Herbicide safeners available as commercial products.

Safener	Crop	Herbicide	Application method
Benoxacor (CGA 154281)	Maize	Metolachlor	Spray as pre-emergence mixture
Cloquintocet-mexyl (CGA 184927)	Wheat	Clodinafop-propargyl	Spray as post-emergence mixture
Cyometrinil (CGA 43089)	Sorghum	Metolachlor	Seed treatment
Dichlormid (DDCA, R25788)	Maize	EPTC, butylate, vernolate	Pre-plant or pre-emergence mixture
Fenchlorazole-ethyl (HOE 70542)	Wheat	Fenoxaprop-ethyl	Spray as post-emergence mixture
Fenclorim (CGA 123407)	Rice	Pretilachlor	Spray as pre-emergence mixture
Flurazole (MON 4606)	Maize	Alachlor	Seed treatment

Safener		Crop	Herbicide	Application method
Fluxofenim (CGA 133205)		Sorghum	Metolachlor	Seed treatment
Furilazole (MON 13900)		Sorghum	Halosulfuron-methyl	Seed treatment
Mefenpyr-diethyl		Wheat, rye, triticale, barley	Fenoxaprop-ethyl	Spray as post-emergence mixture
MG 191		Maize	Thiocarbamates	Spray as pre-emergence mixture
Naphthalic anhydride (NA)		Maize	EPTC, butylate, vernolate	Seed treatment
Oxabetrinil (CGA 92194)		Sorghum	Metolachlor	Seed treatment
AD-67		Maize	ALS inhibitors Dichloroacetamides	Pre-em
Dietholate		Rice and cotton	Clomazone	Seed treatment

(Continued)

Table 4.3 *(Continued)*

Safener		Crop	Herbicide	Application method
Isoxadifen-ethyl		Rice and maize	ACCase and ALS inhibitors	Post-em
Cyprosulfamide		Maize	HPPD and ALS inhibitors Clodinafop-propargyl	Pre- or post-em
Metcamifen		Rice	Clodinafop-propargyl	Pre- or post-em

Sources: adapted from Davies, J. and Caseley, J.C. (1999) Herbicide safeners: a review. *Pesticide Science* **55**, 1043–1058; Kraehmer, H. *et al.* (2014). Herbicides as weed control agents: state of the art: 1. Weed control research and safener technology: the path to modern agriculture. *Plant Physiology* **166**, 1119–1131.

Research in the last decade has helped to unravel how safeners work and contribute to crop selectivity to herbicides. Since the definitive review by Davies and Caseley (1999) there are now about 20 safener molecules used commercially (Table 4.3). Safeners applied as seed treatments assume that the crop is influenced and not the weed. When applied in a pre-emergence tank mix, the safener needs to be crop selective. 'Leaf-active' safeners were developed in the 1980s as co-formulants with post-emergent herbicides, to confer safety of fenoxaprop-ethyl in cereal crops, for example, fenchlorazole-ethyl, since superseded by mefenpyr-diethyl. Subsequently, clodinafop-propargyl was launched with the safener cloquintocct-mcxyl. The patents for the latter two safeners have now expired and they have since been used to safen other ACCase inhibitors, such as pinoxaden and the acetolactate synthase (ALS) inhibitor, pyroxsulam. More recently, isoxadifen-ethyl has become established as a safener for the use of post-emergence treatments of ACCase and ALS inhibitors in rice and maize. Cyprosulfamide is also an active safener when used post-emergence in maize, and can also safen pre-emergence treatments of 4-hydroxyphenylpyruvate dioxygenase (HPPD) and ALS inhibitors in this crop (Table 4.3).

4.3.1 Safener modes of action

It is generally accepted that safeners enhance the metabolism of herbicides or their detoxification in the treated crop, with the net effect that less active ingredient reaches the target site (Figure 4.8). There is increasing evidence that safener protection involves increased expression of genes coding for glutathione transferases, cytochrome P450 mono-oxygenases and many others (Reichers *et al.*, 2010; Rosinger, 2014). In a definitive study, Behringer *et al.* (2011) investigated the safeners isoxadifen-ethyl and mefenpyr-diethyl in *A. thaliana* when applied to leaves, equating to a post-emergence application in the field. Gene expression profiling revealed the induction of 446 genes potentially involved in detoxification processes.

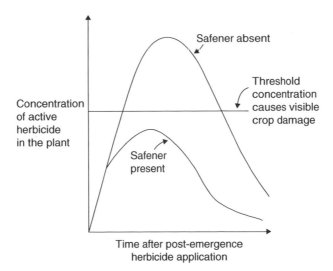

Figure 4.8 Cereal crop selectivity to a theoretical herbicide in the presence or absence of a safener (after Rosinger, 2014).

Genes involved in toxin-catabolism showed a 12-fold enrichment when compared with controls. Also overexpressed were genes responsive to chemical stimuli (including toxins, drugs, salicylic acid, hydrogen peroxide and reactive oxygen species), abiotic stress (high light, heat, oxidative stress) and pathogens. GSTs and ATP-dependent membrane transporters were found to be more than four-fold overexpressed among safener-induced genes. The authors concluded that their observations are consistent with an induction of cellular detoxification machinery at the transcription level after safener treatment of leaves. Further studies are now needed to confirm a similar or different response in other crops following safener application.

Duhoux *et al.* (2017) have observed that since herbicide safeners increase selectivity in crops by increasing the activity of enzymes involved in herbicide metabolism, this is also a mechanism that results in non-target-site-resistance (NTSR) in weeds. Working with populations of ryegrass (*Lolium* sp.), they studied the effect of the safeners cloquintocet-mexyl and mefenpyr-diethyl when applied with the herbicides that inhibit acetolactate synthase. Safener co-application caused an increase in surviving plants and a decreased sensitivity to the herbicides. At the molecular level, such treatment enhanced the expression of NTSR marker genes. They concluded that the evolution of NTSR could be by selection for safener action. It then follows that safeners could be more effective if applied alone to the crop.

In safening rice from clodinafop-propargyl with metcamifen, Brazier-Hicks *et al.* (2019) demonstrated changes in gene expression over a 4 h period that they divided into three phases. The first phase of gene induction (30 min) was dominated by transcription factors and unknown proteins; the second phase was dominated by genes involved in herbicide detoxification (90 min); and the third phase, to 4 h after treatment, was linked to cellular homeostasis, similar to that caused by abscisic acid, salicylic acid and methyl-jasmonate, which are known stress signalling agents in plants. Similar conclusions have been made by Baek *et al.* (2019), who reported the induction of defence genes being following the application of safeners in sorghum.

In the absence of safener, slow rates of herbicide metabolism prolong the active concentration of the herbicide, and so visible crop damage may be observed. In the presence of the safener, there is a more rapid rate of herbicide metabolism, preventing visible damage. Note that the safening effect is only observed in cereals, i.e. monocotyledonous crops.

Almost 250 million kg of herbicides are annually applied in the USA, often with large amounts of safeners to reduce crop damage. Note, however, that the environmental effects of safeners are largely unknown as they are regarded as 'inert ingredients' to the spray mixture. Only recently have studies been done to establish their physicochemical properties, which may be valuable in predicting their distribution and fate in the agricultural environment. While some can be photolysed in ultra-violet light (Kral *et al.*, 2019), others are likely to persist in aquatic environments (Acharya and Weidhass, 2018), but may be removed by treatment with granular activated carbon (Acharya *et al.*, 2020).

4.3.2 Synergists

In contrast to safeners functioning by stimulating herbicide degradation, herbicide synergists inhibit these enzyme systems, so that plant defence mechanisms are overcome and phytotoxicity rapidly ensues. Aminobenzotriazole and tetcyclasis, for example, are able to bind to the haem portion of the cytochrome P450 to prevent herbicide oxidations. Furthermore, tridiphane can inhibit GSTs to synergise atrazine, may form its own phytotoxic glutathione conjugate, and may also inhibit cytochrome P450-linked monooxygenases (Moreland *et al.*, 1989; Figure 4.9).

A generalised summary of safener and synergist action is presented in Figure 4.10.

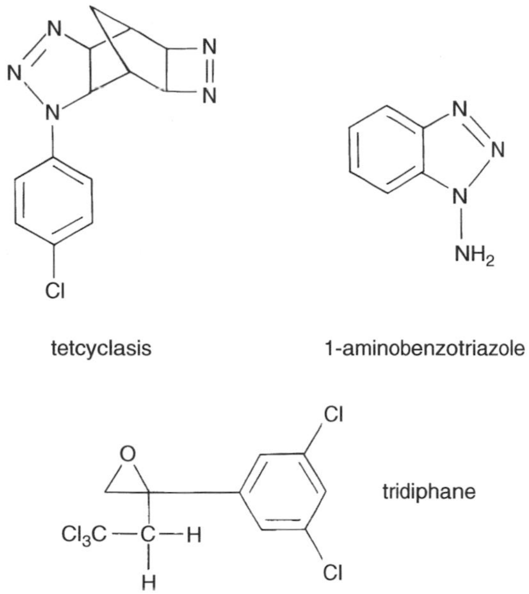

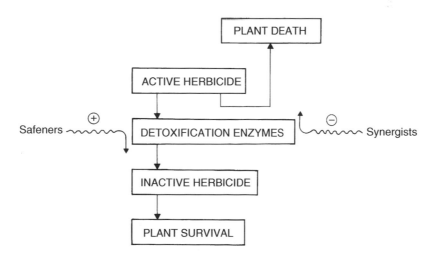

tetcyclasis   1-aminobenzotriazole

tridiphane

Figure 4.9 Structures of tetcyclasis, 1-aminobenzotriazole and tridiphane. Source: based on Moreland, D.E., Novitzky, W.P. and Levi, P.E. (1989) Selective inhibition of cytochrome p450 isozymes by the herbicide synergist tridiphane. *Pesticide Biochemistry and Physiology* **35**, 42–49.

Figure 4.10 A generalised summary of safener and synergist action.

References

Acharya, S.-P. and Weidhaas, J.L. (2018) Solubility, partitioning, oxidation and photodegradation of dichloroacetamide safeners, benoxacor and furilazole. *Chemosphere* **211**, 1018–1024.

Acharya, S.-P, Johnson, J. and Weidhass, J. (2020) Adsorption kinetics of the herbicide safeners, benoxacor and furilazone, to activated carbon and agricultural soils. *Journal of Environmental Sciences* **89**, 23–34.

Baek, Y.S., Goodrich, L.V., Brown, P.J., James, B.T., Moose, S.P., Lambert, K.N. and Reichers, D.E. (2019) Transcriptome profiling and genome-wide association studies reveal GSTs and other defence genes involved in multiple signalling pathways induced by herbicide safener in grain sorghum. *Frontiers of Plant Science*, March; doi: org/10.3389/fpls.2009.

Bak, S., Beisson, F., Bishop, G.J., Hofer, R., Hamberger, B., Paquette, D. *et al.* (2011) Cytochromes P450. *The Arabidopsis Book* 9, e0144. American Society of Plant Physiologists.

Beheringer, C., Bartsch, K. and Schaller, A. (2011) Safeners recruit multiple signalling pathways for the orchestrated induction of the cellular xenobiotic detoxification machinery in *Arabidopsis*. *Plant Cell and Environment* **34**, 1970–1985.

Brazier-Hicks, M. and six others (2020) Chemically induced herbicide tolerance in rice by the safener metcamifen is associated with a phased stress response. *Journal of Experimental Botany* **71**,411–421; doi: org/10.1093/jxb/erz438

Cummins, I., Burnet, M. and Edwards R. (2001) Biochemical characterisation of esterases active in hydrolysing xenobiotics in wheat and competing weeds. *Physiologia Plantarum* **113**, 477–485.

Davies, J. and Caseley, J.C. (1999) Herbicide safeners: a review. *Pesticide Science* **55**, 1043–1058.

Duhoux, A., Pernin, F., Desserre, D., and Delye, C. (2017) Herbicide safeners decrease sensitivity to herbicides inhibiting acetolactate synthase and likely activate non-target-site-resistance pathways in the major grass weed *Lolium sp. Frontiers of Plant Science*, August; doi: org/10.3389/fpls.2017.01310 f

Edwards, R. and Dixon, D.P. (2000) The role of glutathione transferases in herbicide metabolism. In: Cobb, A.H. and Kirkwood, R.C. (eds) *Herbicides and their Mechanisms of Action*. Sheffield: Sheffield Academic Press, Chapter 3.

Gershater, M.C., Cummins, I. and Edwards, R. (2007) Role of a carboxylesterase in herbicide bioactivation in *Arabidopsis thaliana. Journal of Biological Chemistry* **282**, 21460–21466.

Jiang, H.-W., Liu, M.-J., Chen, I.-C., Huang, C.-H., Chao, L.-Y. and Hseih, H.-L. (2010) A glutathione-*S*-transferase regulated by light and hormones participates in the modulation of *Arabidopsis* seedling development. *Plant Physiology* **154**, 1646–1658.

Kraehmer, H. and three others (2014) Herbicides as weed control agents: state of the art: 1. Weed control research and safener technology: the path to modern agriculture. *Plant Physiology* **166**, 1119–1131.

Kral, A.E., Pflug, N.C., McFadden, M.E., LeFevre, G.H., Sivey, J.D. and Cwiertny, D.M. (2019) Photochemical transformations of dichloroacetamide safeners. *Environmental Science and Technology* **53**, 6738–6746.

Leah, J.M., Worrall, T.W. and Cobb, A.H. (1992) Isolation and characterisation of two glucosyltransferases from *Glycine max* associated with bentazone metabolism. *Pesticide Science* **34**, 81–87.

Li, D., Xu, L., Pang, S., Liu, Z., Wang, K. and Wang, C. (2017) Variable levels of glutathione-*S*-transferases are responsible for the differential tolerance to metolachlor between maize shoots and roots. *Journal of Agriculture and Food Chemistry* **65**, 39–44.

Lu, Y.P., Li, Z.S. and Rea, P.A. (1997) AtMRP1 gene of *Arabidopsis* encodes a glutathione *S*-conjugate pump. *Proceedings of the National Academy of Sciences, USA* **94**, 8243–8248.

Moreland, D.E., Novitzky, W.P. and Levi, P.E. (1989) Selective inhibition of cytochrome p450 isozymes by the herbicide synergist tridiphane. *Pesticide Biochemistry and Physiology* **35**, 42–49.

Nelson, D.R. (2013) A world of cytochrome P450s. *Philosophical Transactions of the Royal Society*, *B***368**, 20120430.

Pflugmacher, S. and Sandermann Jr, H. (1998) Taxonomic distribution of plant glucosyltransferases acting on xenobiotics. *Phytochemistry* **49**, 507–511.

Reade, J.P.H., Milner, L.J. and Cobb, A.H. (2004) A role for glutathione *S*-transferases in resistance to herbicides in grasses. *Weed Science* **52**, 468–474.

Reichers, D.E., Kreuz, K. and Zhang, Q. (2010) Detoxification without intoxification: herbicide safeners activate plant defence gene expression. *Plant Physiology* **153**, 3–13.

Rosinger, C. (2014) Herbicide safeners: an overview. *Developments in Herbicides Julius-Kuhn Archiv* **443**, 516–525; doi: org10.5073/jka.2014.443.066

Wen-Sheng, X., Xiang-Jin, W., Tian-Rui, R. and Xiu-Lian, J. (2005) Expression of a wheat cytochrome P450 monooxygenase in yeast and its inhibition by glyphosate. *Pest Management Science* **62**, 402–406.

Werck-Reichhart, D., Hehn, A. and Didierjean, L. (2000) Cytochrome P450 for engineering herbicide tolerance. *Trends in Plant Science* **5**, 116–123.

Chapter 5
Herbicides that Inhibit Photosynthesis

The sun gives spirit and life to the plants and the earth nourishes them with moisture.

Leonardo da Vinci (1452–1519)

5.1 Introduction

Photosynthesis is the process whereby solar energy is converted into chemical energy by plants, algae and some bacteria, and takes place in the chloroplast, an organelle unique to the plant kingdom (Figure 5.1). The generalised and highly simplified equation

$$CO_2 + H_2O \xrightarrow{\text{light}} (CH_2O) + O_2$$

summarises photosynthesis, in which, firstly, water is oxidised within the thylakoid membranes to protons, electrons and oxygen. The movement of protons and electrons through the thylakoid generates ATP and reduced $NADP^+$, and this energy is utilised in the enzymatic reduction of CO_2 to carbohydrates within the chloroplast stroma. No herbicides are known to directly interfere with carbon reduction, but many of the herbicides currently in use act by either blocking or diverting thylakoid electron flow. It is therefore not surprising that considerable research effort has been directed towards the chemical inhibition of photosynthetic electron flow.

5.2 Photosystems

In energy terms, carbon reduction and water oxidation are separated by about 1.2 V. This means that electrons have to move uphill against an energy gradient of 1.2 V in order to reduce CO_2 to carbohydrates. This unique process is driven by two inputs of light energy at huge pigment–protein complexes termed photosystems, which span the thylakoid membrane. The thylakoid itself is now known to contain two functional regions, namely appressed

Herbicides and Plant Physiology, Third Edition. Andrew H. Cobb.
© 2022 John Wiley & Sons Ltd. Published 2022 by John Wiley & Sons Ltd.

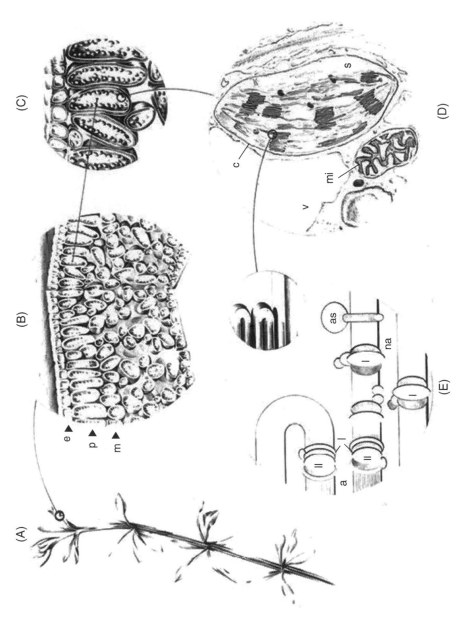

Figure 5.1 Leaf structure in *Galium aparine* (cleavers). (A) Leaf arrangement in whorls. (B) Leaf section to show epidermal (e), palisade (p) and mesophyll (m) cells. (C) Palisade cell detail. (D) Cell ultrastructure featuring chlorplast (c), stroma (s), vacuole (v) and mitochondrion (mi). (E) Detail of thylakoids. Appressed thylakoids (a) are enriched in Photosystem II (II) and light-harvesting complexes (L); non-appressed thylakoids (na) are enriched in Photosystem I (I) and ATP synthase (as).

and non-appressed membranes (Figure 5.1). The appressed region shows close membrane interaction and is enriched in Photosystem II (PS II) and the non-appressed region contains stroma-exposed areas and is enriched in Photosystem I (PS I) and the ATP synthase complex. The relative proportions of appressed vs. non-appressed regions depends on the chemical environment of the chloroplast, and is very sensitive to available photon flux density. Thus, leaves growing in shade conditions will contain chloroplasts with more appressed areas (i.e. granal stacks), whereas leaves in full sun will exhibit proportionately more non-appressed regions and a decreased thylakoid to stroma ratio. Such changes in thylakoid architecture can occur within hours and suggest that this membrane is particularly fluid in nature. Indeed, thylakoids are about 50% lipid with high concentrations of electroneutral galactolipids, and this property permits the movement of electron carriers, such as plastoquinone, and light-harvesting complexes, from appressed to non-appressed areas.

Each photosystem contains specific polypeptides, pigments and electron donors/acceptors, and a unique chlorophyll termed a reaction centre (p680 or p700) at which an electron is moved from low to higher energy. Functionally, the two photosystems operate in series, such that the primary reductant generated from the photolysis of water at the oxidising side of PS II passes electrons through a series of carriers of lower reducing power to PS I. Here a second light reaction transfers electrons to their eventual natural acceptor, $NADP^+$.

When a photon of light is absorbed by the light-harvesting complex and the excitation energy transferred to the PS II reaction centre (p680), a charge separation occurs to $p680^+$ and $p680^-$. The species $p680^+$ is quenched by an electron from water via a tyrosine residue on the D1 protein, and phaeophytin, a chlorophyll molecule lacking magnesium, is the primary electron acceptor from $p680^-$. The first stable electron acceptor is a quinone, Q_A, tightly bound within a particular protein environment. Q_A acts as a single electron carrier and is closely associated with the two-electron carrier, Q_B, located on the D1 protein, which delivers pairs of electrons to the mobile plastoquinone pool (Figure 5.2). Photosystem II may therefore be regarded as a water–plastoquinone oxidoreductase. Electrons then pass via a cytochrome b_6–f complex to the copper-containing protein plastocyanin and to PS I.

The PS I complex contains the reaction centre chlorophyll p700, comprising iron–sulphur centres which act as electron acceptors, several polypeptides and the electron carriers A_0 and A_1. Light excitation causes charge separation at p700 and A_0, a specific monomeric chlorophyll a, is the first acceptor. A_1 is thought to be a phylloquinone (vitamin K_1) which donates an electron to the iron–sulphur centres and hence reduces $NADP^+$ via ferredoxin (Figure 5.3). PS I therefore operates as a plastocyanin–ferredoxin oxidoredutase.

Detailed structures are now known of the light-harvesting and photosystem complexes and the electron transport chain components in the thylakoid membrane. Indeed, some have actually been crystallised and their three-dimensional structures established. Figure 5.4 gives a schematic version of the thylakoid, demonstrating H^+ and e^- flow. The reader is referred to Blankenship (2002) and Lawlor (2001) for more details.

5.3   Inhibition at Photosystem II

The action of photosynthetic inhibitors has traditionally been monitored *in vitro* by measuring the so-called 'Hill Reaction', named after Robin Hill, who first described it in 1937. This is the ability of a thylakoid preparation to evolve oxygen in the presence of a suitable electron acceptor. The natural acceptor, $NADP^+$, is washed free from isolated thylakoids, so artificial acceptors (A) are used experimentally.

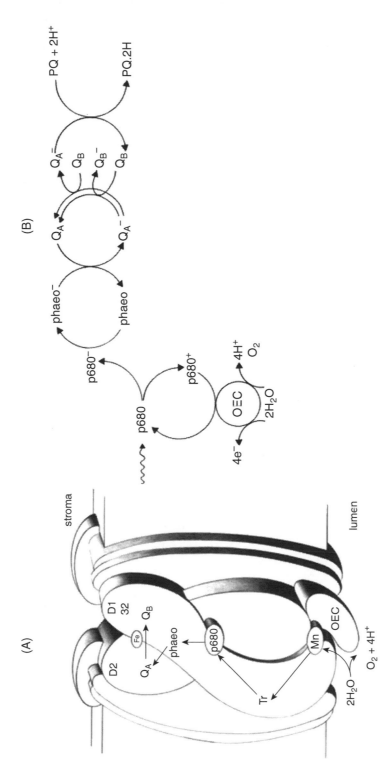

Figure 5.2 Structural (A) and functional (B) models of Photosystem II. Number 32 refers to the molecular mass of the D1 polypeptide in kDa. Tr, A tyrosine residue acting as an electron donor to p680$^+$; the bold arrow shows the direction of electron transfer; OEC, oxygen-evolving complex; p680, PS II reaction centre; phaeo, phaeophytin a; Q_A and Q_B, quinones; PQ, mobile plastoquinone pool. Source: adapted from Rutherford, A.W. (1989) Photosystem II, the water-splitting enzyme. *Trends in Biochemical Sciences* **14**, 227–232.

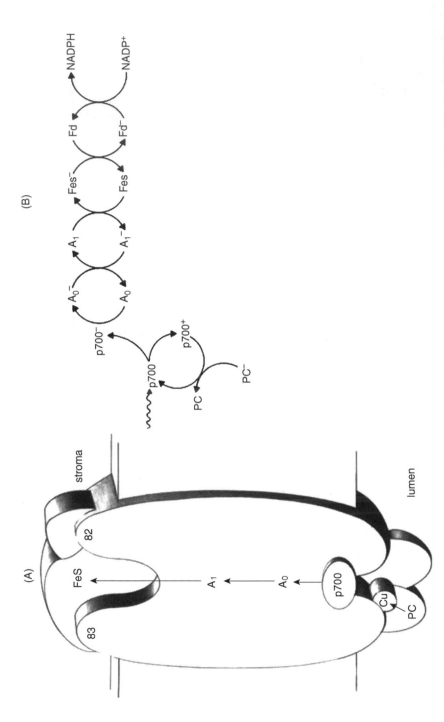

Figure 5.3 Structural (A) and functional (B) models of Photosystem I. Numbers 82 and 83 refer to molecular masses of polypeptides in kDa. PC, plastocyanin (the electron donor to p700⁺); p700, PS I reaction centre; FeS, iron–sulphur centres; Fd, ferredoxin. The final electron acceptor is NADP⁺. Source: adapted from Knaff, D.B. (1988) The Photosystem I reaction centre. *Trends in Biochemical Sciences* **13**, 460–461.

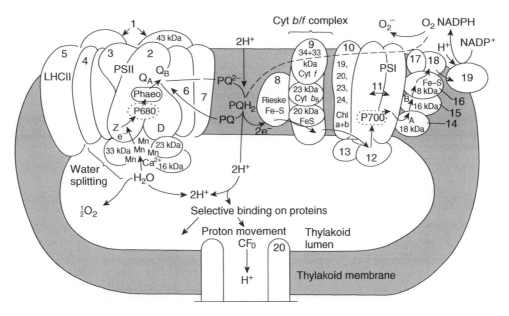

Figure 5.4   Schematic representation of the light-harvesting and photosystem complexes and electron transport chain in the thylakoid membrane (Lawlor, 2001, by permission of Oxford University Press). The number of protein complexes and their relation are shown. The molecular mass of components is indicated. The components are listed as follows. Key: 1, antenna protein–pigment complex; 2, 32 kDa, D1 herbicide binding protein of the reaction centre; 3, 32 kDa, D2 reaction centre protein; 4, cytochrome b_{559}, 9 kDa b_{559} type 1 and 4 kDa b_{559} type 2 proteins; 5, light-harvesting antenna; 6, 10 kDa docking protein; 7, 22 kDa stabilising protein (intrinsic membrane protein); 8, 20 kDa Rieske Fe–S centre; 9, cytochrome b_6–f complex with polypeptides; 10, light-harvesting protein–pigment (chlorophyll a and b) complex of PS I and polypeptides; 11, PS I reaction centre, with two 70 (?) kDa polypeptides; 12, plastocyanin, 10.5 kDa; 13, plastocyanin binding protein (10 kDa); 14, Fe–S protein A, 18 kDa; 15, Fe–S proteins B, 16 kDa; 16, Fe–S protein 8 kDa; 17, ferredoxin binding protein; 18, ferredoxin; 19, ferredoxin, NADP oxidoreductase; 20, coupling factor, CF_0, membrane subunits. Source: Lawlor, D.W. (2001) *Photosynthesis*, 3rd edn. Oxford: BIOS.

$$2H_2O + 2HA \xrightarrow[\text{thylakoids}]{\text{light}} 2AH_2 + O_2$$

A commonly used artificial electron acceptor is potassium ferricyanide, which is reduced to ferrocyanide with the evolution of oxygen. Oxygen evolution can be conveniently monitored using an oxygen electrode. Similarly, the blue dye 2,6-dichlorophenolindophenol (DCPIP) will accept electrons from the functioning thylakoid and the reduced product is colourless. Thus, photosynthetic electron flow can be measured by following DCPIP decolourisation with a spectrophotometer (Figure 5.5).

Such reactions, however, measure the activity of both photosystems rather than specific sites, and are therefore of limited value in identifying precise sites of photosynthetic inhibition. Nowadays the partial reactions of electron flow can be characterised in great detail using highly specific artificial electron donors and acceptors (e.g. Table 5.1; Trebst, 1980). Indeed, herbicides have proved invaluable in unravelling the flow of electrons in the thylakoid.

It was by using DCPIP as an artificial electron acceptor that Wessels and van der Veen (1956) first demonstrated that the urea herbicide diuron could reversibly inhibit photosynthetic electron flow at micromolar concentrations. Subsequent studies in the following decades

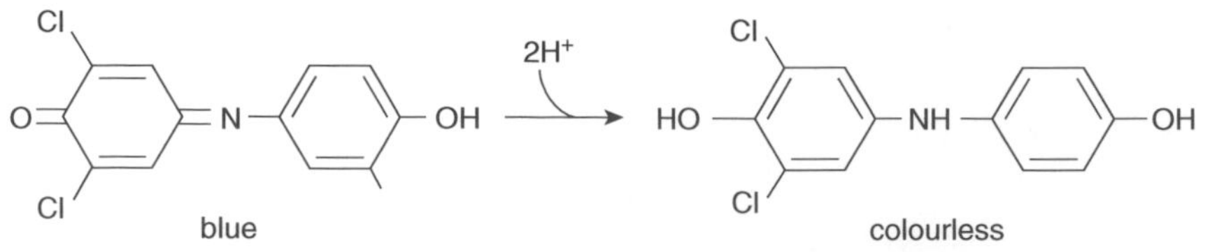

Figure 5.5 DCPIP decolourisation in the presence of functioning thylakoids.

Table 5.1 Examples of the measurement of photosystem activity *in vitro* by oxygen exchange.

Donor	Acceptor	Inhibitor	Reactions measured	Oxygen
H_2O	Ferricyanide	Absent	PS II and PS I ($H_2O \rightarrow$ FeS centres)	Evolution
H_2O	Dimethyl-benzoquinone	Absent	PS II only ($H_2O \rightarrow$ plastoquinone)	Evolution
H_2O	Silicomolybdate	Absent	PS II only ($H_2O \rightarrow$ phaeophytin)	Evolution
Ascorbate + DCPIP	Methyl viologen	Diuron	PS I only (plastocyanin $\rightarrow$ ferredoxin)	Uptake

established that diuron was acting at the reducing side of PS II in the vicinity of Q_A and Q_B. This was deduced from the observations that (a) PS I activity was insensitive to diuron, (b) electron flow at the cytochrome b_6–f complex was similarly unaffected and (c) diuron had no influence on the photolysis of water, nor on charge separation at p680, but electron flow to plastoquinone was potently inhibited.

Several classes of herbicide including the ureas, triazines and phenols are now known to inhibit PS II activity by displacing plastoquinone from the Q_B site and so preventing electron flow from Q_A^-. Although Q_A is tightly bound to the D2 protein, Q_B is not firmly bound to D1, and so herbicides successfully compete with Q_B for this site. Inhibition by the ureas and triazines is characteristically reversible and competitive. Furthermore, it was found that there was only one binding site for each PS II reaction centre, and that binding/dissociation constants were very similar to inhibition constants, hence occupancy of this site is required for photosynthetic inhibition.

Studies in the 1970s established that thylakoids treated with trypsin became insensitive to diuron, suggesting a stromal-facing proteinaceous binding site (Renger, 1976). This protein was further characterised using radiolabelled herbicides as a rapid turnover polypeptide (molecular mass 32 kDa) that was encoded in the chloroplast *psb*A genome (Mattoo *et al.*, 1981). Conclusive evidence that this protein played a dominant role in herbicide binding came from its analysis in plants showing resistance to PS II herbicides. Hirschberg *et al.* (1984) cloned the *psb*A gene from atrazine-resistant and susceptible biotypes of *Solanum nigrum* and *Amaranthus retroflexus*, and detected a single base substitution from serine to glycine at amino acid 264 to be the basis of resistance to the herbicide. Resistance is therefore achieved by the substitution of one amino acid in a protein containing 353 amino acid residues!

Figure 5.6 presents the interaction between the quinones and atrazine at protein D1. A separate binding site specific to the phenols (such as dinoseb) and the hydroxybenzonitriles (e.g. ioxynil) has been proposed with a molecular mass of 41–47 kDa that is less

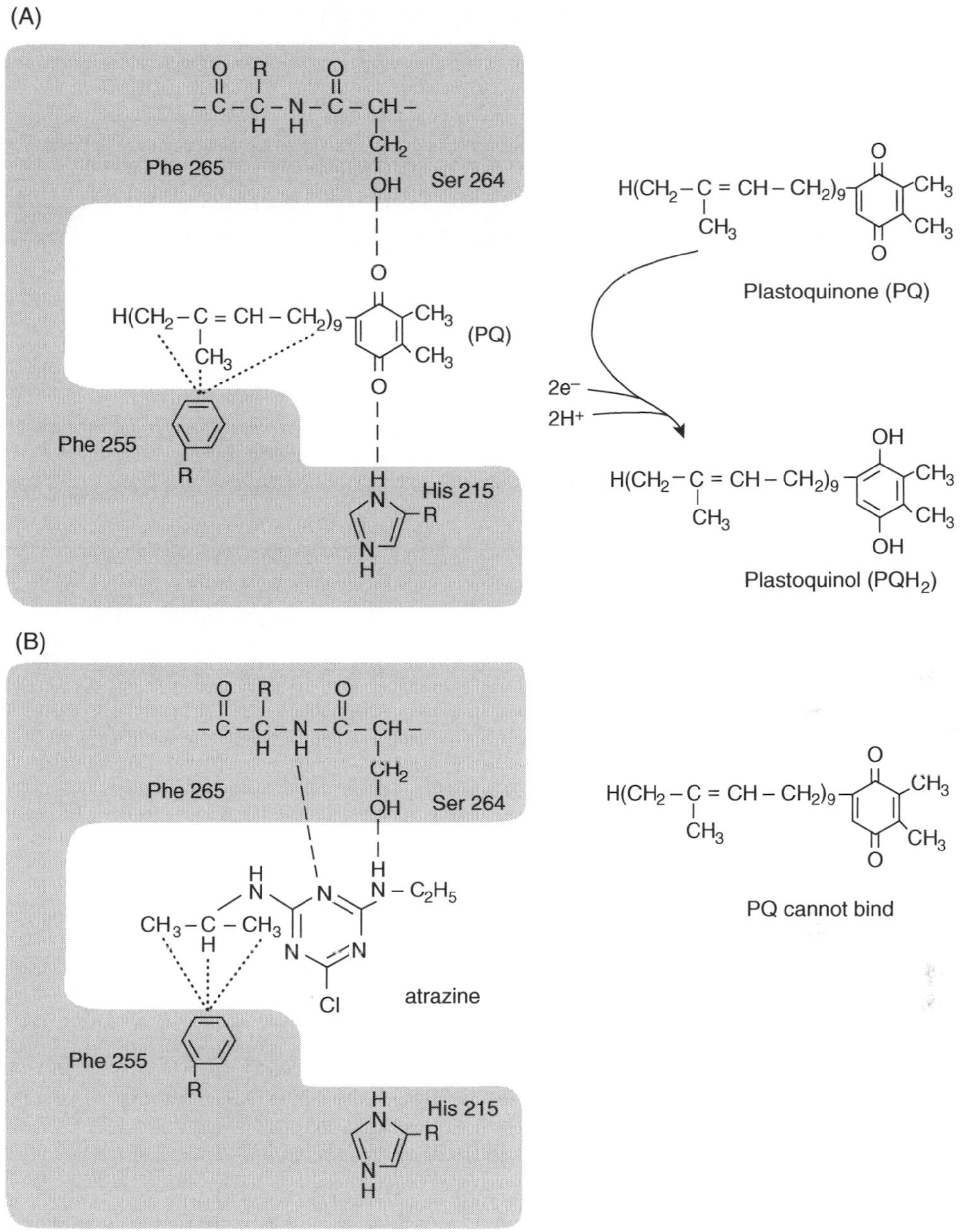

Figure 5.6 Proposed interaction between plastoquinone (A) and atrazine (B) with the Q$_B$ site of protein D1. Dashes represent hydrogen bonds and dots represent hydrophobic interactions (modified from Fuerst and Norman, 1991, from the journal *Weed Science*, courtesy of the Weed Science Society of America). (A) PQ binds to the D1 protein, accepts two electrons and two protons, and is released as PQH$_2$. (B) Atrazine binding to the D1 protein prevents the binding of PQ. Source: modified from Fuerst, E.P. and Norman, M.A. (1991) Interactions of herbicides with photosynthetic electron transport. *Weed Science* **39**, 458–464.

susceptible to trypsin digestion, and therefore more deeply located in the thylakoid. It is argued that these compounds bind to a different site on the D1 protein.

Our understanding of this protein was vastly enhanced when the reaction centre of the photosynthetic bacterium *Rhodopseudomonas viridis* was first crystallised and then its structure resolved by Michel and Deisenhofer in 1984 (see Michel and Deisenhofer, 1988). This seminal work, for which the Nobel Prize for Chemistry was awarded in 1988, pointed to clear similarities and sequence homologies between the bacterial reaction centre and PS II, and provided three-dimensional detail of the amino acids involved in this quinone-binding herbicide niche on the D1 protein. Trebst (1987) utilised this X-ray data and, with information gained from photoaffinity labelling of herbicides and site-directed mutagenesis, proposed a detailed model of the herbicide binding niche. This model shows that the D1 polypeptide contains five transmembrane helical spans (1–5) and two parallel helices (A and B), and that helices 4, 5 and B specifically participate in herbicide binding (Figure 5.7a). The model predicts the orientation of residues at the quinone niche (Figure 5.7b) and envisages Q_B binding via two hydrogen bridges at His 215 and close to Ser 264.

Following closer examination of the Q_B binding niche, Trebst (1987) suggested that inhibitors with a carbonyl or equivalent group (e.g. ureas, triazines, triazinones) were orientated towards the peptide bond close to Ser 264 and could form a hydrogen bridge to this peptide bond. However, the phenol group of inhibitors cannot form this bridge and are therefore thought to bind towards His 215, where they are bound more strongly to the membrane. In this way, Trebst envisages two families of PS II herbicides, namely the serine and the histidine families.

By 2002, additional mutants were identified with altered amino acids in the stroma-exposed β-helix, particularly in algae, and other substitutions noted that produce herbicide resistance (Table 5.2). Substitutions at position 264 from serine to glycine yield atrazine resistance, as demonstrated in Table 13.1, while serine to alanine or threonine produces additional resistance to diuron. This observation has led to the proposal that a serine hydroxyl group may be essential for atrazine binding. Changes at nearby residues 255 and 256 also cause triazine resistance, although changes at positions 219 and 275 give resistance to diuron and also to bromoxynil in the latter example. Resistance may then be due to conformational changes in the binding niche, an absence of a specific residue so that a herbicide is unable to bind or the introduction of steric hindrance to prevent herbicide access to the binding niche.

A new group of triazine herbicides, the 2-(4-halogenobenzyl amino)-4-methyl-6-trifluoromethyl-1,3,5-triazines, were shown in 1998 to inhibit photosynthetic electron flow at the D1 protein. Interestingly, they have shown activity against atrazine-resistant species, such as *Chenopodium album* L. and *Solanum nigrum*, presumably by binding to different amino acids in the D1 niche (Kuboyama *et al.*, 1999; Kohno *et al.*, 2000). Figure 5.8 shows the structure of one example, BW 314, with the structure of atrazine also presented for comparison.

5.4   Photodamage and repair of Photosystem II

Plants are remarkable organisms in their ability to absorb light energy and convert it into chemical energy. Unlike animals, however, plants cannot move away from unfavourable environmental conditions and may not be able to process all of the excitation energy to which they are exposed. This excess energy will lead to the generation of active oxygen species and

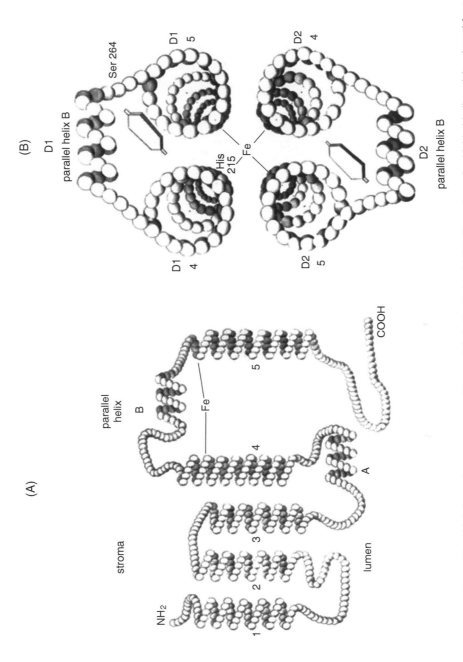

Figure 5.7 The D1 polypeptide. (A) Schematic arrangement of the protein in the thylakoid. (B) The quinone/herbicide binding niche viewed from above. Source: adapted from Trebst, A. (1987) The three-dimensional structure of the herbicide binding niche on the reaction centre polypeptides of Photosystem II. *Zeitschrift für Naturforschung* **42c**, 742–750.

Table 5.2 Characteristics of mutants with amino acid substitutions in D1.

Mutation	Organism	Primary resistances	Negative cross-resistance
Phe$_{211}$→Ser	Cyanobacteria	Atrazine ×9	
Val$_{219}$→Ile	Alga + cyano	Metribuzin ×200	Ketonitrile × 0.6
		Metabenzthiazuron ×62	
	Poa annua	Metribuzin and/or diuron	
Tyr$_{237}$→Phe	Cyanobacterium	Diuron/ioxynil ×5	BNT ×0.2
Lys$_{238}$→Val	Cyanobacterium	Ioxynil ×2.3	BNT ×0.4
Ile$_{248}$→Thr	Cyanobacterium	Metribuzin ×28	
Ala$_{250}$→Arg	Alga	Phenmedipham ×6.3	Bromoxynil ×0.3
Ala$_{250}$→Asn	Alga	Metamitron ×5	Bromoxynil ×0.2/ atrazine ×0.3
Ala$_{250}$→Asp	Alga	Phenmedipham ×5	-oxynils ×0.2/ atrazine ×0.25
Ala$_{250}$→His	Alga	Phenmedipham ×10	
Ala$_{250}$→Ile	Alga	Phenmedipham ×2.5	Metribuzin ×0.25
Ala$_{250}$→Tyr	Alga	Phenmedipham ×20	Bromoxynil ×0.4
Ala$_{251}$→Cys	Alga	Metamitron/bromoxynil ×6.3	
Ala$_{251}$→Gly	Alga	Metamitron ×10	
Ala$_{251}$→Ile	Alga		Diuron ×0.6
Ala$_{251}$→Leu	Alga	Metribuzin ×108	
		Bromacil ×26	
Ala$_{251}$→Val	Alga/cyano	Metribuzin ×1000	Ketonitrile ×0.5
Phe$_{255}$→Tyr	Alga/cyano	Cyanoacrylate ×39	Metamitron ×0.3
Gly$_{256}$→Asp	Alga	Bromacil ×10	
Arg$_{257}$→Val	Cyanobacterium	Atrazine/diuron ×34	BNT ×0.3
Ala$_{263}$→Pro	Cyanobacterium	Atrazine ×2000/ metribuzin ×1600	
Ser$_{264}$→Ala	Alga/cyano/ *Euglena*	Metribuzin ×>3000 Chloroxuron ×480/ atrazine variable	
Ser$_{264}$→Gly	Weeds/cyano	Most s- and as-triazines ×>500	-oxynils/pyridate ×<0.5
Ser$_{264}$→Asn	Tobacco cells	Terbutryn	
Ser$_{264}$→Pro	Cyanobacterium	Atrazine ×10,000	
Ser$_{264}$→Thr	Plant cells/*Euglena* *Portulaca oleracea*	Atrazine ×>50 Linuron	Dinoseb/-oxynils/ BNT ×<0.3
Asn$_{266}$→Asp	Cyanobacterium	Ioxynil ×2.5	
Asn$_{266}$→Thr	Cyanobacterium	Bromoxynil ×15	
Ser$_{268}$→Pro	Soybean cells	Atrazine ×50	
Arg$_{269}$→Gly	Alga	Terbutryn ×8	Ioxynil ×0.2
Leu$_{275}$→Phe	Alga	Metamitron ×63	

Source: modified from Gressel, J. (2002) *Molecular Biology of Weed Control*. London: Taylor & Francis.

membrane protein damage if it is not dissipated effectively. Such photodamage appears to be mainly confined to the D1 protein in the core Photosystem II reaction centre complex and inactivation can be rapidly detected following minutes of exposure to high light energies.

Fortunately, a damage-repair cycle (Figure 5.9) is present in which the PS II reaction centre is disassembled, a new D1 protein is synthesised and the complex reassembled and

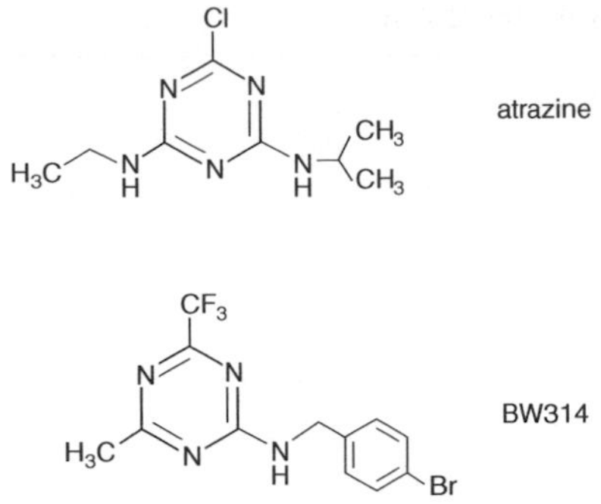

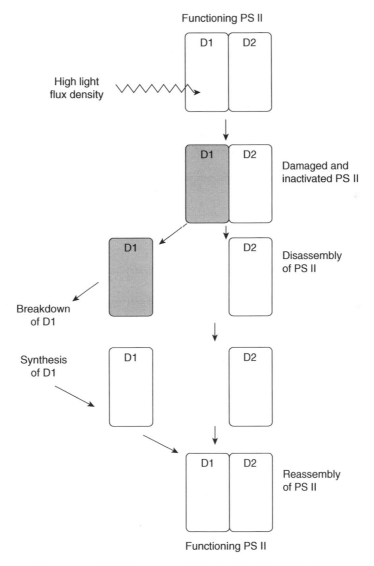

Figure 5.8 The structures of atrazine and BW314.

Figure 5.9 Damage–repair cycle in Photosystem II. Source: Based on Melis, A. (1999) Photosystem II damage and repair cycle in chloroplasts: what modulates the rate of photodamage *in vivo*? *Trends in Plant Science* **4**, 130–135.

activity is restored (Melis, 1999). The precise mechanism of photodamage remains unclear. Perhaps an over-reduction of the quinone acceptor is the cause, although the presence of two β-carotene molecules in the reaction centre appears to be crucial to the process. Without these molecules, the breakdown of protein D1 is favoured. While the precise role of these pigments is uncertain, it is clear that the inhibition of carotenoid biosynthesis by lycopene cyclase or phytoene desaturase inhibitors leads initially to an inhibition of photosynthetic electron flow and later to pigment bleaching (Fedtke *et al.*, 2001).

Herbicidal action as a consequence of binding to the D1 protein is now well known. Activity is effective and well characterised both in crops and weeds. One disadvantage, however, is the relatively high application rates required for phytotoxicity, typically 2–4 kg ha^{-1} with atrazine. Duff *et al.* (Fabbri *et al.*, 2005; Duff *et al.*, 2007) at Monsanto have suggested that the inhibition of D1 protein biosynthesis could be a better target for herbicide action at gram rather than kilogram doses. They have discovered that a carboxy-terminal processing protease, CtpA, is a low-abundance thylakoid enzyme that catalyses the conversion of pre-D1 into the active form by cleaving the 9-C terminal residues. They expressed a recombinant form of the enzyme in *Escherichia coli* (rCtpA), purified it and used it in a high throughput screen for CtpA inhibitors. Lead compounds were studied *in vitro* and most were shown to be competitive inhibitors with K_i values in the range of 2–50 µM. These authors conclude that CtpA inhibitors could become a new generation of effective herbicides, although none have yet been marketed.

As mentioned above, many thylakoid structures involved in photosynthetic electron transfer are present as large pigment–protein complexes, which need to be continually assembled and maintained. Many of the proteins are encoded in the nucleus and must be imported into the chloroplast to function. These proteins have to be imported in an unfolded form, which is then refolded in the chloroplast. If not correctly folded, they must be detected and removed in order to protect the thylakoid from damage, especially from reactive oxygen species in the presence of high light stress. Normally, heat-shock proteins, chaperones and proteases ensure that the proteins are correctly folded. However, when exposed to high light intensities, or when a protease ClpP1 is absent, the system can become overwhelmed, resulting in an unfolded protein response (cpUPR), which causes the production of specialist proteins in the nucleus that migrate to the chloroplast to help protect and repair the thylakoid complexes.

How the chloroplast and nucleus coordinate this response is unclear. Perlaza *et al.* (2019) identified a gene coding for a protein termed MARS1 (mutant affected in chloroplast to nucleus retrograde signalling) that is essential for activating cpUPR in the single cell alga, *Chlamydomonas reinhardtii*. It is located in the cytoplasm, has a kinase domain, and is involved in sending the cpUPR signal from the chloroplast to the nucleus. Cells with defective MARS1 were more vulnerable to thylakoid damage. Thus, activation of cpUPR via MARS1 mitigates photooxidative stress and delays or prevents thylakoid photobleaching. The search is now on to identify more components of the cpUPR pathway, to further understand this protective pathway and so become targets to enhance crop yield or to develop new herbicides.

5.5   Structures and uses of Photosystem II inhibitors

Observations made in recent decades have shown that the quinone–herbicide binding domain in PS II is highly conserved throughout the plant kingdom. Since this domain involves the interaction of many amino acids in the D1 protein, it is not too surprising that a large number of chemically dissimilar molecules have been found to bind at this site, thus preventing Q_B reduction. Many have become highly successful herbicides and so detailed quantitative structure–activity relationships have been attempted with the PS II inhibitors. Trebst *et al.* (1984) have suggested that the ureas and triazines, now classified in the 'serine' family, all have a lipophilic group in close association with an *sp²* hybrid and an essential positive charge (Figure 5.10). Examples of herbicide classes in the 'serine' family are shown in Figure 5.11.

Trebst *et al.* (1984) considered that the essential features of the phenol-type 'histidine' herbicides are those shown in Figure 5.12. Examples of herbicides of this type are shown in Figure 5.13.

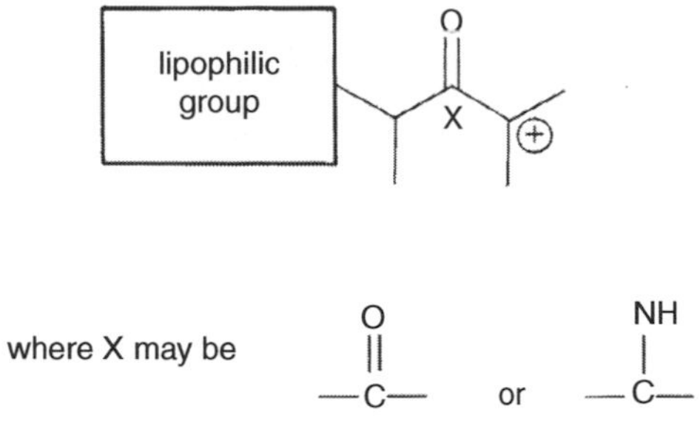

Figure 5.10   Detail of the environment surrounding the lipophilic group in the 'serine' family of herbicides. Source: based on Trebst, A., Donner, W. and Draber, W. (1984) Structure–activity correlation of herbicides affecting plastoquinone reduction by Photosystem II: electron density distribution in inhibitors and plastoquinone species. *Zeitschrift für Naturforschung* **39c**, 405–411.

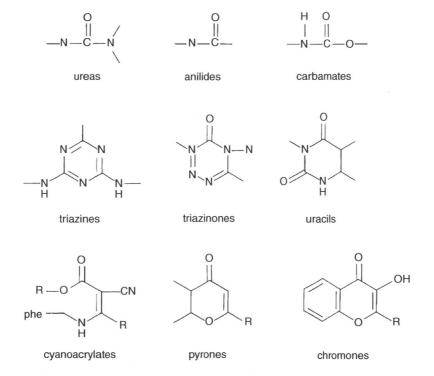

Figure 5.11   Examples of herbicide classes in the serine family.

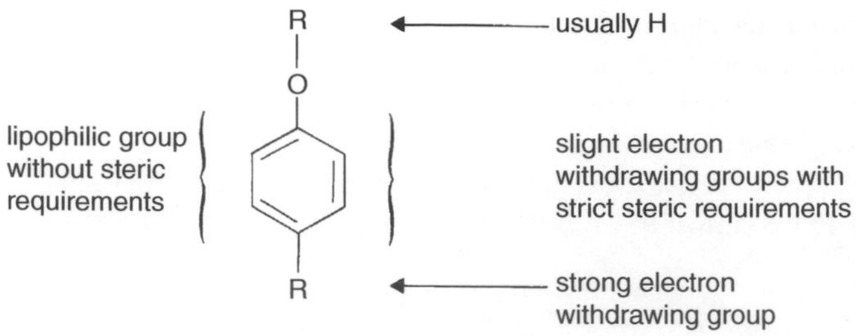

Figure 5.12 Essential features of the phenol-type 'histidine' herbicides. Source: based on Trebst, A., Donner, W. and Draber, W. (1984) Structure–activity correlation of herbicides affecting plastoquinone reduction by Photosystem II: electron density distribution in inhibitors and plastoquinone species. *Zeitschrift für Naturforschung* **39c**, 405–411.

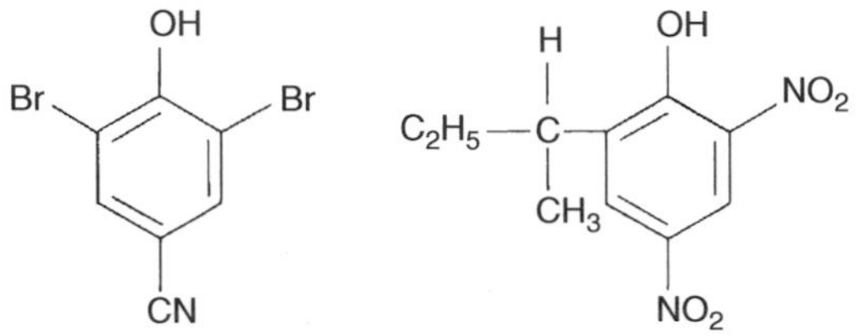

Bromoxynil   Dinoseb

Figure 5.13 Structures of the 'histidine' family of herbicides: the hydroxybenzonitriles, bromoxynil and the nitrophenol, dinoseb.

A few major examples of PS II herbicides in commercial use today are presented in Figure 5.14. It is noteworthy that about 50% of all the pesticides cited in *The Pesticide Manual* (Tomlin, 2000) are inhibitors of photosynthetic electron transport, although recent legislation referred to in Chapter 2 may reduce this value in years to come.

5.6   Interference with electron flow at Photosystem I

The bipyridinium compounds paraquat and diquat are well known as potent, total herbicides with a contact action. Paraquat was known in the 1930s as methyl viologen and was used as an oxidation–reduction indicator, but it was not until the mid-1950s that its herbicidal properties were discovered, and it was commercialised in 1959. Paraquat and diquat, with redox potentials of −0.446 and −0.349 V respectively, exist as di-cations which can accept an electron from one of the iron–sulphur proteins near ferredoxin (−0.420 V) on the stromal side of PS I to form a stable free radical.

This radical is re-oxidised by molecular oxygen to produce active oxygen species, and the paraquat is then free to accept another electron from near ferredoxin (Figure 5.15). In this way a futile cycle operates, whereby the di-cation is constantly regenerated in the light and active oxygen species kill the plant (see Section 5.8).

Although many molecules have been screened, only paraquat and diquat have found commercial use. They appear to possess unique properties for bipyridinium-type action, namely:

1 they are readily soluble in water and are stable at physiological pH ranges;
2 the herbicidal di-cation can be tightly bound by plants, organic matter or soil clay minerals, and so is rapidly inactivated and unavailable to plants;
3 they have suitable negative oxidation–reduction potentials to accept a single electron from PS I;
4 the reduced di-cation is stable and readily oxidised by molecular oxygen; and
5 the molecule is always available to accept further electrons from PS I.

Paraquat and diquat are contact, non-selective, total herbicides which are not metabolised in treated, susceptible plants. They are only active in the light. Phytotoxicity results from the rapid accumulation of active oxygen species causing lipid peroxidation, irreversible membrane destruction and cell death. On a more cautionary note, the bipyridinium herbicides are toxic to animals including mammals, as well as plants. Consequently, their use is now banned

Figure 5.14   A selection of herbicides that inhibit Photosystem II electron flow.

UREAS	USES
Diuron	General weed control in non-crop areas. Selective, pre-emergent control of annual weeds in alfalfa, maize, cotton, pineapple, sugar cane and sorghum. Soil active and residual. Readily absorbed by roots and leaves and rapidly translocated.
Isoproturon	Control of annual grasses and broadleaf weeds in barley, rye and wheat. Soil active and residual. Can leach to groundwater.
Chlorotoluron	As isoproturon.
Linuron	Selective, pre-emergent control of annual weeds in potatoes, carrots, peas, beans, cotton, maize, soybean and winter wheat. Soil active and residual.

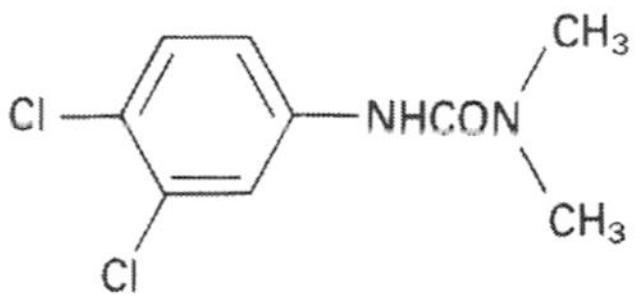

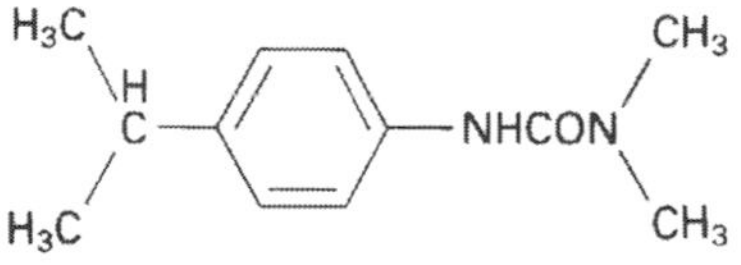

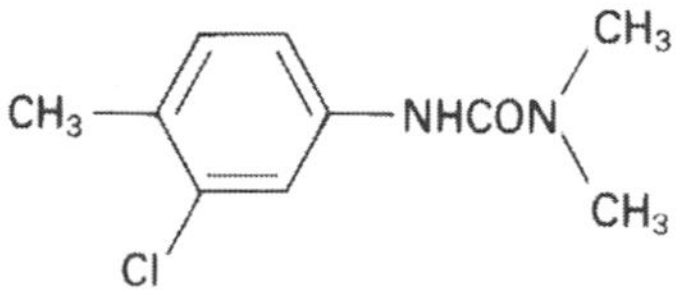

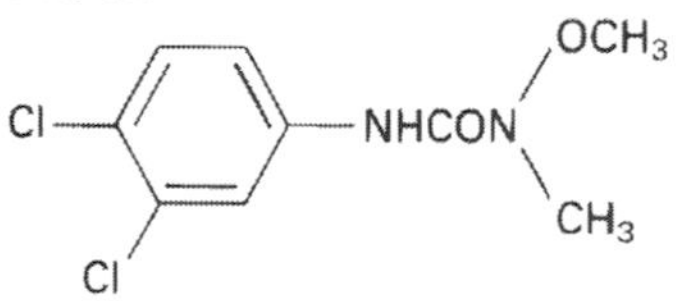

(Continued)

Figure 5.14　(*Continued*)

TRIAZINES	USES
Atrazine	Selective pre- and post-emergent control of annual weeds in maize, sorghum, sugar cane, raspberries and roses. Residual, can persist in soils at high soil pH. Can leach to groundwater.
Simazine	Total weed control in non-crop areas. Pre-emergent or early post-emergent control of annual weeds in beans and maize, fruit bushes, canes and trees.
Prometryn	Pre-emergent control of annual weeds in cotton, peas and carrots. Post-emergent weed control in many vegetable crops and sunflowers.
Terbutryn	Pre-emergent control of blackgrass and broadleaf weed seedlings in winter wheat, barley, sugar cane and sunflowers.
TRIAZINONES	
Metribuzin	Pre- or post-emergent control of annual weeds in lucerne, potatoes, soybean, tomatoes, asparagus and sugar beet. Readily absorbed by roots and leaves and rapidly translocated.
Metamitron	Pre- or post-emergent control of annual weeds in sugar beet and fodder beet.

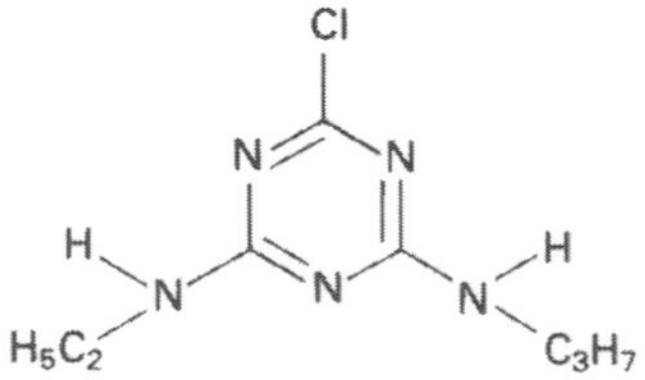

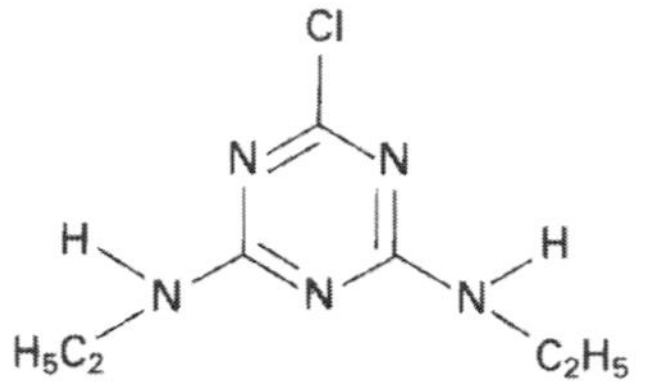

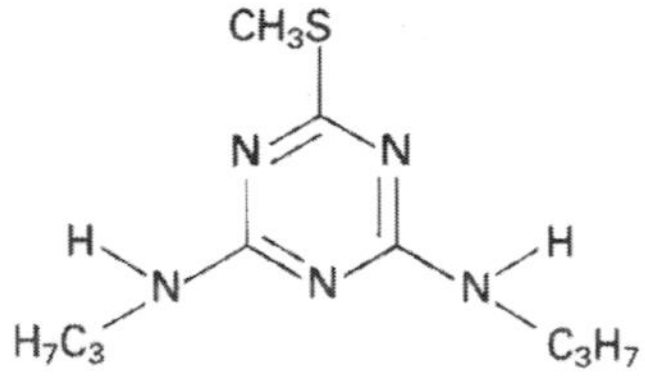

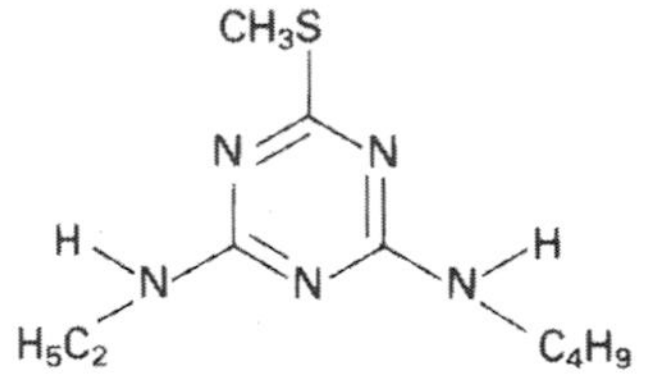

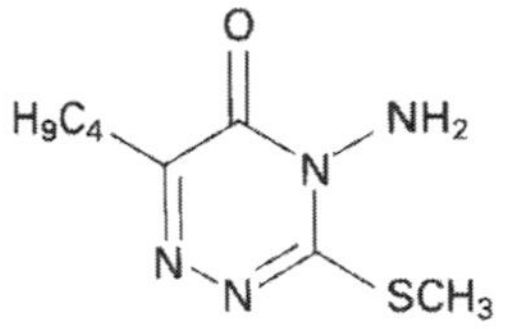

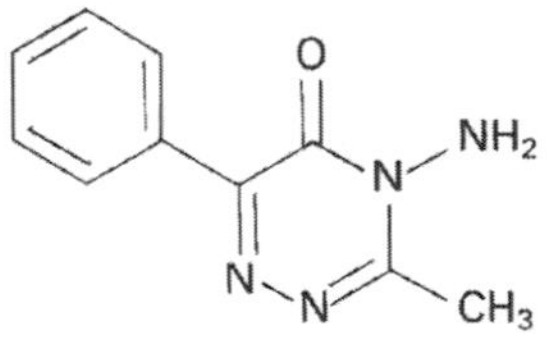

Figure 5.14   (*Continued*)

	USES
URACILS Lenacil	

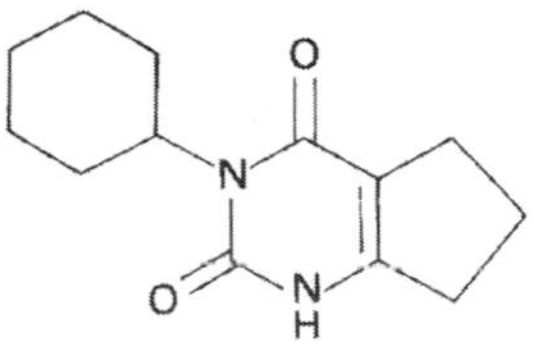

Pre-emergent control of annual weeds in sugar beet and fodder beet. Readily absorbed by roots and leaves and rapidly translocated.

Terbacil

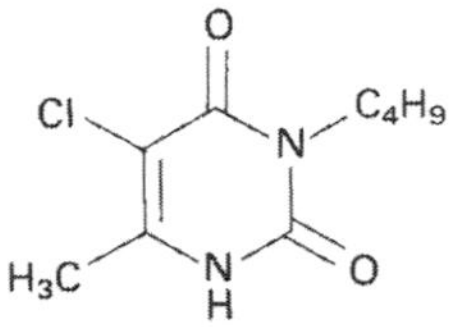

Soil applied for weed control in sugar cane, strawberry, peach, citrus and apple.

ANILIDES
Propanil

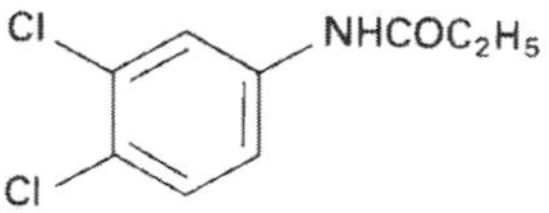

Post-emergent control of barnyard grass, sedges and some broadleaf weeds in rice, maize and wheat. Contact action.

Pentanochlor

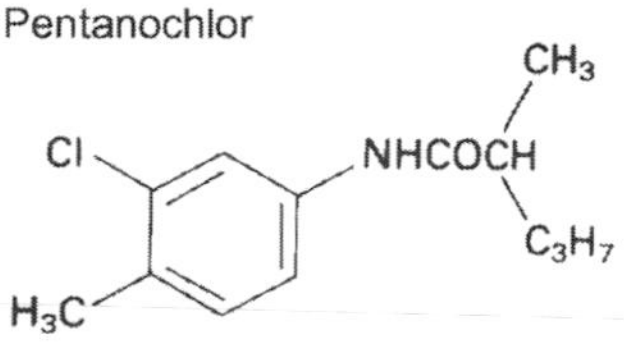

Controls annual weeds in carrots, celery, strawberries, tomato and soft fruit.

PHENYLCARBAMATES
Phenmedipham

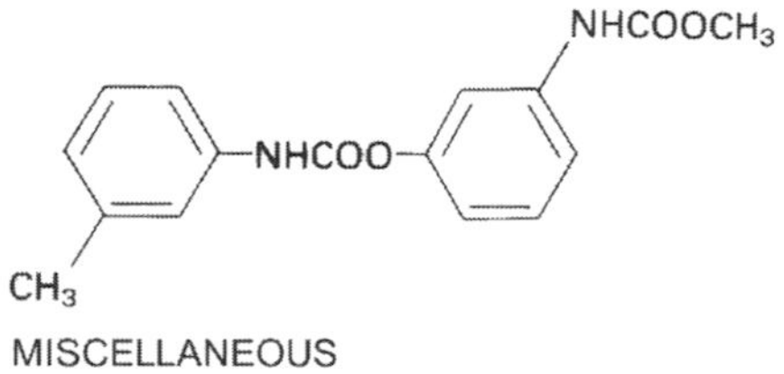

Post-emergent selective control of most broadleaf weeds in sugar beet. Readily absorbed by foliage but poorly translocated.

MISCELLANEOUS
Bentazone

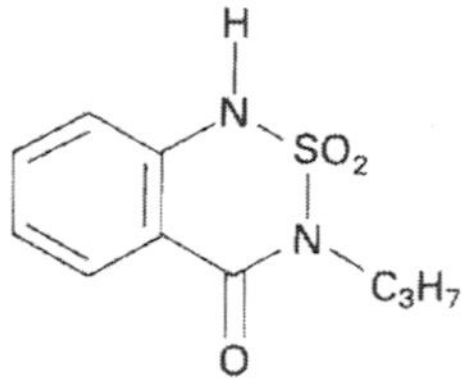

Post-emergent selective control of broadleaf weeds in wheat, barley, maize, rice, peas, beans and soybean. Absorbed by leaves and rapidly transported in the transpiration stream.

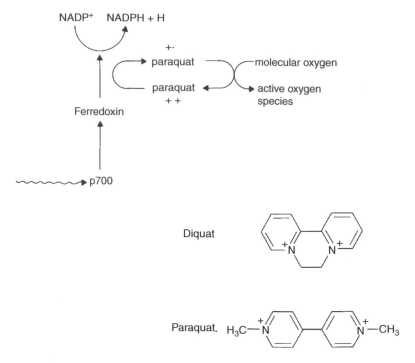

Figure 5.15 How paraquat and diquat act at Photosystem I to give rise to active oxygen species.

in some countries and becoming increasingly limited in others. Although the manufacturers have added bright colourants and powerful emetics to the formulation, careful use of these molecules is called for.

There have been many attempts to develop new herbicides based on this site of action, of which a few are noted here. Itoh and Iwaki (1989) have proposed a herbicide-binding site in PS I, designated the Q_ψ or phylloquinone-binding site. They found that phylloquinone (A_1) could be reversibly extracted from PS I with diethyl ether, leaving the photochemical charge separation almost intact. Reconstitution experiments identified potential inhibitors at this site, which appears to be more hydrophobic than equivalent quinone sites in PS II, and diuron and atrazine, for example, bind weakly to it.

A report by Smith *et al.* (2005) demonstrated impressive *in vitro* activity of indolizine-5,8-diones, a novel class of quinone-like compounds, although volatility and photo-instability conspired to reduce foliar persistence, and so these compounds have not been developed commercially.

Other PS II inhibitors continue to be discovered and are noted in the literature. Three are included below (Figure 5.16).

1 *Amicarbazone*. This triazolinone was first registered for use on maize in 2005 and was approved in 2012 for use on golf courses and turf grass. It can be absorbed by both leaves and roots, while selective in grasses. It has a good selectivity profile and is a more potent herbicide than atrazine, which enables it to be applied at lower rates than traditional PS II inhibitors (Dayan *et al.*, 2009).

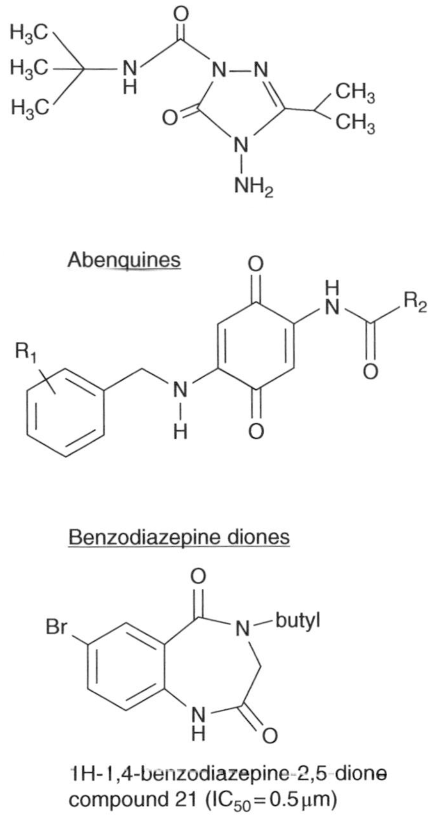

Abenquines

Benzodiazepine diones

1H-1,4-benzodiazepine-2,5-dione
compound 21 (IC$_{50}$ = 0.5 μm)

Figure 5.16 Structures of new Photosystem II herbicides.

2 *Abenquines.* These are natural compounds isolated from *Streptomyces* species growing in the Atacama desert in Chile (Nain-Perez *et al.*, 2017). They are *N*-acetyl aminobenzoquinones attached to an amino acid. Four analogues have IC$_{50}$ values comparable with diuron (0.1–0.4 μM) and the D1 protein is the most likely target site.

3 *Benzodiazepine diones.* These molecules were first reported by Singh *et al.* in 1996, sharing the D1 protein as a common binding site with atrazine. Intriguingly, they are also active in animals, with anxiolytic, anti-convulsant and anti-tumour properties in humans (Praba and Velmurugan, 2007).

The structure and function of Photosystem II has been the subject of detailed investigation for over six decades and herbicides have been important probes in our understanding of how PS II works. On the other hand, the assembly of PS I is relatively poorly understood and there are no commercialised herbicides that bind to PS I. Unlike PS II, its abundance is not controlled by environmental conditions. Recent progress has come from the studies of mutants and the protein CGL71.

CGL 71 is a thylakoid protein that is part of a family termed the 'GreenCut' proteins, of which about 600 are currently known, although over half have unknown functions. Many of the rest are thought to have key roles in photosynthesis and plastid maintenance. The GreenCut is a conserved gene set, shared by all green algae and plants, termed the Viridiplantae. Heinnickel *et al.* (2016) have studied a *Chlamydomonas reinhardtii* mutant lacking CGL 71 and propose that it has a role in PS I assembly and by protecting it from oxidative damage. It follows that inhibitors to CGL 71 may act as potent herbicides. Furthermore, the same team has demonstrated the involvement of CPLD 38 in the accumulation and physiological action of the cytochrome b6f complex, another potential target site for herbicide development, as it catalyses the slowest step in photosynthetic electron flow.

5.7   RuBisCo activase

While there are no known herbicides that interfere with photosynthetic carbon reduction or oxidation, a promising candidate enzyme target has emerged in recent years: RuBisCo activase.

Ribulose 1,5-bisphosphate carboxylase-oxygenase (E.C. 4.1.1.39), or RuBisCo, is one of the most important and probably the most abundant soluble proteins in Nature. It is located in the chloroplast stroma and is responsible for the net incorporation of carbon dioxide into products. It is a large and complex protein, consisting of eight large and eight small subunits with a total molecular mass of over 500 kDa. The large subunits, containing the active site, are encoded in the chloroplast genome, while the small subunits, which may confer stability or in some way enhance catalytic efficiency, are nuclear encoded (see Gutteridge and Gatenby, 1995 for further details).

For RuBisCo to be catalytically active, a lysine residue at the active site becomes modified by carbamylation. This is achieved by the reaction of a molecule of CO_2 with the free ε-amino lysine group. Subsequent binding of an atom of magnesium then renders RuBisCo active (Figure 5.17).

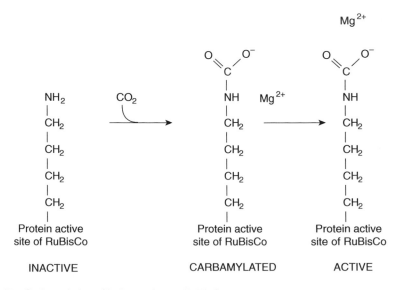

Figure 5.17   Carbamylation of lysine activates RuBisCo.

It appears, however, that the substrate RuBP is able to bind very tightly to the non-carbamylated form of RuBisCo, preventing activation of the enzyme and inhibiting carbon fixation. There are other natural RuBP analogues, such as carboxyarabinitol-1-phosphate and xylulose 1,5-bisphosphate, which are equally able to inhibit the enzyme, leaving it in the inactive state. This inhibition is overcome by RuBisCo activase which alters the conformation of RuBisCo protein, increasing the rate at which the inhibitors dissociate from the decarbamylated enzyme, thus allowing carbamylation, magnesium attachment and a competent enzyme.

RuBisCo activase is a nuclear-encoded tetrameric protein of 47 kDa subunits which may account for 1–2% (tobacco) or 5% (*Arabidopsis*) of the total soluble protein in leaves. Enzyme activity requires ATP hydrolysis to alter the conformation of RuBisCo. The activase is an abundant stromal protein because it is a relatively slow enzyme. Decreasing the amount of activase protein by antisense methodology indicates an important role in regulating CO_2 assimilation.

The ratio of activase to RuBisCo concentration and activity appears important in the regulation of carbon fixation, with the ratio greater in fluctuating light than in constant bright light (Mott and Woodrow, 2000). Clearly, the herbicidal inhibition of RuBisCo activase would favour the inactive form of the enzyme, carbon fixation would be prevented and the cessation of plant growth would ensue.

It is now known that most plant species have two forms of RuBisCo activase. A short beta-isoform (42 kDa) and a longer alpha-form (47 kDa). In wheat, two genes code for three isoforms, which differ in activity at different temperatures. Recently, Degan *et al.* (2020) demonstrated that substitution of a methionine residue by an isoleucine in the 2 beta-isoform improved the thermo-tolerance of the enzyme without affecting enzyme activity, thus permitting more photosynthesis at higher temperatures, especially in sub-Saharan Africa.

5.8   How treated plants die

Following root absorption of PS II herbicides, initial symptoms of chlorosis between leaf veins on lower leaves are observed. This is followed by the death of leaf tips and margins. After foliar absorption, leaves contacted by the spray become chlorotic and then necrotic. In both cases, there is limited herbicide movement to other leaves and weed death follows within days after treatment. This is influenced by temperature and photon flux density after treatment.

All of the herbicides mentioned so far in this chapter kill plants by the photo-peroxidation of thylakoid and other cell membrane lipids. Thylakoids are particularly susceptible to lipid peroxidation because they contain a high concentration of the unsaturated fatty acids, linoleic (18:2) and linolenic (18:3) acids, and molecular oxygen is always being produced by the photolysis of water, which can be used to generate active oxygen species.

5.8.1   *Active oxygen species*

Electrons usually exist as pairs in an atomic orbital and a free radical is defined as any species with one or more unpaired electrons (Halliwell and Gutteridge, 1984). By this definition oxygen itself is a free radical since it possesses two unpaired electrons, each located in a different π^* anti-bonding orbital, with the same parallel spin quantum number (Figure 5.18), which renders the molecule relatively unreactive. This is because, when

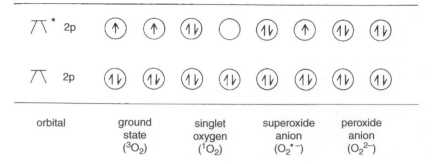

Figure 5.18 Bonding orbitals in diatomic oxygen. Source: Halliwell, B. and Gutteridge, J.M.C. (1984) Oxygen toxicity, oxygen radicals, transition metals and disease. *Biochemical Journal* **219**, 1–14. Reproduced with permission of The Biochemical Society.

another molecule is oxidised, by accepting a pair of electrons from it, both new electrons must be of parallel spin to fit into the vacant spaces in the π^* orbitals.

However, such a pair would be expected to have anti-parallel spins ($\downarrow\uparrow$) which slow and prevent many potential oxidations. Spin restriction can be overcome in biological systems by the presence of transition metals, especially iron, found at the active site of many oxygenases. The metals can do this by their ability to accept or donate single electrons. The reactivity of oxygen is also increased by moving one of the unpaired electrons to overcome the spin restriction and form singlet oxygen. Singlet oxygen has no unpaired electrons and is therefore not a radical, but can exist for long enough (2–4 µs) to react with many biological molecules. It is commonly generated by excitation energy transfer from photosensitisers such as chlorophylls, porphyrins, flavins and retinal (Figure 5.19).

The superoxide radical $\left(O_2^{\cdot-}\right)$ is formed when a single electron is accepted by ground state oxygen and the addition of a second electron, either enzymically by superoxide dismutase or non-enzymically, yields the peroxide ion $\left(O_2^{2-}\right)$.

$$2O_2^{\cdot-} + 2H^+ \longrightarrow H_2O_2 + O_2$$

In addition, hydroxyl free-radicals (·OH) are formed in the presence of peroxide (H_2O_2) and ferrous salts in what is termed the Fenton Reaction (Figure 5.19). The reactivity of ·OH is so great that it reacts immediately with whatever biological molecules are in the vicinity to produce secondary radicals.

$$H_2O_2 \underset{Fe^{2+}}{\overset{Fe^{3+}}{\longrightarrow}} \cdot OH + OH^-$$

All four active oxygen species ($^1O_2$, $O_2^{\cdot-}$, .OH and H_2O_2) are naturally generated at the thylakoid and are normally quenched by a series of scavengers or antioxidants within the stroma or dissipated as heat.

Chloroplast superoxide dismutase is a metalloprotein (apparently existing as Cu–Zn, Mn or Fe forms) partially bound to the thylakoid. Its product, hydrogen peroxide, can inhibit several stromal enzymes involved in photosynthetic carbon reduction, and is removed by

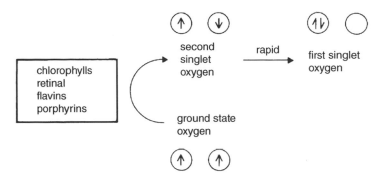

Figure 5.19   Generation of singlet oxygen by photosensitisers.

the enzymes of the ascorbate–glutathione cycle (Figure 5.20). Stromal ascorbate and glutathione concentrations fluctuate seasonally and millimolar amounts are commonly measured in the summer months.

α-Tocopherol (vitamin E) is a most effective antioxidant and can react rapidly with singlet oxygen and lipid peroxide radicals. The carotenoid pigments, particularly β-carotene, are also important quenchers of triplet chlorophyll in addition to singlet oxygen, and lead to the dissipation of excess excitation as heat. More recent studies have suggested that polyamines and some flavonols may also act as natural photoprotectants.

Plant survival therefore depends on the balance between photo-oxidative stress and the effectiveness of natural antioxidant protection systems, which have the ability to 'mop up' oxygen free radicals (Figure 5.21).

In the presence of PS II inhibitors, excitation energy generated by p680 cannot be dissipated by normal electron flow beyond Q_A^- and so fluorescence yield is dramatically enhanced and activated oxygen species generated. Similarly, paraquat will divert electrons at higher energy from PS I and rapidly generate oxygen free radicals, as previously described. Under these conditions the natural protective mechanisms are rapidly overloaded, especially at increased temperatures and photon flux densities, and lipid peroxidation is initiated in thylakoids by hydrogen abstraction (Figure 5.22).

The free-radical attacks unsaturated membrane fatty acids and is quenched by hydrogen atom abstraction. Since a hydrogen atom has only one electron, it leaves behind an unpaired electron on a carbon atom. This carbon (lipid) radical rapidly reacts with oxygen to yield a hydroperoxy radical, which is itself able to abstract hydrogen atoms from other unsaturated lipid molecules, thus initiating a chain reaction of lipid peroxidation. Eventually, the unsaturated fatty acids of the thylakoid are totally degraded to malondialdehyde and ethane, and the appressed thylakoid structure progressively opens up and disintegrates (Figure 5.23). Finally, cell membranes and tissues disintegrate from this chain reaction of free-radical attack.

Investigations in recent years have expanded our understanding of reactive oxygen species to the conclusion that they should no longer be regarded as features of chloroplast metabolism alone. We now know that they are formed in all plant organelles with high electron flux (including chloroplasts, mitochondria and peroxisomes), and also in the plasmalemma, cell wall, extracellular matrix, endoplasmic reticulum and the nucleus. They are regarded as signalling molecules in diverse metabolic pathways and control gene expression in response to stress. Indeed, the view is now taken that maintaining a basal level of

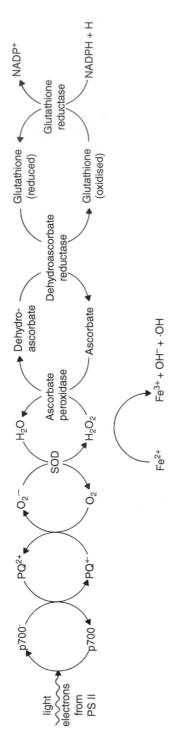

Figure 5.20 Superoxide generation and detoxification at Photosystem I. PQ, paraquat; SOD, superoxide dismutase. Source: modified from Shaaltiel, Y. and Gressel, J. (1986) Multienzyme oxygen radical detoxifying system correlated with paraquat resistance in *Conyza bonariensis*. *Pesticide Biochemistry and Physiology* **26**, 22.

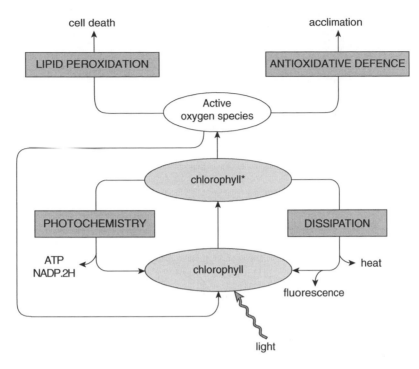

Figure 5.21   The fates of solar energy absorbed by the thylakoid light-harvesting complexes. * An excited state.

reactive oxygen species is essential for optimal plant cell function, although high concentrations, as noted above, can be fatal. The reader is referred to the fascinating article by Mittler (2017), and the review by Caverzan *et al.* (2019).

Mittler (2017) presents a convincing case that active oxygen species are also involved in metabolic signalling, and are necessary for the progression of cell processes, including cell proliferation and differentiation. Indeed, cell death, which was previously considered to be due to the direct effect of active oxygen species, is now considered to be due to active oxygen species triggering a programmed pathway for cell death. Mittler suggests that active oxygen species are therefore beneficial to plant cell function, and that a basal level of active oxygen species is essential for plant metabolism.

5.9   Chlorophyll fluorescence

The analysis of chlorophyll fluorescence has become a valuable experimental technique in recent years to investigate the effects of environmental stress, including herbicides, on PS II activity. Devices that are relatively simple to use and portable have become available, which enable rapid measurements in the laboratory, glasshouse and field. Indeed, chlorophyll fluorescence has become a useful screening technique for PS II inhibitors.

As shown in Figure 5.21, chlorophyll in an excited state can be used in one of three ways: photochemistry to drive photosynthesis, dissipation as heat or re-emission as fluorescence. The three processes are essentially competitive and environmentally sensitive. Under low

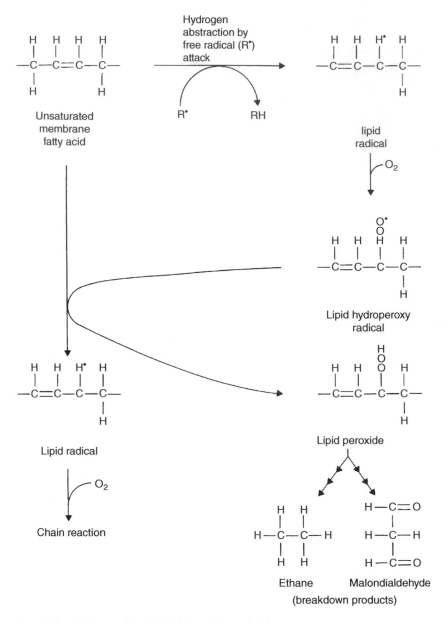

Figure 5.22   The chain reaction of thylakoid lipid peroxidation.

light exposure about 97% of the absorbed photons are used in photochemistry, 2.5% are transformed to heat and 0.5% emitted as fluorescence. On the other hand, if all of the PS II reaction centres are closed by high light intensity or the presence of PS II inhibitors, 95–97% of the absorbed energy may be dissipated as heat and as much as 3–5% via fluorescence.

Kautsky *et al.* (1960) were the first to report the pattern of fluorescence induction when dark-adapted leaves are exposed to light. An example of this fluorescent transient or Kautsky curve is presented in Figure 5.24.

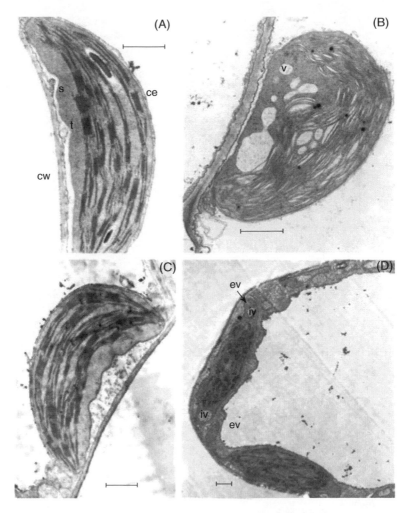

Figure 5.23 Ultrastructural symptoms following treatment with photosynthetic inhibitors. (A) Mesophyll cell chloroplast from an untreated leaf of *Tripleurospermum maritimum* subsp.*inodora* (scentless may-weed). Note appressed and non-appressed thylakoids (t), cell wall (cw), stroma (s) and chloroplast envelope (ce). (B) As (A), but 7 days after treatment with ioxynil-sodium at a rate equivalent to 560 g active ingredient (a.i.) ha^{-1}. Note that the chloroplast is swollen and that vesicles (v) are present in both the stroma and the thylakoids. Such inter-granal vacuolation is a typical symptom of photoperoxidative damage. (C, D) Mesophyll cell chloroplasts from *Galium aparine* leaves 3 h after treatment with (C) water and (D) 100 μM of the diphenyl ether herbicide, acifluorfen. Note the invaginations (iv) and evagi-nations (ev) of the chloroplast envelope. Bar, 1 μM. Sources: A and B, Sanders, G.E. and Pallett, K.E. (1986) Studies into the differential activity of the hydroxybenzonitrile herbicides. *Pesticide Biochemistry and Physiology* **26**(2), 116–127; C and D, Derrick, P.M., Cobb, A.H. and Pallett, K.E. (1988). Ultrastructural effects of the diphenyl ether herbicide acifluorfen and the experimental herbicide M&B 39279. *Pesticide Biochemistry and Physiology* **32**(2), 153–163; doi: 10.1016/0048-3575(88)90008-9. Reproduced with permission of Elsevier.

An initial rise from a dark-adapted low value (Fo) to a maximum value (Fm) is observed in less than a second (fast phase), followed by a slower phase, lasting minutes, before a constant value is reached. The fast phase is related to PS II photochemistry, while the slower phase is determined by thylakoid function such as electron transfer away from PS II

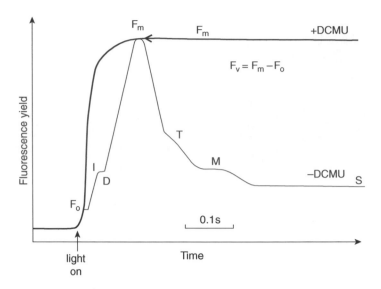

Figure 5.24 A characteristic fluorescent transient or Kautsky curve induced when a dark-adapted leaf is exposed to light in the presence and absence of DCMU (diuron). Source: adapted from Bolhar-Nordenkampf, H.R. and Oquist, G. (1993) Chlorophyll fluorescence as a tool in photosynthesis research. In: Hall, D.O., Scurlock, J.M.O., Bolhar-Nordenkaempf, H.R., Leegood, R.C. and Long, S.P. (eds.) *Photosynthesis and Production in a Changing Environment*. London: Chapman and Hall, pp. 193–206.

and carbon metabolism in the stroma. Specifically, at Fo all of the reaction centres at PS II are fully oxidised and ready to accept electrons. At Fm, all quinone carriers, especially Q_A, are reduced and unable to accept another electron until they have passed the first one on to the next carrier, Q_B. During this time the reaction centres are thought to be closed. Thus, when reaction centres are closed, photochemical efficiency is reduced and fluorescence yield is increased. Note that in the presence of PS II-inhibiting herbicides that prevent electron flow between QA and QB, photochemistry is reduced to zero and Fm is sustained over a prolonged period.

Variable fluorescence, Fv, (Fm − Fo) is a valuable parameter that reflects the efficiency with which incident light energy is used in, and downstream of, the photosystems. It is also a valuable measure of how a plant may tolerate a herbicide, since metabolism in a crop plant will detoxify the herbicide and the Fm value will decline.

Kaiser *et al.* (2013) report measuring the maximum efficiency of PS II in treated and control plants of *Alopecurus myosuroides* in pot tests and found that it was possible to distinguish between resistant and susceptible populations to fenoxaprop-P-ethyl and a mixture of mesosulfuron and iodosulfuron. For both ACCase and acetolactate synthase (ALS) inhibitors, all resistant populations could be identified by chlorophyll fluorescence imaging, irrespective of the mode of resistance. Thus, Fv/Fm was reduced. The authors considered that herbicides cause a collapse of proton gradients and peroxidation of polyunsaturated fatty acids, leading to an accumulation of active oxygen species, and electron flow from PS II to PS I will be disrupted. On the other hand, the reason why ALS inhibitors are able to affect PS II remains to be explained, and any possible role of branch-chain amino acids in this process.

Zhang *et al.* (2016) have also used leaf fluorescence to detect resistant plants of *Echinochloa* species at the four to five leaf stage, so enabling alternative methods of weed control to be used. Differences in Fv/Fm in resistant and tolerant plants varied with time after treatment, depending on the mode of action of the herbicide used. Discrimination after treatment with PS II herbicides took 64 h, a much shorter time than plants treated with ACCase or ALS inhibitors, which required 168 and 192 h, respectively.

As a note of caution in the value of this technique in herbicide research, the reader is referred to the study of Olesen and Cedergreen (2010). During their study of the effects of glyphosate on barley, they recorded decreases in carbon fixation one day after treatment, yet without significant changes in chlorophyll fluorescence. This observation was interpreted as an indirect effect of glyphosate on photosynthesis owing to indirect effects on sink processes, such as the inhibition of flavonoid metabolism. In this case, we may conclude that chlorophyll fluorescence is not always a sufficiently sensitive biomarker for glyphosate phytotoxicity. For further information, the reader is referred to the definitive review on the measurement and analysis of chlorophyll fluorescence by Murchie and Lawson (2013).

5.10   Inhibition of photosynthetic carbon reduction in C4 plants

RuBisCo is the sole enzyme in green plants, algae and cyanobacteria that catalyses carbon fixation and is one of the most abundant proteins in the biosphere. Yet it is an inefficient carboxylating enzyme that has defied evolutionary attempts to improve it. Since it is a rate-limiting step, its improvement would lead to a direct effect on crop productivity. C_4 plants have an additional carboxylator, phospho-enol pyruvate carboxylase (PEPC), that is far more efficient than RuBisCo. Since many of the worst weeds in the world are C_4 plants, it would be valuable to be able to selectively control this enzyme.

RuBisCo comprises eight large and eight small subunits, of 50 and 15KDa, respectively, and its assembly requires several chaperones. Feiz *et al.* (2012) identified a chloroplast

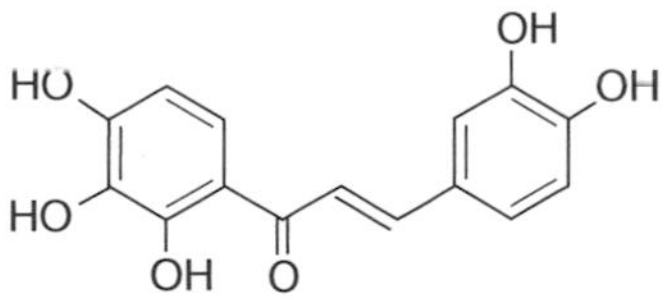

2′,3′,4′,3,4-pentahydroxychalcone (also known as okanin, a natural pentachalcone from the Asteracae family)

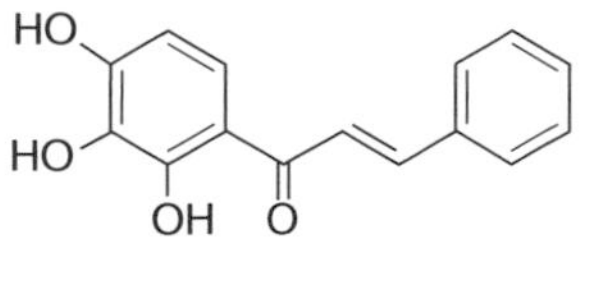

2′,3′,4-trihydrochalcone

Figure 5.25   Structures of two chalcone derivatives as potential inhibitors of PEP carboxylase in C_4 plants.

protein, termed RAF1 (RuBisCo accumulation factor 1) which they found to be expressed in the bundle sheath cells in maize, acting in the assembly of the large subunits. Whitney *et al.* (2015) have since expressed RAF1 in *Arabidopsis* with a doubling of photosynthetic rate caused by a doubling of RuBisCo production. If may follow that targeting inhibitors to RAF 1 could lead to less growth of C_4 weeds.

Nguyen *et al.* (2016), have demonstrated that 2′,3′,4′,3,4-penta-hydroxy chalcone (IC_{50}, 600 nM) and 2′,3′,4-trihydroxychalcone (IC_{50}, 4.2 μM) are potent inhibitors of PEPC in C_4 plants, but inactive on PEPC in C_3 plants over the same concentration range. The compounds (Figure 5.25) bind to PEPC at the same site as malate and aspartate, the natural feedback inhibitors of the C_4 pathway, and caused growth inhibitory effects in *A. retroflexus*, an important C_4 weed. Catechins and quinoxalines are also selective inhibitors of PEPC with IC_{50} values around 100 μM. The quantity and position of hydroxyl groups influences the potency and selectivity of the chalcones. These studies may generate new chemical leads in the development of selective herbicides against C_4 weeds.

References

Blankenship, R.E. (2002) *Molecular Mechanisms of Photosynthesis*. Oxford: Blackwell Science.

Bolhar-Nordenkampf, H.R. and Oquist, G. (1993) Chlorophyll fluorescence as a tool in photosynthesis research. In: Hall, D.O., Scurlock, J.M.O., Bolhar-Nordenkaempf, H.R., Leegood, R.C. and Long, S.P. (eds.) *Photosynthesis and Production in a Changing Environment*. London: Chapman and Hall, pp. 193–206.

Caverzan, A., Piasecki, C., Chavarria, G., Stewart, C.N. and Vargas, L. (2019) Defenses against ROS in crops and weeds: The effects of interference and herbicides. *International Journal of Molecular Sciences* **20**, 1086–1106; doi: org/10.3390/ijms20051086

Dayan, F.E., Trindade, M.L.B. and Velini, E.D. (2009) Amicarbazone, a new photosystem II inhibitor. *Weed Science* **57**, 579–583.

Degan, G.E., Worrall, D. and Carmo-Silva, E. (2020) An isoleucine residue acts as a thermal and regulatory switch in wheat Rubisco activase. *The Plant Journal*; doi: org/10.1111/tpj.14766

Derrick, P.M., Cobb, A.H. and Pallett, K.E. (1988) Ultrastructural effects of the diphenyl ether herbicide Acifluorfen and the experimental herbicide M & B 39279. *Pesticide Biochemistry and Physiology* **32**, 153–163.

Duff, M.G.S., Chen, Y.-C.S., Fabbri, B.J. *et al.* (2007) The carboxyterminal processing protease of D1 protein: herbicidal activity of novel inhibitors of the recombinant and native spinach enzymes. *Pesticide Biochemistry and Physiology* **88**, 1–13.

Exposito-Rodriguez, M., Laissue, P.-P., Yvon-Durocher, G., Smirnoff, N. and Mullineaux, P.M. (2017) Photosynthesis-dependent H_2O_2 transfer from chloroplasts to nuclei provides a high-light signalling mechanism. *Nature, Communications* **8**, article number 49.

Fabbri, B.J., Duff, M.G.S., Remsen, E.E., Chen, Y.-C.S., Anderson, J.C. and Cajacob, C.A. (2005) The carboxyterminal processing protease of D1 protein: Expression, purification and enzymology of the recombinant and native spinach proteins. *Pest Management Science* **61**, 682–690.

Fedtke, C.B., Depka, O., Schallner, K., *et al.* (2001) Mode of action of new diethylamines in lycopene cyclase inhibition and in photosystem II turnover. *Pesticide Management Science* **57**, 278–282.

Feiz, L. and five others (2012) Ribulose-1,5-bisphosphate carboxylase/oxygenase accumulation factor 1 is required for holoenzyme assembly in maize. *The Plant Cell* **24**, 3435–3446.

Fuerst, E.P. and Norman, M.A. (1991) Interactions of herbicides with photosynthetic electron transport. *Weed Science* **39**, 458–464.

Gressel, J. (2002) *Molecular Biology of Weed Control*. London: Taylor & Francis.

Gutteridge, S., and Gatenby, A.A. (1995) Rubisco synthesis, assembly, mechanism and regulation. *Plant Cell* **7**, 809–819.

Halliwell, B. and Gutteridge, J.M.C. (1984) Oxygen toxicity, oxygen radicals, transition metals and disease. *Biochemical Journal* **219**, 1–14.

Heinnickel, M.L. and six others (2016) Tetratricopeptide repeat protein protects photosystem I from oxidative disruption during assembly. *Proceedings of the National Academy of Science* **113**, 2774–2779.

Hirschberg, J., Bleeker, D.J., Kyle, L., McIntosh, L. and Arntzen, C.J. (1984) The molecular basis of triazine-herbicide resistance in higher-plant chloroplasts. *Zeitschrift für Naturforschung* **39c**, 412–420.

Itoh, S. and Iwaki, M. (1989) Vitamin k_1 (phylloquinone) restores the turnover of FeS centers in the ether-extracted spinach Photosystem I particles. *FEBS Letters* **243**, 47–52.

Kaiser, Y.I., Menegat, A. and Gerhards, R. (2013) Chlorophyll fluorescence imaging: a new method for rapid detection of herbicide resistance in *Alopecurus myosuroides*. *Weed Research* **53**, 399–406.

Kautsky, H., Appel, W. and Amman, H. (1960) Chlorophyllfluorescenz und kohlensaureassimilation. *Biochemische Zeitschrift* **322**, 277–292.

Knaff, D.B. (1988) The Photosystem I reaction centre. *Trends in Biochemical Sciences* **13**, 460–461.

Kohno, H., Ohki, A., Ohki, S., Koizumi, K., Van den Noort, M.E., Van Rensen, J.J. and Wakabayashi, K (2000) Low resistance against novel 2-benzylamino-1,3,5-triazine herbicides in atrazine resistant *Chenopodium album* plants. *Photosynthesis Research* **65**, 115–120.

Kuboyama, N., Koizumi, K., Ohki, A., Ohki, S., Kohno, H. and Wakabayashi, K. (1999) Photosynthetic electron transport inhibitory activity of 2-aralkylamino-4-methyl-6-trifluoromethyl-1,3,5-triazine derivatives. *Journal of Pesticide Science* **24**, 138–142.

Lawlor, D.W. (2001) *Photosynthesis*, 3rd edn. Oxford: BIOS.

Matoo, A.K., Pick, U., Hoffman-Falk, H. and Edelman, M. (1981) The rapidly metabolised 32,000 dalton polypeptide of the chloroplast is the 'proteinaceous shield' regulating Photosystem II electron transport and mediating diuron herbicide sensitivity. *Proceedings of the National Academy of Sciences, USA* **78**, 1572–1576.

Melis, A. (1999) Photosystem II damage and repair cycle in chloroplasts: what modulates the rate of photodamage *in vivo*? *Trends in Plant Science* **4**, 130–135.

Michel, H. and Deisenhofer, I. (1988) Relevance of the photosynthetic reaction center from purple bacteria to the structure of Photosystem II. *Biochemistry* **27**, 1–7.

Mittler, R. (2017) ROS are good. *Trends in Plant Science* **22**, 11–19; doi: org/10.1016/j.tplants.2016.08.002

Mott, K.A. and Woodrow, I.E. (2000) Modelling the role of rubisco activase in limiting nonsteady-state photosynthesis. *Journal of Experimental Botany* **51**, 399–406.

Murchie, E.H. and Lawson, T. (2013) Chlorophyll fluorescence analysis: a guide to good practice and understanding some new applications. *Journal of Experimental Botany* **64**, 3983–3998; doi: org/10.1093/jxb/ert208

Nain-Perez, A., Barbosa, L.C.A., Maltha, C.R.A., Gilberti, S. and Forlani, G. (2017) Tailoring natural abenquines to inhibit photosynthetic electron transport through association with the D1 protein of photosystem II. *Journal of Agriculture and Food Chemistry* **65**, 11304–1131.

Nguyen, G.T.T., Erlenkamp, G., Jack, O., Kuberl, A., Bott, M., Fiorani, F., *et al.* (2016) Chalcone-based selective inhibitors of a C4 plant key enzyme as novel potential herbicides. *Nature, Scientific Reports* **6**, 27333.

Olesen, C.F. and Cedergreen, N. (2010) Glyphosate uncouples gas exchange and chlorophyll fluorescence. *Pest Management Science* **66**, 536–542 (2010)

Perlaza, K., Toutkoushian, H., Boone, M., Lam, M., Jonikas, M.C., Walter, P. and Ramundo, S. (2019) The Mars1 kinase confers photoprotection through signalling in the chloroplast unfolded protein response. *eLife*, October; doi: org/10.7554/eLife.49577

Praba, G.O. and Velmurugan, D. (2007) Quantitative structure–activity relationships of some pesticides. *Indian Journal of Biochemistry and Biophysics* **44**, 470–476.

Renger, G. (1976) Studies on the structural and functional organisation of system II photosynthesis. The use of trypsin as a structurally selective inhibitor at the outer surface of the thylakoid membrane. *Biochimica et Biophysica Acta* **440**, 287–300.

Rutherford, A.W. (1989) Photosystem II, the water-splitting enzyme. *Trends in Biochemical Sciences* **14**, 227–232.

Sanders, G.E. and Pallett, K.E. (1986) Studies into the differential activity of the hydroxybenzonitrile herbicides. *Pesticide Biochemistry and Physiology* **26**, 116–127.

Shaaltiel, Y. and Gressel, J. (1986) Multienzyme oxygen radical detoxifying system correlated with paraquat resistance in *Conyza bonariensis. Pesticide Biochemistry and Physiology* **26**, 22.

Singh, B., Szamosi, I.T., Dahlike, B.J., Karp, G.M. and Shaner, D.L. (1996) Benzodiazepinediones: a new class of photosystem II inhibiting herbicides. *Pesticide Biochemistry and Physiology* **56**, 62–68.

Smith, S.C., Clarke, E.D., Ridley, S.M., Bartlett, D., Greenhow, D.T., Glithro, H. *et al.* (2005) Herbicidal indolizine-5,8-diones: photosystem I redox mediators. *Pest Management Science* **61**, 16–24.

Tomlin, C.D.S. (ed.) (2000) *The Pesticide Manual*, 12th edn. Farnham: British Crop Protection Council.

Trebst, A. (1980) Inhibitors in electron flow: tools for the functional and structural localisation of carriers and energy conservation sites. *Methods in Enzymology* **69**, 675–715.

Trebst, A. (1987) The three-dimensional structure of the herbicide binding niche on the reaction centre polypeptides of photosystem II. *Zeitschrift für Naturforschung* **42c**, 742–750.

Trebst, A., Donner, W. and Draber, W. (1984) Structure–activity correlation of herbicides affecting plastoquinone reduction by photosystem II: electron density distribution in inhibitors and plasto-quinone species. *Zeitschrift für Naturforschung* **39c**, 405–411.

Wessels, J.S.C. and van der Veen, R. (1956) The action of some derivatives of phenylurethan and of 3-phenyl-1,1-dimethyl urea in the Hill reaction. *Biochimica et Biophysica Acta* **19**, 548–549.

Whitney, S.M., Birch, R., Kelso, C., Black, J.L. and Kapralov, M.V. (2015) Improving recombinant Rubisco biogenesis, plant photosynthesis and growth by co-expressing its ancillary RAF1 chaperone. *Proceedings of the National Academy of Science* **112**, 3564–3569.

Zhang, C.-J., Lim, S.-H., Kim, J.W., Nash, G., Fischer, A. and Kim, D.S. (2016) Leaf chlorophyll fluorescence discriminates herbicide resistance in *Echinochloa* species. *Weed Research* **56**, 424–433.

Chapter 6
Inhibitors of Pigment Biosynthesis

More energy is expended for the weeding out of man's crops than for any other single human task.

LeRoy Holm (1971)

6.1 Introduction: structures and functions of photosynthetic pigments

Photosynthesis relies on the unique light-harvesting abilities of the chlorophyll and carotenoid pigments to trap solar energy in the chloroplast thylakoids and transfer excitation energy to the reaction centres. The principal light-harvesting pigment is chlorophyll *a* which contains a light-absorbing 'head', centred on a magnesium ion, and a hydrophobic 'tail' attached to proteins and lipids, and embedded in the thylakoid membrane. Light is absorbed at the rate of about one photon per chlorophyll molecule per second and an electron is boosted to an excited state for each photon absorbed.

The light-harvesting complexes contain two other types of pigment in addition to chlorophyll *a*, namely chlorophyll *b* and the carotenoids. These so-called accessory pigments absorb light of shorter (i.e. higher-energy) wavelengths than does chlorophyll *a*, so they increase the width of the spectrum available for photosynthesis. Energy transfer occurs in the sequence: carotenoids, chlorophyll *b*, chlorophyll *a*, reaction centre.

Chlorophylls *a* and *b* are found in the leaves of all higher plants and in green algae, typically at a concentration of 0.5 g m^{-2} of leaf area. Chlorophyll *b* differs from chlorophyll *a* in having a formyl group (–CHO) instead of a methyl group (–CH$_3$) on one of the N-containing rings that make up the head group, as indicated in Figure 6.1. The carotenoids are abundant in green tissues and over 600 have been found in Nature.

Herbicides that interfere with the biosynthesis of chlorophylls or carotenoids are generally termed 'bleaching herbicides' because bleaching (or whitening, i.e. a lack of typically green pigmentation) is a principal symptom in treated plants. Bleaching herbicides were the most frequently patented class of herbicides in the 1980s and the 1990s.

Herbicides and Plant Physiology, Third Edition. Andrew H. Cobb.
© 2022 John Wiley & Sons Ltd. Published 2022 by John Wiley & Sons Ltd.

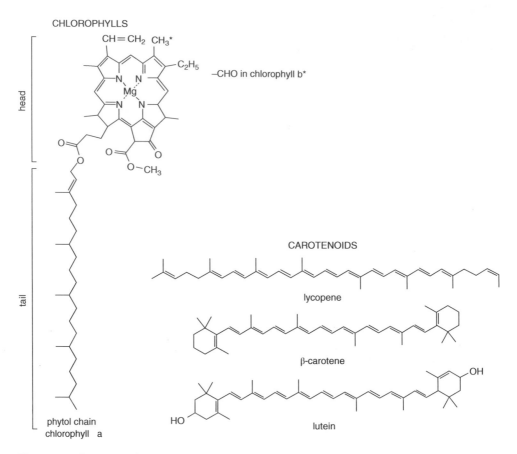

Figure 6.1   Structures of the major chlorophylls and carotenoids.

6.2   Inhibition of chlorophyll biosynthesis

Chlorophyll is the principal pigment in photosynthesis. In addition to a light-harvesting function, chlorophyll is located within reaction centres, and so it plays a pivotal role in the movement of electrons from low energy (H_2O) to high energy ($NADP^+$). The biosynthesis of porphyrins and tetrapyrroles begins with the formation of δ-aminolaevulinic acid (δ-ALA) and its conversion to porphobilinogen (PBG) (Figure 6.2). In animals, δ-ALA is formed from glycine and succinyl-CoA and in plants δ-ALA is synthesised *via* glutamate and a pyridoxal phosphate-linked transaminase. Gabaculine inhibits the transaminase by covalent binding to the pyridoxal phosphate cofactor, whereas 4-amino-5-fluropentanoic acid (Gardner *et al.*, 1988) can attack the transaminase active site to form a stable complex that inactivates the enzyme. Laevulinic acid, dioxovaleric acid and dioxoheptanoic acid all compete with δ-ALA to inhibit the synthesis of PBG. A succinyl moiety ($COOH–(CH_2)_2–C=O$) appears to be an essential prerequisite for this competitive inhibition of δ-ALA dehydratase (Figure 6.2).

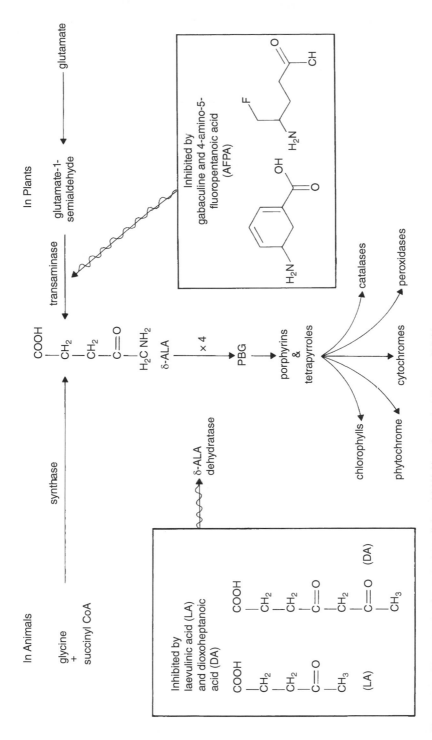

Figure 6.2 Biosynthesis of δ-aminolaevulinic acid (ALA) and porphobilinogen (PBG).

Since these two reactions are common to the biosynthesis of all tetrapyrroles, their inhibition also results in the cessation of phytochrome, cytochrome, peroxidase and catalase synthesis. The inhibitors just described clearly show some potential as *non-selective* herbicides, but since terapyrroles have a central role in the metabolism of all organisms, a lack of mammalian toxicity would need to be conclusively demonstrated.

The chlorophyll biosynthesis inhibitors are also referred to as 'peroxidising herbicides' because their action results from the peroxidising effect of active oxygen species. The pioneering examples are the *p*-nitrophenyl ethers (Table 6.1) and the cyclic imides (Table 6.2), developed between 1965 and 1980, and the most recent examples (such as fluthiacet-methyl, Table 6.2) have been developed in the last decade to be active at very low rates of $1–10$ g ha^{-1}, in a wide range of crops, when applied pre-plant or post-emergence.

The diphenyl ethers and cyclic imides are an important group of broad-spectrum herbicides that have been widely used for the selective control of annual grasses and broadleaf weeds in major world crops, including soybean, peanut, cotton and rice. Successful active ingredients include acifluoren, oxyflurorfen and bifenox (Table 6.1), and the generation of singlet oxygen produces phytotoxicity. It was initially thought that the phytotoxic species was generated in the thylakoid – as is the case with the inhibitors of photosynthetic electron flow – but research has conclusively demonstrated that photosynthesis is not directly involved in diphenyl ether action, and that the chloroplast envelope is an initial site of action (see Figure 5.14; Derrick *et al.*, 1988).

Observations that diphenyl ether action was particularly sensitive to low-energy blue light led to the suggestion that carotenoids could mediate singlet oxygen generation. However, Matringe and Scalla (1988) have shown typical phytotoxic symptoms in carotenoid-free cell lines and instead reported the accumulation of protoporphyrin IX in treated plants. These authors have since conclusively demonstrated that protoporphyrinogen oxidase is the target enzyme (Matringe *et al.*, 1989), since abbreviated to PROTOX.

It is currently envisaged that protoporphyrinogen IX synthesis and its oxidation to protoporphyrin IX are chloroplast envelope, membrane-bound reactions and that protoporphyrinogen IX spontaneously and non-enzymatically oxidises to form the potent photosensitiser, protoporphyrin IX, often termed the protogen. This molecule cannot be further

Table 6.1　Structure of some diphenyl ether peroxidising herbicides.

R$_1$	R$_2$	Common name
CF$_3$	COOH	acifluorfen
CF$_3$	OC$_2$H$_5$	oxyfluorfen
Cl	H	nitrofen
Cl	COOH	bifenox

Table 6.2 Structures of some PROTOX inhibitors.

	Dose and use
Fluthiacet-methyl	5–15 g ha^{-1} post-emergence in maize and soybean
Cinidon-ethyl	30–50 g ha^{-1} post-emergence in winter wheat and winter barley
Pentoxazone	150–450 g ha^{-1} pre- and early post-emergence in rice
Sulfentrazone	Pre-emergence and preplant. Controls sedges, broad-leaf and grass weeds in turf, 250–500 g ha^{-1}
Carfentrazone	Post-emergence control of broad-leaf and sedge weeds in cereals, 20–30 g ha^{-1}
Azafeniden	Controls weeds in fruit crops. Post-emergence, residual, 100–240 g ha^{-1}

(*Continued*)

Table 6.2 (*Continued*)

	Dose and use
Flumioxazin	Controls weeds pre-emergence in soybean, peanuts and a range of vegetable and fruit crops, 50–100 g ha^{-1}
Flumiclorac	Post-emergence control of broad-leaf weeds in soybean and maize, 40–100 g ha^{-1}

metabolised to porphyrins and so accumulates, possibly in the stroma, where it generates toxic singlet oxygen and lipid photoperoxidation ensues (Figure 6.3).

The diphenyl-ethers, first introduced in the 1960s, became lead compounds for more active successors that were discovered, developed and commercialised in more recent decades. It is thought that the relatively low rates required for herbicidal activity are far below the doses required to adversely affect porphyrin metabolism in animals and humans as a result of exposure on application or in food. They are now known to be potent inhibitors of protoporphyrinogen IX oxidase, with concentrations as low as 4 nM aciflurofen-methyl inhibiting this enzyme by 50% in corn etioplasts (Matringe *et al.*, 1989). However, this enzyme from either mammalian and yeast sources appears equally sensitive, which suggests that the toxicological properties of these herbicides may need to be closely re-examined.

Peroxidising herbicides inhibit protoporhyrinogen oxidase, commonly abbreviated to PROTOX (E.C. 1.3.3.4). Recent studies suggest that this activity is located in several parts of the cell in addition to the plastids, including at the mitochondrial inner membrane for haem production, the endoplasmic reticulum and possibly at the plasmalemma, in addition to soluble forms in the chloroplast stroma.

Lipid peroxidation results in a rapid loss in cell membrane integrity and function, especially at the thylakoid, chloroplast envelope, tonoplast and plasmalemma. This leads to ion leakage and water loss from the cell. Further effects are the inhibition of photosynthesis, ethylene formation, evolution of ethane from the products of membrane lipid peroxidation, bleaching of pigments and eventual plant death (Derrick *et al.*, 1988).

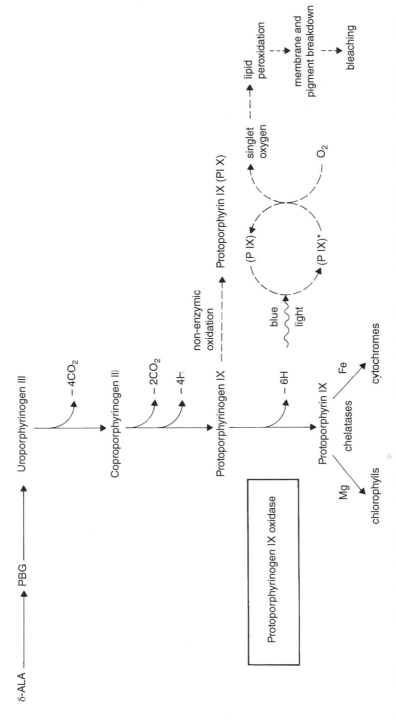

Figure 6.3   An overview of porphyrin biosynthesis in the presence (dashed line) and absence (solid line) of diphenyl ether and cyclic imide herbicides.

The membrane-bound PROTOX activity shows very high sensitivity to peroxidising herbicides with IC_{50} values as low as 10^{-10} M. Since all eukaryotic protox enzymes are sensitive to these inhibitors, their safety in animals and humans has been questioned. Some authors cite evidence that these compounds can alter porphyrin metabolism in animals, with an accumulation of porphyrin intermediates. The literature suggests, however, that no health problems associated with the consumption of crops treated with protox inhibitors have been reported and that these herbicides are rapidly metabolised in animals.

Peroxidising herbicides act as rapid, reversible, competitive inhibitors of protox activity and the most potent structures show close resemblance to the geometric shape and electronic characteristics of one-half of the protogen molecule. Since this molecule is a photodynamic pigment in its own right, it generates highly active oxygen species in the stroma and the cytoplasm.

Protox catalyses the six-electron oxidation of protoporphyrinogen IX to protoporphyrin IX, a known and potent photosensitiser. Flavoproteins are closely involved in the activity. Photoaffinity labelling of radio-ligands has been used to study herbicide binding, with one herbicide binding site suggested for each flavin adenine dinucleotide (FAD) molecule, with the active site at the *C*-terminal domain of the protein. The enzyme protein appears highly resistant to protease degradation and it is thought that acylation stabilises its conformation.

Genes involved in protox activity have now been isolated from bacteria, yeasts and higher plants. The *Arabidopsis* protox gene encodes a protein of 537 amino acid residues with little structural homology to the enzymes encoded in human or mouse genes. There have been two cDNA sequences found in tobacco and other plants, which share only 27% sequence similarity, one located in the chloroplast and the other in the mitochondrion. These two isoforms have molecular masses of 60 and 55 kDa, respectively.

Several approaches have been used to obtain plants that are resistant to the protox inhibitors. This is an active area of herbicide research and development, and new protox inhibitors continue to emerge (Matsumoto, 2002). For example, Grossmann *et al.* (2010) have announced saflufenacil for the pre-emergent control of dicot weeds in several crops, including maize. The mode of action of saflufenacil, a pyrimindinedione, was discovered using a physionomics approach, which demonstrated that its physiological profiling was very similar to the diphenyl-ether, bifenox. It inhibited PROTOX *in vitro* with an IC_{50} of 0.4 nM. Saflufenacil is absorbed by both roots and leaves, and is mainly translocated in the xylem. As shown in Figure 6.4, the side-chain ensures that it is metabolised in tolerant crops, making it an effective, selective, pre-emergence herbicide.

Mutant cell cultures have been obtained by several groups by incorporating the herbicides into the growth medium, followed by later studies to identify the gene sequence that confers resistance. This strategy has generated resistance to pyraflufen-ethyl in tobacco. A mutation of Val 389 to Met in the plastidic isoform has also generated resistance in *Chlamydomonas reinhardtii*. Expression of the bacterial enzyme in plants has generated tolerance to the diphenyl ether oxyfluorfen. Overexpression of the *Arabidopsis* plastidic form has resulted in resistance to acifluorfen. Clearly, several companies are now pursuing the development of peroxidising herbicide resistance in crops and further details are anticipated.

Readers are referred to the multi-authored text *Peroxidising Herbicides*, edited by Böger and Wakabayashi (1999) for a detailed treatise on these important herbicides, including their chemistry, physiology, mode of action and toxicology.

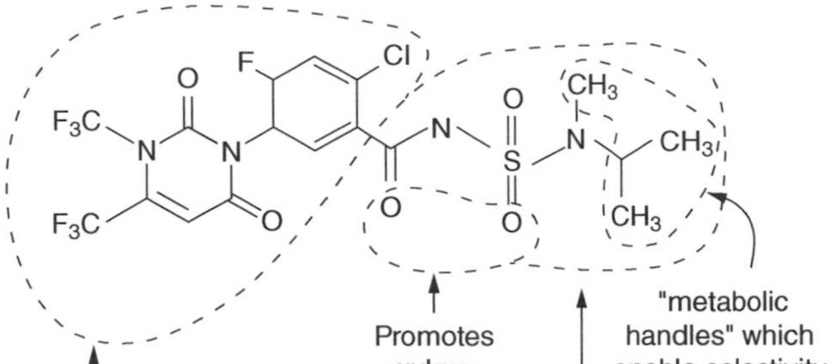

Figure 6.4 Structure of saflufenacil, indicating structure-activity relationships. Source: After Grossmann et al., 2011

6.3 Inhibition of carotenoid biosynthesis

In addition to their light-harvesting role, carotenoids protect chlorophyll from attack by active oxygen species by quenching both triplet chlorophyll and singlet oxygen, dissipating this energy as heat (Figure 6.5). How thermal energy dissipation is achieved has challenged researchers for several decades. In 2000, Li *et al.* reported a breakthrough using a mutant of the plant *Arabidopsis thaliana* that contained normal concentrations of zeaxanthin but was unable to dissipate thermal energy. They used molecular and genetic markers to show that a Photosystem II (PS II) protein known as CP22 (see Figure 5.4, a chlorophyll-binding protein with a relative molecular mass of 22,000) was absent in the mutant. On reintroduction into the mutant plant of a normal copy of the gene encoding CP22, the ability to dissipate thermal energy was regained. CP22 is one of over 30 proteins thought to be involved in light harvesting and its precise function remains unclear. Li and colleagues imply that a conformational change in the thylakoid takes place via CP22 to achieve the dissipation, which requires zeaxanthin and a *trans*-thylakoid pH gradient. Interestingly, these workers have also suggested that CP22 synthesis may increase in leaves exposed to excess light on a daily basis, implying an important physiological role for this PS II protein.

Surplus reductive capacity in the chloroplasts (i.e. production of NADP.2H in excess of that required for carbon fixation and other biosynthetic reactions) is dissipated by the xanthophyll cycle (Figure 6.6), which involves three different carotenoids: violaxanthin (a di-epoxide), antheraxanthin (a mono-epoxide) and zeaxanthin (epoxide-free). These three carotenoids are reversibly interconvertible by the addition or removal of an epoxide group, as shown in Figure 6.6. In strong light, violaxanthin is converted to zeaxanthin via antheraxanthin. This conversion, which is catalysed by a de-epoxidase enzyme, is optimal at low pH (around 5.1). The conversion of violaxanthin to zeaxanthin takes place within a few minutes in the presence of high-intensity light. During irradiation of chloroplasts, the de-epoxidase enzyme is activated by the drop in pH within the thylakoid owing to photosynthetic electron transport. Under the

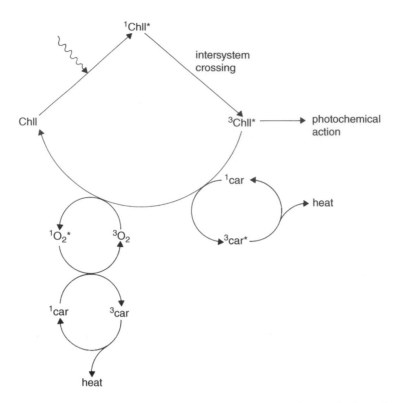

Figure 6.5 Protection of chlorophyll (Chll) by carotenoids (car). Source: Britton, G., Barry, P. and Young, A.J. (1989) Carotenoids and chlorophylls: herbicide inhibition of pigment biosynthesis. In: Dodge, A.D. (ed.) *Herbicides and Plant Metabolism*. Cambridge: Cambridge University Press, pp. 51–72. Reproduced with permission of Cambridge University Press.

conditions of a proton gradient across the thylakoid membrane, violaxanthin is reduced to zeaxanthin with the participation of the redox systems glutathione/oxidized glutathione and ascorbic acid/dehydroascorbic acid. Reconversion of zeaxanthin to violaxanthin is catalysed by an epoxidase enzyme, requires NADP.2H and uses oxygen. This reaction occurs rapidly in low-intensity light or darkness and is optimal at higher pH (around 7.5; see Figure 6.6). Thus, under high-intensity light conditions, the xanthophyll cycle favours the production of zeaxanthin, thereby increasing the capacity for dissipating light energy as heat and protecting against photooxidative damage.

The atomic structure of the major light-harvesting antenna protein, LHC II, has been determined by X-ray crystallography. It can exist in different reversible states, operating in either light-harvesting or in energy-dissipation mode. In addition to chlorophyll molecules, the LHC II also contains the carotenoid lutein. The formation of the quenched antenna state is controlled by the carotenoids of the xanthophyll cycle. In this way, de-epoxidation of violaxanthin to zeaxanthin, which stimulates energy dissipation, is thought to modulate structural changes in LHC II. Pascal *et al.* (2005) consider that the LHC II molecule behaves as a 'natural nanoswitch' to control the emission or transfer of incoming light quanta.

It follows from the above that if carotenoids biosynthesis is inhibited, free-radical attack and lipid peroxidation will rapidly ensue under high light-intensity conditions, as previously described.

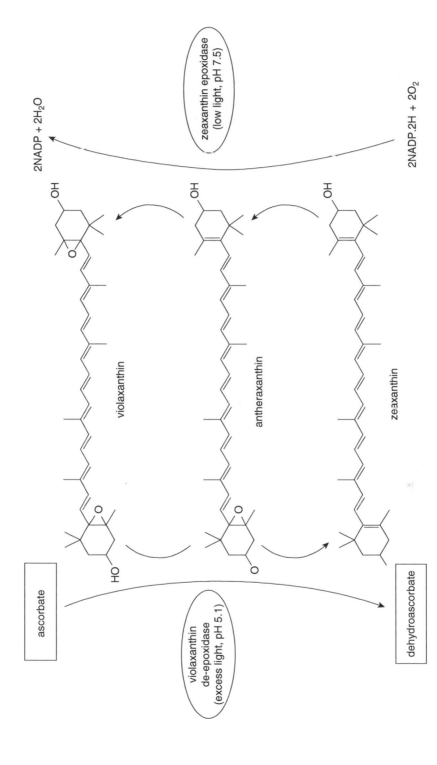

Figure 6.6 The xanthophyll cycle.

6.3.1 Inhibition of phytoene desaturase

Carotenoids are synthesised in the chloroplast via the mevalonic acid pathway (Figure 6.7). In essence, this pathway involves the condensation of five carbon units (isopentenyl pyrophosphate) to yield many molecules of both physiological (e.g. natural plant growth

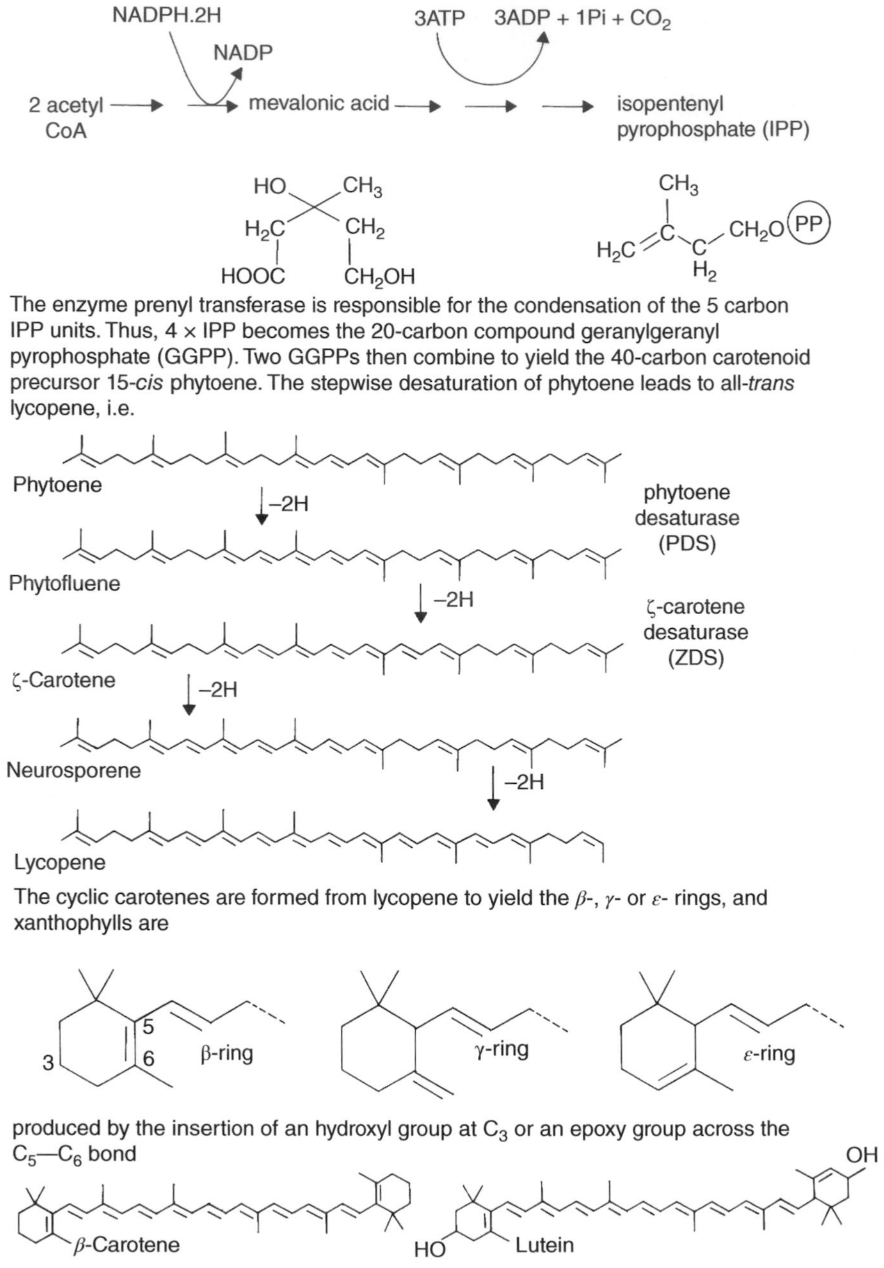
NADPH.2H
NADP
3ATP
3ADP + 1Pi + CO2
2 acetyl CoA
mevalonic acid
isopentenyl pyrophosphate (IPP)
PP
The enzyme prenyl transferase is responsible for the condensation of the 5 carbon IPP units. Thus, 4 × IPP becomes the 20-carbon compound geranylgeranyl pyrophosphate (GGPP). Two GGPPs then combine to yield the 40-carbon carotenoid precursor 15-cis phytoene. The stepwise desaturation of phytoene leads to all-trans lycopene, i.e.
Phytoene
−2H
Phytofluene
−2H
ζ-Carotene
−2H
Neurosporene
−2H
Lycopene
phytoene desaturase (PDS)
ζ-carotene desaturase (ZDS)
The cyclic carotenes are formed from lycopene to yield the β-, γ- or ε- rings, and xanthophylls are
5
6
3
β-ring
γ-ring
ε-ring
produced by the insertion of an hydroxyl group at C3 or an epoxy group across the C5—C6 bond
β-Carotene
HO
Lutein
OH

Figure 6.7 Carotenoid biosynthesis.

regulators) and biotechnological (e.g. terpenes) importance. In carotenoid biosynthesis two molecules of the 20-carbon geranylgeranyl pyrophosphate combine to yield the carotenoid precursor 15-*cis*-phytoene which undergoes a series of desaturation steps to all-*trans*-lycopene. Cyclisation followed by ring hydroxylation yields the major higher plant carotenoids, namely β-carotene and lutein. The inhibition of isopentenyl pyrophosphate or geranylgeranyl pyrophosphate biosynthesis would affect the production of all plant terpenoids and could present a target for future herbicide action, although no such inhibitors are currently known. The desaturation steps between phytoene and lycopene, however, have been successfully exploited agrochemically. These enzymes are present as a multi-enzyme complex situated at the thylakoid membrane.

15-*cis*-Phytoene desaturases (PDS) add two double-bonds into their colourless substrate, phytoene, by dehydrogenation, and isomerise two neighbouring double bonds from *trans* to *cis*. Subsequent actions by three further enzymes produce the red-coloured lycopene. The activity of PDS requires the reduction of a non-covalently bound co-factor, FAD. The electrons involved in the reaction are subsequently transferred to plastoquinone and the plastid terminal oxidase.

Any enzyme involved in the reaction sequence to β-carotene is a potential herbicide target to induce bleaching symptoms, although all commercial examples are phytoene desaturase inhibitors, with *in vitro* IC_{50} values of 0.01–0.1 μM. In plants, both PDS and ζ-carotene desaturase (ZDS) (Figure 6.7) catalyse the desaturation sequence from phytoene to lycopene, proceeding via hydrogen abstraction forming double bonds with NAD and NADP as potential hydrogen acceptors, although plastoquinone has a 20-fold higher affinity than NADP. This role of plastoquinone in phytoene desaturation also explains why the presence of inhibitors of plastoquinone biosynthesis leads to phytoene accumulation.

While the herbicidal inhibition of phytoene desaturase is non-competitive with respect to the substrate phytoene, competition has been observed for the cofactors.

Phytoene desaturase inhibitors are numerous and have been intensively studied during the last two decades. Commercial examples include norflurazon, flurochloridone and diflufenican (Figure 6.8), which are used as pre-planting or pre-emergent herbicides. The persistent activity of diflufenican has been effectively exploited in long-term weed control in paths and drives. Since germinating weeds lack photoprotective carotenoids, lipid peroxidation rapidly follows seedling emergence; weeds appear bleached and quickly die. Fewer compounds are known to interfere with ζ-carotene desaturase or lycopene cyclase, and none appears to have been commercially developed as yet.

Efficient bleaching herbicides possess a central five- or-six-membered heterocycle, carrying one or two substituted phenyl rings, implying a common binding site at the phytoene desaturase protein (Sandmann, 2002).

Fedtke *et al.* (2001) have reported the inhibition of lycopene cyclase by some novel diethylamines. Lycopene accumulates in treated plants with marked reductions in the concentrations of neoxanthin, violaxanthin, lutein and β-carotene. Those workers demonstrated that an observed inhibition of photosynthesis was due to interference with the turnover of the D1 protein in the PS II reaction centre. The reassembly process requires the continuous biosynthesis of two reaction-centre β-carotene molecules, without which protein D1 disappears, especially at high light-flux densities. Interestingly, this depletion of PS II precedes the bleaching process, which may imply a new mechanism of herbicidal activity shared by both the lycopene cyclase and phytoene desaturase inhibitors.

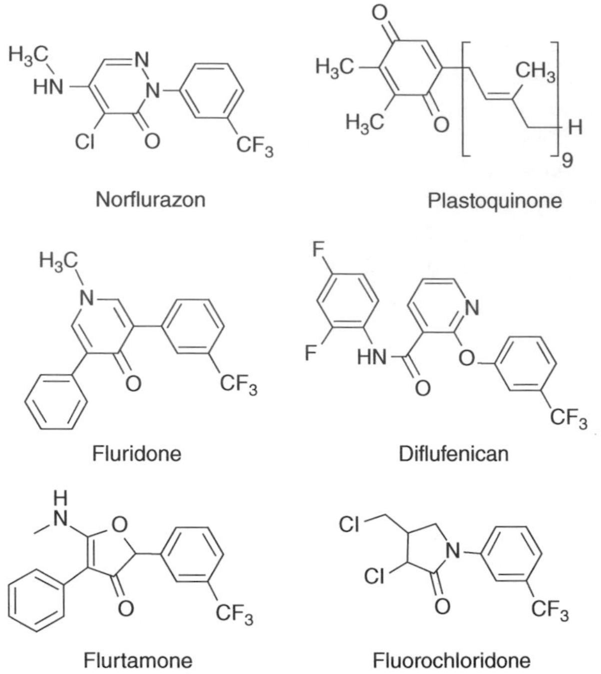

Figure 6.8 Structures of five commercialised inhibitors of phytoene desaturase (PDS) and its electron acceptor plastoquinone. Source: Brausemann, A., Gemmecker, S., Koschmieder, J., Ghisla, S., Beyer, P. and Einsle, O. (2017) Structure of phytoene desaturase provides insights into herbicide binding and reaction mechanisms involved in carotene desaturation. *Structure* **25**, 1222–1232; doi: org/10.1016/j. str.2017.06.002.

Brausemann *et al.* (2017) propose that phytoene and plastoquinone bind in a long hydrophobic tunnel in the enzyme, and that the herbicidal inhibitors bind at the plastoquinone site, preventing the re-oxidation of reduced FAD. These authors note the structural similarity of the PDS inhibitors to plastoquinone (see Figure 6.8).

The chloroplast is also the site of the biosynthesis of 5-*n*-alkyl resorcinols. These phenolic lipids play a key role in plant defence against fungal pathogens. Magnucka *et al.* (2007) have demonstrated that the metabolism of resorcinolic lipids is also affected by the PDS inhibitor norflurazone, such that their concentrations increased in plants grown in the light and in the dark. They conclude that seedlings of monocotyledonous plants which survive the herbicide treatment may become less susceptible to infection by microbial pathogens.

Intriguingly, Kaya and Yigit (2014) found that while sunflower plants treated with flurochloridone had lower chlorophyll and carotenoid content, pre-treatment with salicylic acid alleviated these responses, by a mechanism unknown.

These latter publications suggest that PDS inhibition leads to other physiological responses in plants. Indeed, Qin *et al.* (2007) suggest that inhibition of PDS can induce changes in at least 20 metabolic pathways, and that dwarfing effects in a mutant lacking the PDS gene can be partially overcome by the addition of exogenous gibberellic acid, GA3.

6.4   Inhibition of plastoquinone biosynthesis

Plastoquinone is an electron acceptor in carotenoid biosynthesis in addition to its key role in photosynthetic electron transport (Figure 6.9). Inhibitors of plastoquinone biosynthesis are also herbicides and cause typical bleaching symptoms. Plants synthesise plastoquinone from the aromatic amino acid tyrosine via the intermediate homogentisic acid (Figure 6.10).

Natural molecules and herbicides that inhibit the activity of 4-hydroxyphenylpyruvate dioxygenase (HPPD) are shown in Figure 6.11.

Isoxaflutole is a pro-herbicide and its metabolic byproduct, a diketonitrile, acts as the inhibitor. Similarly, a metabolite of the herbicide pyrazolate, termed detosyl-pyrozolate,

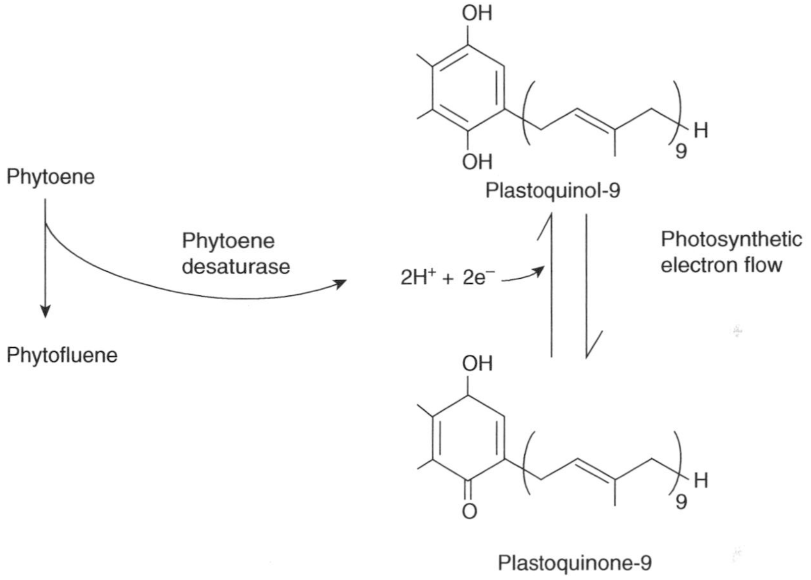

Figure 6.9   The role of plastoquinone in carotenoid biosynthesis and its regeneration by photosynthetic electron flow.

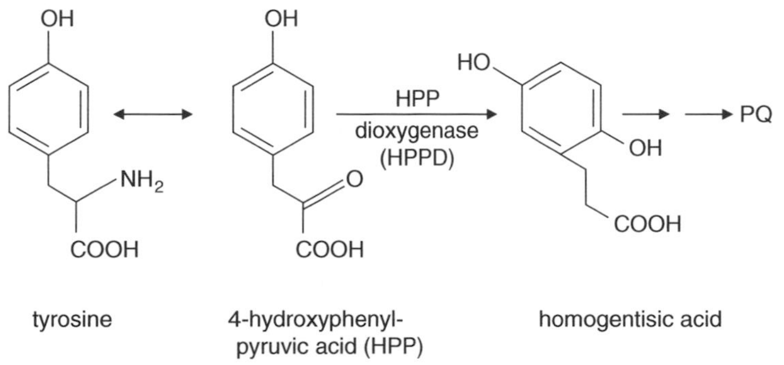

Figure 6.10   Synthesis of homogentisic acid from tyrosine.

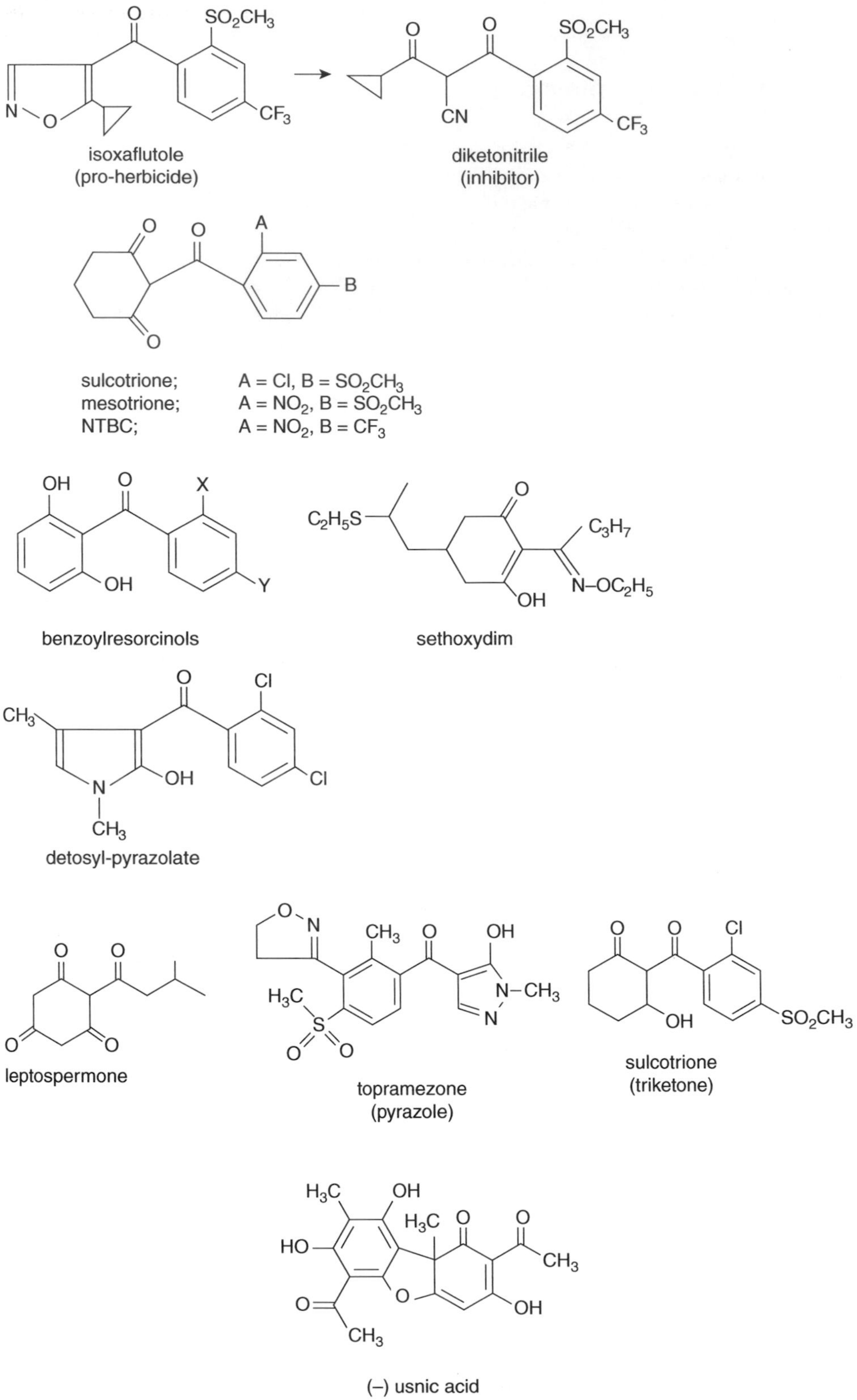

Figure 6.11 Structures of HPPD inhibitors referred to in the text

is a potent HPPD inhibitor. Intriguingly, inhibition of HPPD activity by the graminicide sethoxydim has also been claimed, in addition to its inhibitory activity against acetyl-CoA carboxylase (Lin and Young, 1999).

The benzoyl-cyclohexanediones, sometimes termed the herbicidal triketones, were first patented in 1985 as a new group of bleaching herbicides. They are particularly effective when applied pre-emergence in maize at doses as low as 60–100 g ha^{-1} and provide an effective low-dose alternative to atrazine.

Barta and Böger (1996) isolated HPPD from maize and demonstrated potent competitive inhibition by several experimental benzoyl-cyclohexanediones with IC_{50} values in the range of 3–23 nm. Viviani *et al.* (1998) performed a detailed kinetic study of HPPD inhibition by diketonitrile and other triketones and concluded that the herbicides acted as tightly binding inhibitors that dissociate extremely slowly from the enzyme–inhibitor complex.

HPPD (E.C. 1.13.11.27; E.C. 1.14.2.2) is a monomeric polypeptide of molecular mass 43 kDa (Barton and Böger, 1996) in maize, whereas it behaves as a homodimer of 48 kDa subunits in cultured carrot cells. Amino acid sequences are known from plant, animal and microbial HPPDs. Highly conserved regions at the *C*-terminus suggest an involvement of this region in the catalytic process, perhaps including sites for the binding of the substrate and an iron atom at specific histidine and glutamate residues.

The HPPD inhibitors have become important additions to the agrochemical armoury in recent years owing to their impressive crop selectivity in major crops, low application rates and broad-spectrum weed control, including those resistant to other herbicides, and applications that can often be both pre- and post-emergence. Indeed, their market share is estimated at 6.6% of all herbicides with collective sales of almost US$2 billion, and growing. Each of the major agrochemical companies has representatives in this herbicide class (Ndikuryayo *et al.*, 2017). Zhang *et al.* (2019) report the development of a new HPPD inhibitor, QYM201, that gives the additional control of grass weeds in winter wheat in China, when applied post-emergence at a dosage of 90–180 g active ingredient (a.i.) ha^{-1}.

Inhibition of HPPD prevents the conversion of tyrosine to plastoquinone and tocopherol. The former is important for electron flow and carotenoid biosynthesis in the thylakoids and the latter is an effective protectant against oxidative stress in the chloroplast. Over three decades of study of HPPD, including X-ray crystallography, have shown that the enzyme has a six-stage catalytic mechanism, involving the formation of an HPPD–HPPA (4-hydroxyphenylpyruvic acid) complex, dioxygen addition, decarboxylation, the cleavage of an 0–0 bond, acetic acid chain migration and C1 hydroxylation (Ndikuryayo *et al.*, 2017).

HPPD can be inhibited by natural products, such as leptospermone (secreted by *Callistemon citrinus*) and usnic acid (a lichen metabolite, commonly found in the genus *Usnea*), and three families of synthetic compounds, the isoxazoles, triketones and pyrazoles (Table 6.3 and Figure 6.11). It is thought that the inhibitors bind progressively to the catalytic site and compete with HPPA with time, leading to irreversibility. Thus, (−)-usnic acid inhibits HPPD at an IC_{50} of 50 nm (Romagni *et al.*, 2000).

The pyrazoles were first introduced for selective weed control in rice, although they have since been superseded by the sulphonylureas. The triketones have been successfully used in maize and sugarcane for the control of broad-leaf weeds and some grasses.

As indicated by Pallett (2000), the prospects of further HPPD inhibitors are enhanced by the resolution of the crystal structure of the enzyme. Similarly, other enzymes in the biosynthesis of quinones and tocopherols should be evaluated as potential targets for herbicide development.

Table 6.3 Representative synthetic HPPD inhibitors.

Class	Active ingredient	Year	Crop	Application
Diketonitrile	Isoxaflutole	1998	Maize, soybean	Pre-emergence
Pyrazoles	Benzofenap	1987	Rice	Pre- and post-emergence
	Pyrasulfotole	2008	Wheat, barley	Post-emergence
	Pyrazoxyfen	1985	Rice	Pre- and post-emergence
	Pyrazolynate	1980	Rice	Pre- and post-emergence
	Topramezone	2006	Maize	Post-emergence
Triketones	Tefuryltrione	2010	Rice	Post-emergence
	Tembotrione	2007	Maize	Post-emergence
	Mesotrione	2001	Maize	Pre- and post-emergence
	Bicyclopyrone	2015	Maize	Pre- and post-emergence
	Sulcotrione	1991	Maize	Post-emergence
	Benzobicyclon	2001	Rice	Pre- and post-emergence

Source: Modified from Ndikuryayo, F. *et al*. (2017) 4-Hydroxyphenypyruvate dioxygenase inhibitors: From chemical biology to agrochemicals. *Journal of Agriculture and Food Chemistry*, **65**, 8523–8537.

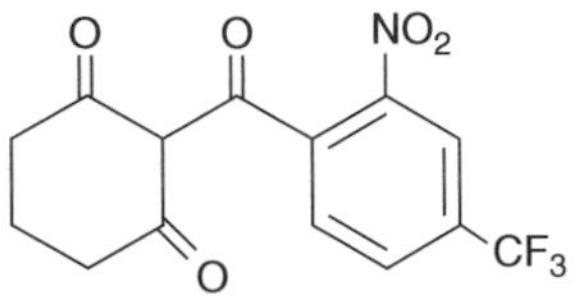

Nitisinone (NTBC)

Figure 6.12 Nitisinone.

Alkaptonuria is a rare genetic condition in humans that can lead to severe joint pain and, in some cases, death. It is also referred to as 'black bone disease' as it causes bones and ligaments to turn black, produces black spots in the eyes, and even makes the ears turn blue. It is caused by a mutation in the gene coding for homogentisate 1,2-dioxygenase (E.C. 1.13.11.5), so the blood and tissues accumulate homogentisic acid, and the breakdown of tyrosine and phenylalanine is prevented. It is estimated that 30,000 persons have this condition, affecting 1 in every 250,000 people in the EU.

Progress of the condition can be halted by nitisinone (Figure 6.12) which arose via herbicide discovery, as a reversible inhibitor of HPPD.

6.5   How treated plants die

The *p*-nitrophenyl ethers, cyclic imides and HPPD inhibitors induce rapid wilting and browning of shoots, leading to pigment bleaching, retardation of growth and plant death. They are applied pre- or post-emergence and are typically poorly translocated in treated leaves.

HPPD inhibitors cause sensitive plants to die by producing rapid wilting and browning of shoots, followed by the depletion of the plastoquinone pool, leading to the loss of tocopherols and carotenoids. Membrane damage ensues with an eventual bleaching of leaf tissues.

6.6 Selectivity and metabolism

The contact activity of the peroxidising herbicides may lead to poor selectivity. Cinidon-ethyl, however, selectively controls broadleaf weeds in cereals, with selectivity resulting from increased metabolism in wheat.

The basis of isoxaflutole selectivity in weeds and maize also appears to be differential rates of metabolism. The herbicide is rapidly taken up and translocated following both soil and foliar applications. In maize the herbicide is rapidly metabolised, so that 6 days after a root application 59% of recoverable activity was found in a benzoic acid derivative and 29% in the active diketonitrile. Conversely, in the susceptible weed *Abutilon theophrasti*, after the same period, 82% remained as the active diketonitrile, with only 12% of recoverable activity found in the benzoic acid metabolite (Pallet *et al.*, 1998). The pathway of isoxaflutole metabolism is shown in Figure 6.13.

Sulfentrazone shows selective pre-emergence activity in soybeans and peanuts at doses of 125–500 g a.i. ha⁻¹. Its selective action is explained by oxidative metabolism (Figure 6.14), specifically demethylation. Interestingly, the tolerant weed sicklepod (*Cassia obtusifolia*) also metabolises the herbicide in the same way (Dayan *et al.*, 1996).

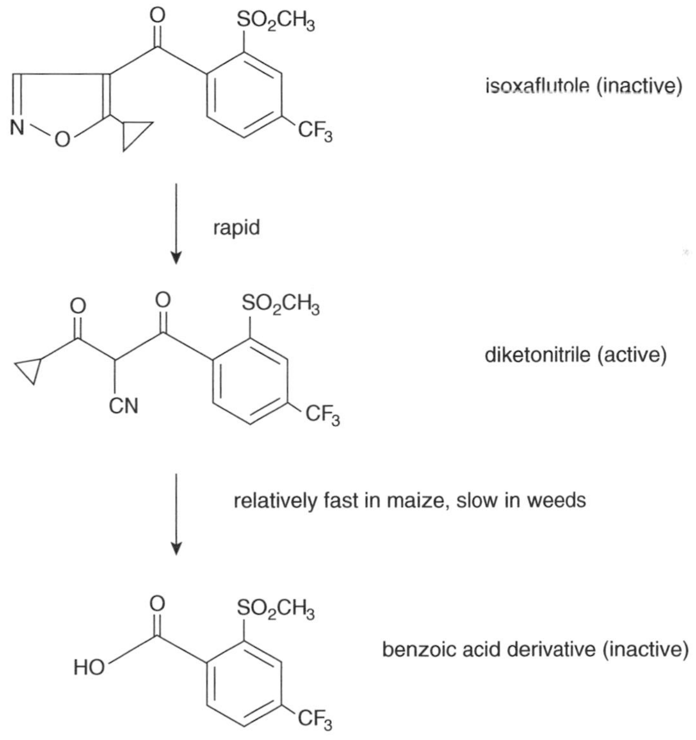

Figure 6.13 Metabolism of isoxaflutole in crops and weeds.

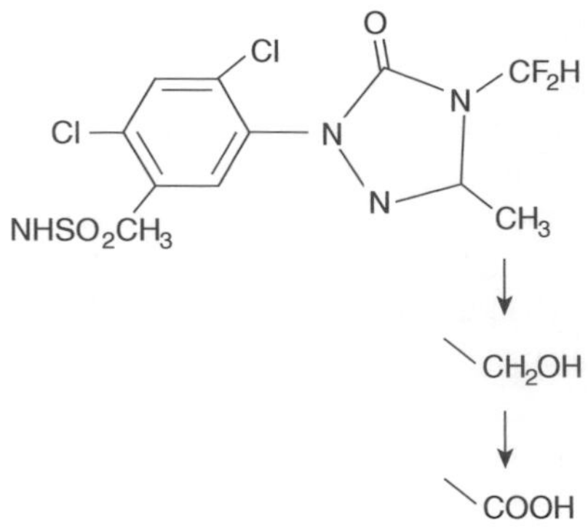

Figure 6.14 Metabolism of sulfentrazone in soybean. Source: modified from Dayan, F.E., Weete, J.D. and Hancock, H.G. (1996) Physiological basis for differential sensitivity to sulfentrazone by sicklepod (*Cassia obtusifolia*) and coffee senna (*Senna occidentalis*). *Weed Science* **44**, 12–17.

The diphenyl ether herbicides have shown commercially successful selectivity in soybean and rice. Soybean contains homoglutathione (γ-glutamyl-cysteinyl-β-alanine) instead of glutathione and uses this alternative thiol in herbicide metabolism. As examples, acifluorfen and fomesafen are rapidly detoxified by homoglutathione conjugation in soybean. Interestingly, diphenyl ether herbicides can increase the expression of glutathione *S*-transferases (GSTs) in soybean. The activity of the GST in question (*GM GSTU1-1*) was selectively enhanced by homoglutathione rather than by glutathione (Skipsey *et al.*, 1997).

Tolerance of peas to the diphenyl ether fluorodifen is due to rapid conjugation with glutathione and the GST responsible has been further characterised (Edwards, 1996).

Tolerance and selectivity in maize to the triketones is due to a rapid 4-hydroxylation of the cyclohexanedione ring. In the absence of this ring, e.g. the pyrazolones, *N*-demethylation may also confer tolerance.

Waterhemp (*Amaranthus tuberculatus*) has evolved resistance to several HPPD inhibitors in dicots owing to rapid oxidative metabolism of the parent herbicide in resistant populations. Furthermore, this species exhibits multigenic, complex inheritance patterns for the HPPD inhibitors. Lygin *et al.* (2018) consider that such multiple herbicide resistance in this species may result from a potential to evolve metabolic mechanisms that may differ from tolerant crops. Further findings are awaited.

6.7 Summary

In summary, bleaching herbicides are classified into four groups (Figure 6.15), with target sites related to either carotenoid or plastoquinone biosynthesis. Most target PDS or HPPD. Note also that clomazone inhibits 1-deoxy-D-xylulose-5-phosphate synthase, while cyclopyrimorate and haloxydine inhibit homogentisate solanyltransferase (Figure 6.16).

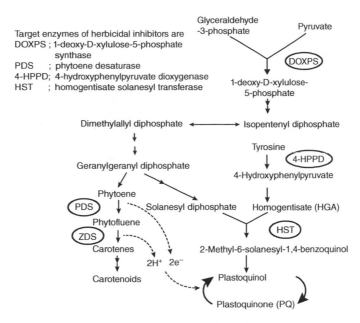

Target enzymes of herbicidal inhibitors are
DOXPS ; 1-deoxy-D-xylulose-5-phosphate synthase
PDS ; phytoene desaturase
4-HPPD; 4-hydroxyphenylpyruvate dioxygenase
HST ; homogentisate solanesyl transferase
Glyceraldehyde -3-phosphate
Pyruvate
DOXPS
1-deoxy-D-xylulose-5-phosphate
Dimethylallyl diphosphate
Isopentenyl diphosphate
Tyrosine
4-HPPD
4-Hydroxyphenylpyruvate
Geranylgeranyl diphosphate
Phytoene
PDS
Solanesyl diphosphate
Homogentisate (HGA)
Phytofluene
ZDS
HST
Carotenes
2-Methyl-6-solanesyl-1,4-benzoquinol
2H+ 2e−
Carotenoids
Plastoquinol
Plastoquinone (PQ)

Figure 6.15 Pathways of carotenoid and plastoquinone biosynthesis

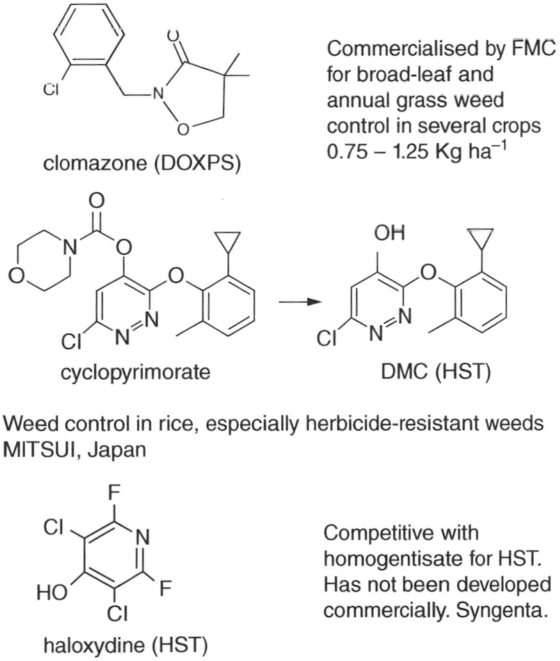

clomazone (DOXPS)
Commercialised by FMC for broad-leaf and annual grass weed control in several crops 0.75 – 1.25 Kg ha−1
cyclopyrimorate
DMC (HST)
Weed control in rice, especially herbicide-resistant weeds MITSUI, Japan
haloxydine (HST)
Competitive with homogentisate for HST. Has not been developed commercially. Syngenta.

Figure 6.16 Structures of herbicides that inhibit 1-deoxy-D-xylulose-5-phosphate synthase (DOXPS) and homogentisate solanyltransferase (HST).

References

Barta, I.C. and Böger, P. (1996) Purification and characterisation of 4-hydroxyl phenylpyruvate diox-ygenese from maize. *Pesticide Science* **48**, 109–116.

Bőger, P. and Wakabayashi, K. (eds) (1999) *Peroxidising Herbicides*. Berlin: Springer.

Brausemann, A., Gemmecker, S., Koschmieder, J., Ghisla, S., Beyer, P. and Einsle, O. (2017) Structure of phytoene desaturase provides insights into herbicide binding and reaction mechanisms involved in carotene desaturation. *Structure* **25**, 1222–1232; doi: org/10.1016/j.str.2017.06.002

Britton, G., Barry, P. and Young, A.J. (1989) Carotenoids and chlorophylls: herbicide inhibition of pigment biosynthesis. In: Dodge, A.D. (ed.) *Herbicides and Plant Metabolism*. Cambridge: Cambridge University Press, pp. 51–72.

Dayan, F.E., Weete, J.D. and Hancock, H.G. (1996) Physiological basis for differential sensitivity to sulfentrazone by sicklepod (*Cassia obtusifolia*) and coffee senna (*Senna occidentalis*). *Weed Science* **44**, 12–17.

Derrick, P.M., Cobb, A.H. and Pallett, K.E. (1988) Ultrastructural effects of the diphenyl ether herbicide Acifluorfen and the experimental herbicide M & B 39279. *Pesticide Biochemistry and Physiology* **32**, 153–163.

Edwards, R. (1996) Characterisation of glutathione transferases and glutathione peroxidases in pea (*Pisum sativum*). *Physiologia Plantarum* **98**, 594–604.

Fedtke, C., Depka, B., Schallner, O., Tietjen, K., Trebst, A., Wollweber, D. and Wroblowsky, H.J. (2001) Mode of action of new diethylamines in lycopene cyclase inhibition and in photosystem II turnover. *Pest Management Science* **57**, 278–282.

Gardener, G., Gorton, H.L. and Brown, S.A. (1998) Inhibition of phytochrome synthesis by the transaminase inhibitor, 4-amino-5-fluoropentanoic acid. *Plant Physiology* **87**, 8–10.

Grossmann, K., Niggeweg, R., Christiansen, N., Looser R. and Ehrhardt, T. (2010) The herbicide saflufenacil (Kixor TM) is a new inhibitor of protoporphyrinogen IX activity. *Weed Science* **58**, 1–9.

Grossmann, K., Hutzler., Caspar, G., Kwiatkowski, J. and Brommer, C.I. (2011) Saflufenacil (KixorTM): Biokinetic properties and mechanism of selectivity of a new protoporphyrinogen IX oxidase inhibiting herbicide. *Weed Science* **59**, 290–298; doi: org/10.1614/WS-D-10-00179.1

Kaya, A. and Yigit, E. (2014) The physiological and biochemical effects of salicylic acid on sunflow-ers (*Helianthus annus*) exposed to flurochloridone. *Ecotoxicology and Environmental Safety* **106**, 232–238.

Lin, S.W. and Young, D.Y. (1999) Inhibition of 4-hydroxyphenylpyruvate dioxygenase by sethoxy-dim, a potent inhibitor of acetyl-coenzyme A carboxylase. *Bioorganic and Medicinal Chemistry Letters* **9**, 551–554.

Lygin, A.V., Kaundun, S.S., Morris, J.A., McIndoe, E., Hamilton, A.R. and Riechers, D.E. (2018) Metabolic pathways of topramezone in multiple-resistant waterhemp (*Amaranthus tuberculatus*) differs from naturally tolerant maize. *Frontiers of Plant Science*, 21 November; doi: org/10.3389/fpls.2018.01644

Magnucka, E.G., Suzuki, Y., Pietr, S.J., Kozubeck, A. and Zarnowski, R. (2007) Effect of norflurazon on resorcinolic lipid metabolism in rye seedlings. *Zeitschrift für Naturforschung* **62c**, 239–245.

Matringe, M. and Scalla, R. (1988) Studies on the mode of action of acifluorfen-methyl in non-chlorophyllous soybean cells: accumulation of tetrapyrroles. *Plant Physiology* **86**, 619–622.

Matringe, M., Camadro, J.M., Labbe, P. and Scalla, R. (1989) Protoporphyrinogen oxidase as a molecular target for diphenyl-ether herbicides. *Biochemical Journal* **260**, 231–235.

Matsumoto, H. (2002) Inhibitors of protoporphyrinogen oxidase: a brief update. In: Bőger, P. Wakabayashi, K. and Hirai K. (eds) *Herbicide Classes in Development*. Berlin: Springer, Chapter 8.

Ndikuryayo, F. *et al.* (2017) 4-Hydroxyphenypyruvate dioxygenase inhibitors: From chemical biol-ogy to agrochemicals. *Journal of Agriculture and Food Chemistry*, **65**, 8523–8537.

Pallett, K.E. (2000) The mode of action of isoxaflutole: a case study of an emerging target site. In: Cobb, A.H. and Kirkwood, R.C. (eds) *Herbicides and their Mechanisms of Action*. Sheffield: Sheffield Academic Press.

Pallett, K.E., Little, J.P., Sheekey, M. and Veerasakaran, P. (1998) The mode of action of isoxaflutole: 1. Physiological effects, metabolism and selectivity. *Pesticide Biochemistry and Physiology* **62**, 113–124.

Pascal, A.A., Liu, Z., Broess, K., *et al*. (2005) Molecular basis of photoprotection and control of photosynthetic light harvesting. *Nature* **436**, 134–137.

Qin, G., Gu, H., Ma, L., Peng, Y., Deng, X.W., Chen, Z. and Qu, L-J. (2007) Disruption of phytoene desaturase gene results in albino and dwarf phenotypes in *Arabidopsis* by impairing chlorophyll, carotenoid and gibberellin biosynthesis. *Cell Research* **17**, 471–482.

Romagni, J.G., Meazza, G., Dhammika, N.P and Dayan, F.E. (2000) The phytotoxic lichen metabolite, usnic acid, is a potent inhibitor of plant *p*-hydroxyphenylpyruvate dioxygenase. *FEBS Letters* **480**, 301–305.

Sandmann, G. (2002) Bleaching herbicides: action mechanism in carotenoid biosynthesis, structural requirements and engineering of resistance. In: Bőger, P., Wakabayashi, K. and Hirai, K. (eds) *Herbicide Classes in Development*. Berlin: Springer, Chapter 2.

Shino, M., Hamada, T., Shigematsu, Y., Hirase, K. and Banba, S. (2018) Action mechanism of bleaching herbicide cyclopyrimorate, a novel homogentisate solanesyltransferase inhibitor. *Journal of Pesticide Science* **43**, 233–239; doi: org/10.1584/jpestics.D18-008

Skipsey, M., Andrews, C.J., Townson, J.K., Jepson, I. and Edwards, R. (1997) Substrate and thiol specificity of a stress-inducible glutathione transferase from soybean. *FEBS Letters* **409**, 370–374.

Viviani, F., Little, J.P. and Pallett, K.E. (1998) The mode of action of isoxaflutole: 2. Characterisation of the inhibition of the carrot 4-hydroxyphenyl pyruvate dioxygenase by the diketonitrile derivative of isoxaflutole. *Pesticide Biochemistry and Physiology* **62**, 125–134.

Zhang, F., Bai, S., Wang, H, Liu, W. and Wang, J. (2019) Greenhouse and field evaluation of a novel HPPD-inhibiting herbicide, QYM201, for weed control in wheat. *Nature, Scientific Reports* **9**, 1625.

Chapter 7
Auxin-type Herbicides

A weed is a plant that has mastered every survival skill, except for learning how to grow in rows.
Doug Larson (1926–2017)

7.1   Introduction

The phenoxyacetic acids MCPA and 2,4-D were discovered apparently independently in the UK and the USA in 1941. Templeman and colleagues at ICI and Nutman and collaborators at Rothamsted Experimental Station first demonstrated the herbicidal activity of MCPA, while in the same year in the USA, Porkorny synthesised 2,4-D, and its growth regulatory properties were characterised by Zimmerman and Hitchcock in 1942 (see Kirby 1980 for further details). Both compounds were developed in secret during the war years as potential chemical warfare agents. Fortunately, they were not used in this context at that time and their agricultural potential was soon realised with the marketing of 2,4-D by the American Chemical Paint Company as 'Weedone' in 1945 and by the launch of MCPA by ICI as 'Agroxone' in 1946. Since these molecules could kill many broadleaf weeds in narrowleaf crops, they were the first truly selective, non-toxic organic herbicides that were effective at low doses. They were also cheap to produce and became available at a time when maximum food production was essential and labour on farms was very scarce. Consequently, they completely altered traditional approaches to weed control and so provided a chemical replacement for the hoe. Furthermore, their success stimulated the chemical industry to invest in the research that led to the discovery of the wide range of herbicides now available. Further information on the discovery and development of the phenoxyalkanoic acid herbicides can be found in Kirby (1980).

In the following decades many structural analogues were developed to broaden the spectrum of weed control and selectivity, and these herbicides remain some of the most widely used pesticides in the world. Indeed, estimates indicate that more than 32 million kg were produced annually in the USA by six manufacturers, leading to more than 1500 formulated products and 35 different esters and salts as active ingredients (Ware, 1983).

Since the registration of 2,4-D for herbicide use in 1945, there has been worldwide acceptance and widespread use of the phenoxy herbicides. Even now, they are probably the most widely used family of herbicides in the world and play a major role in weed

Herbicides and Plant Physiology, Third Edition. Andrew H. Cobb.
© 2022 John Wiley & Sons Ltd. Published 2022 by John Wiley & Sons Ltd.

management when used either alone or in combination with other herbicides. They have been used for over 75 years with few, if any, reports of acute or chronic toxicity to humans, and have an outstanding record of environmental safety.

In recent years, 2,4-D has satisfactorily completed all re-registration and re-evaluation data requirements in North America and Europe. It has been approved and re-registered for use in 74 countries, and in each continent. It is not a reproductive toxicant, nor a neuro-toxin, nor a developmental immunotoxin, and there is no evidence that it is a carcinogen or an endocrine disruptor in humans. The reader may wish to refer to Peterson *et al.* (2016) for a comprehensive review of this pioneering herbicide, which remains one of the most important compounds in the herbicide armoury. Indeed, the auxin herbicides have become the standard for broad-leaf weed control for almost 80 years.

Burnside and colleagues (1996) have estimated the benefits of phenoxy herbicide use in the USA. They found that these herbicides were used in over 65 crops and in many non-crop situations, equivalent in 1992 to a value of US$171 million. They estimated that if phenoxy-herbicides were banned in the USA, this would result in an annual loss of US$2559 million: 37% of this loss would be due to increased costs of weed control from the use of more expensive herbicides or other control methods, 36% would result from decreased crop yield and the remaining 27% would arise from higher retail commodity prices.

Atwood and Paisley-Jones (2017) have compiled data for 2012 of the most commonly used pesticides in the USA for the Agricultural Market sector (Table 7.1A) and the Home and Garden Market sector (Table 7.1B). Note the dominance of glyphosate and how the auxin-type herbicides remain very popular in both sectors.

Morphological symptoms produced by these herbicides are indicative of an exaggerated auxin response, leading to disorganised growth and death in susceptible species. Their mode of action is unique among herbicides, in that they activate processes, rather than inhibit them. Although they have been used for over 75 years and have generated a copious amount of scientific literature, the molecular basis of their activity and selectivity has until recently remained obscure. This ignorance is now being remedied as our understanding of the molecular biology of natural auxin improves.

7.2   Structures and uses of auxin-type herbicides

The auxin-type herbicides currently in use are categorised into six groups (Table 7.2). These are the phenoxyalkanoic acids, benzoic acids, aromatic carboxymethyl derivatives (which includes natural auxin), pyridine carboxylic acids, the quinoline carboxylic acids and the arylpicolinates. All active auxins appear to possess a free carboxyl group, which suggests that a negatively charged group is essential for activity, although other common chemical characteristics are less obvious. The presence and position of halogens has a profound effect on both activity and selectivity since, for example, 2,3-D and 3,5-D have no auxin activity but 2,4-D does. Furthermore, the addition of an extra chlorine creates 2,4,5-T, which can control woody plants that can tolerate 2,4-D.

Table 7.3 provides a brief overview of the weed spectrum controlled by auxin-type herbicides. Although only a few examples are included, Table 7.3 clearly illustrates the reasons for the development of these herbicides and the introduction of mixtures to give an even broader spectrum of weed control. Thus, typical MCPA-susceptible weeds include

Table 7.1 Most commonly used herbicide active ingredients in the USA Agricultural Market sector in 2012 (A) and in the Home and Garden sector (B), their rankings and estimates of usage, as millions of kilograms.

Active ingredient	Rank	Range of use (millions of kilograms)
(A) Agricultural Market in the USA in 2012		
Glyphosate	1	122–132
Atrazine	2	29–34
Metolachlor-*S*	3	15–20
2,4-D	5	14–18
Acetochlor	7	13–17
Pendimethalin	11	3–7
Metolachlor	15	2–4
Propanil	17	1–3
Dicamba	18	1–3
(B) Home and Garden Market in the USA in 2012		
2,4-D	1	3–4
Glyphosate	2	2–3
MCPP	3	1–2
Dicamba	8	1
MCPA	9	1

Source: data from Atwood, D. and Paisley-Jones, C. (2017) Pesticides industry sales and usage 2008–2012 estimates. Office of Pesticide Programs. US Environmental Protection Agency.

charlock, shepherd's purse and fat hen. Mecoprop was introduced to obtain control of cleavers and chickweed and dichlorprop was developed to control *Polygonum* spp. In the following decade clopyralid became available for the additional control of scentless mayweed and creeping thistle and, more recently, soil-applied quinmerac has been developed for further control of speedwells, red deadnettle and cleavers. Mixtures first appeared in the 1960s and dicamba, for example, proved a useful addition for the control of mayweed. More recently, the addition of clopyralid, benazolin and the hydroxybenzonitriles ioxynil and bromoxynil has widened even further the spectrum of weeds controlled, and so these herbicide 'cocktails' have become widely and routinely used, particularly in cereals.

New uses for auxin-type herbicides continue to be reported. For example, in 2005 Brinkworth *et al.* reported that the mixture of two pyridine carboxylic acids, fluroxypyr and aminopyralid, was effective for the long-term selective control of annual and perennial broadleaf weeds in grassland.

The pyridine carboxylic acids, also known as the picolinic carboxylic acids, are characterised by a pyridine ring, chlorine functional groups and a carboxyl group. They are mainly used in non-crop areas, including pastures, and industrial areas, owing to their selectivity, low dose rates and soil persistence. Some also offer selective weed control in crops. Their herbicidal activities differ, possibly owing to variable sensitivity at the target site (Sperry *et al.*, 2020).

Table 7.2   Structures of auxin-type herbicides.

(A) Phenoxyalkanoic acids

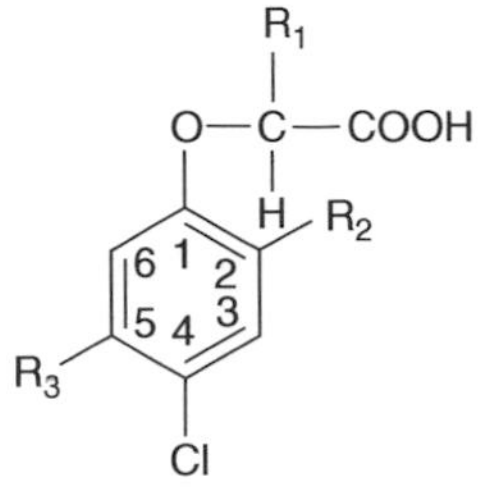

R_1	R_2	R_3	Common name
H	CH_3	H	MCPA
H	Cl	H	2,4-D
H	Cl	Cl	2,4,5-T
CH_3	CH_3	H	Mecoprop
CH_3	Cl	H	Dichlorprop
CH_3	Cl	Cl	Fenoprop

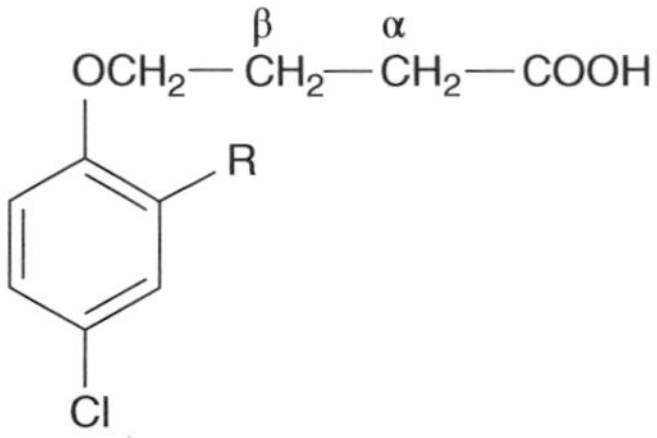

R	Common name
CH_3	MCPB
Cl	2,4-DB

(B) Benzoic acids

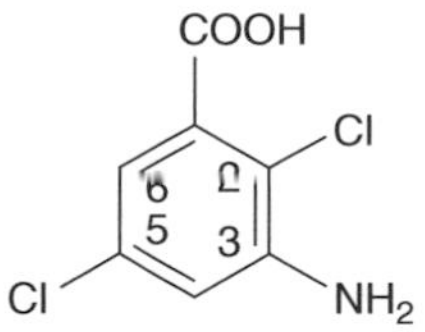

chloramben 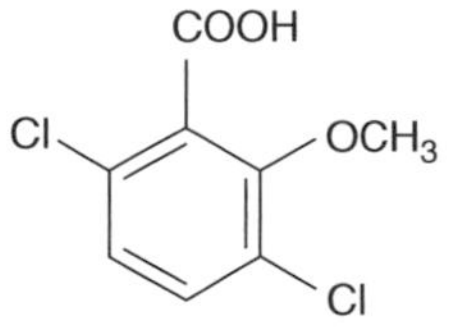   dicamba 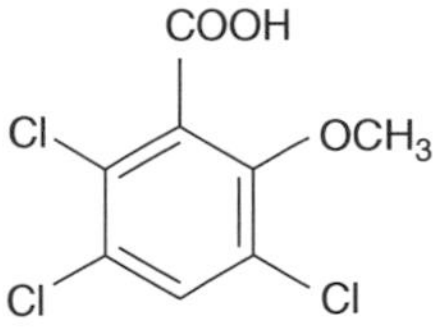   tricamba 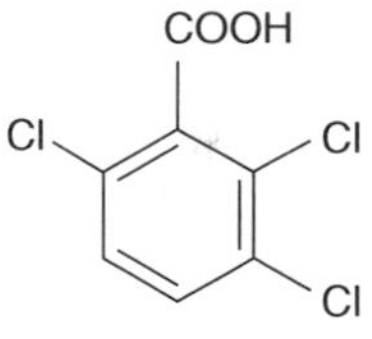   2,3,6-TBA

(C) Aromatic carboxymethyl derivatives

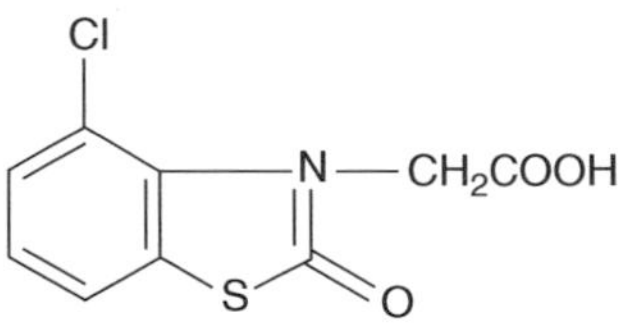

benazolin 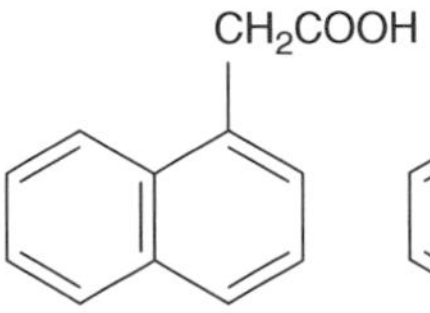   naphthylacetic acid   indol-3yl-acetic acid
                       (NAA)   (IAA)

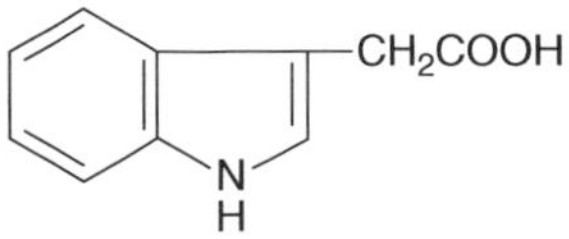

(Continued)

Table 7.2   *(Continued)*

(D) Pyridine derivatives

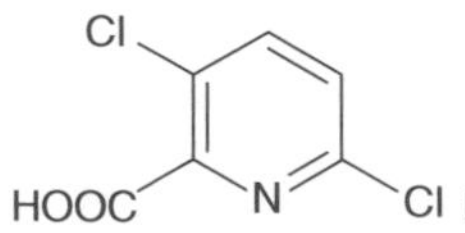

clopyralid

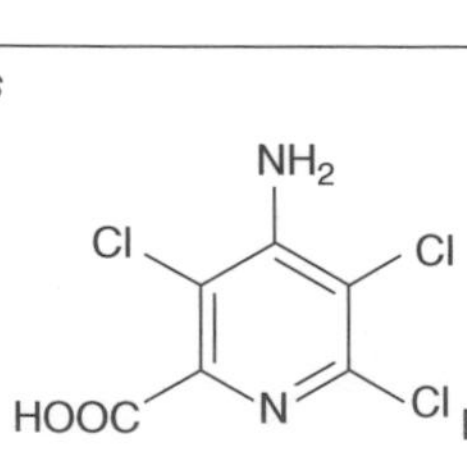

picloram

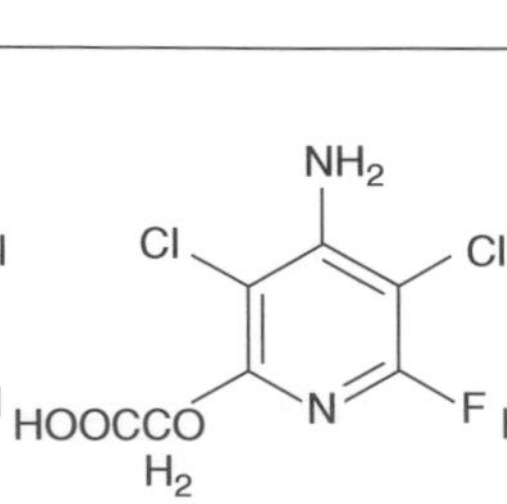

fluroxypyr

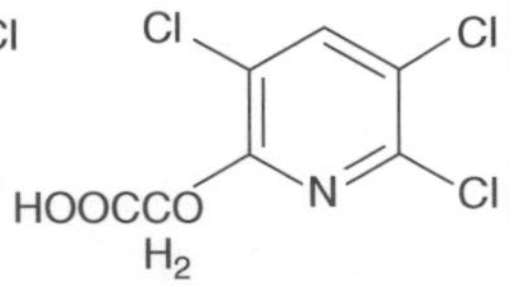

triclopyr

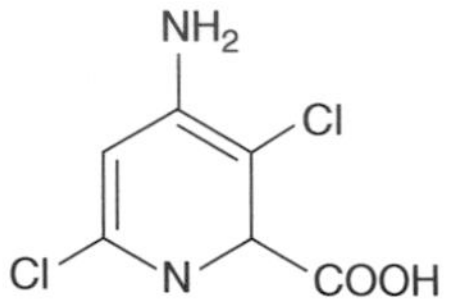

aminopyralid

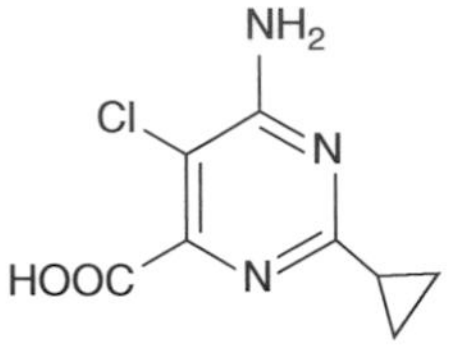

Aminocyclopyrachlor (2010)

(E) Quinoline Carboxylic Acids

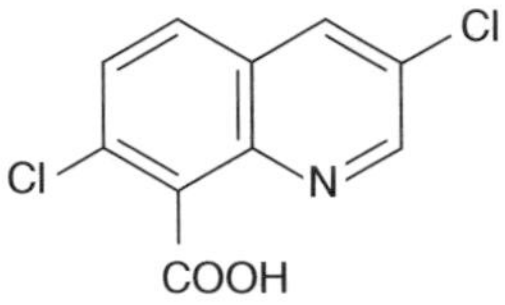

quinclorac

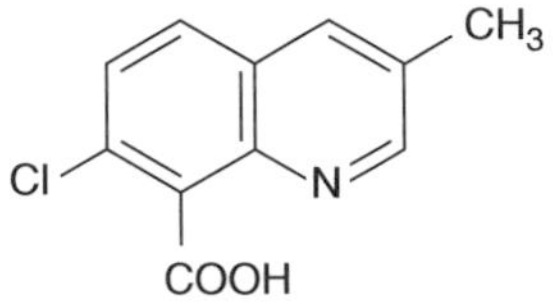

quinmerac

(F) Arylpicolinates

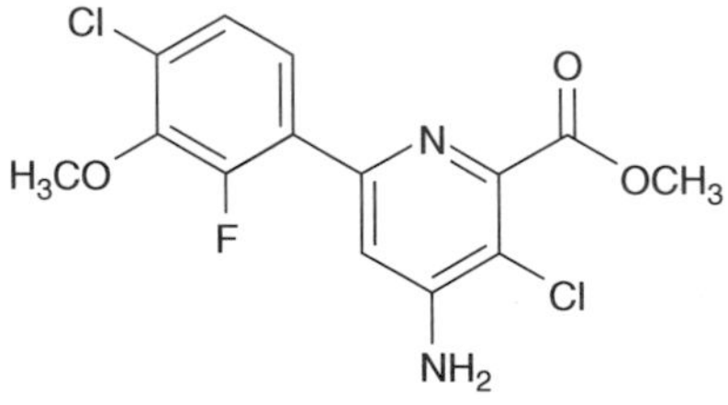

Halauxifen-methyl (2015)

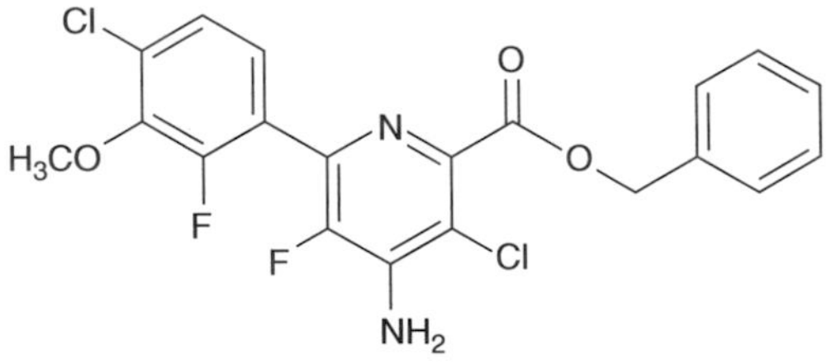

Florpyrauxifen-benzyl (2018)

Recent additions have been aminocyclopyrachlor (2010), halauxifen-methyl (2015) and flopyrauxifen-benzyl (2018). The former controls broad-leaf weeds in turfgrass and pasture grasses (Lewis *et al.*, 2013), although damage may occur in sensitive dicots, including Norway Spruce. Halauxifen-methyl was launched by Dow in 2015 as the first member of a new class of chemistry, the 6-aryl picolinates, for the post-emergent control of a wide spectrum of broad-leaf weeds in cereals, and gives classic symptoms of auxin overdose in sensitive plants. It is a low-dose herbicide (less than 10 g active ingredient (a.i.) per hectare) and

Table 7.3 Some examples of weed seedlings controlled by auxin-type herbicides.

Weed	MCPA (1945)	Mecoprop (1957)	Dichlorprop (1961)	Clopyralid (1975)	Dicamba + mecoprop + MCPA	Benazolin + clopyralid	Quinmerac (1985)
Sinapis arvensis (charlock)	S	S	S		S	S	
Capsella bursa-pastoris (shepherd's purse)	S	S	S		S	S	
Chenopodium album (fat hen)	S	S	S		S	S	
Galium aparine (cleavers)	R	S	S		S	S	S
Stellaria media (chickweed)	R	S	S		S	S	
Polygonum lapathifolium (pale persicaria)	R	R	S		S	S	
Polygonum persicaria (redshank)	R	R	S		S	S	
Bilderdykia convolvulus (black bindweed)	R	R	S		S	S	
Tripleurospermum maritimum (scentless mayweed)	R	R	R	S	S	S	
Cirsium arvense (creeping thistle)	R	R	S	S	R	S	
Veronica hederifolia (ivy-leaved speedwell)	R	R	R		R	R	S
Lamium purpureum (red deadnettle)	R	R	R		R	S	S

R, Resistant; S, Susceptible.
Source: modified from Martin, T.J. (1987) Broad versus narrow-spectrum herbicides and the future of mixtures. *Pesticide Science* **20**, 289–299.

can be mixed with many cereal agrochemicals. Formulations with florasulam and fluroxypyr have been developed. Its discovery is described by Schmitzer *et al.* (2015). Flopyrauxifen-benzyl was commercialised in 2018 for the control of weeds resistant to acetolactate synthase (ALS)- and auxin-type herbicides, in particular resistant barnyard grass in rice. It is also relevant to note that some pyridine derivatives can be phytotoxic to non-target plants, such as grapevines.

Robatscher *et al.* (2019) found that the pyridine-based fungicide, fluopyram, produced growth disorders in grapevines, *Vitis vinifera*, causing significant losses at harvest. These scientists discovered that it was not the parent fungicide that was phytotoxic, but a metabolite, namely 3-chloro-5-trifluoromethylpyridine-2-carboxylic acid, which was structurally almost identical to the herbicide clopyralid, but with an added trifluoromethyl group on the pyrimidine structure.

The auxin-type herbicides have stood the test of time and remain a very effective means of weed control, more than 75 years after their introduction. Those in use today are generally of low persistence, are environmentally benign and are considered unlikely to result in major problems with weed resistance in the foreseeable future.

7.3   Auxin, a natural plant growth regulator

Auxin, or indol-3-yl-acetic acid (IAA), is an endogenous plant growth regulator that plays a crucial role in the division, differentiation and elongation of plant cells. At the organ and whole plant level it has a profound influence on many aspects of plant physiology, including seedling morphology, geotropism, phototropism, apical dominance, leaf senescence and abscission, flowering, fruit setting and ripening. It is synthesised via tryptophan-dependent and tryptophan-independent pathways in meristematic, actively growing tissues and is found throughout the plant body in concentrations ranging from 1 to 100 µg IAA kg^{-1} fresh weight. Young seedlings and tissues that are rapidly growing and elongating contain relatively higher concentrations of auxin than mature tissues and it is believed that younger tissues are the most sensitive to this growth regulator.

During recent decades, many studies have investigated the effects of exogenous auxins on plant growth, with the general conclusions that (a) auxins can inhibit as well as stimulate plant growth in a concentration-dependent manner and (b) different tissues show differential sensitivity to applied auxin. Growth inhibition caused by supra-optimal auxin concentrations is largely attributable to auxin-induced ethylene evolution. Thus, once a critical level of auxin is reached which is tissue-specific, ethylene is produced, and this causes relative inhibition of growth (Figures 7.1 and 7.2).

Klaus Grossmann and colleagues at BASF have proposed a link between hydrogen peroxide production and tissue damage in *Galium aparine* when treated with auxin-type herbicides (Grossmann *et al.*, 2001). They envisage that as a consequence of auxin-herbicide treatment, ethylene synthesis is stimulated, accompanied by an increase in the biosynthesis of the hormone abscisic acid (ABA). Ethylene induces senescence, while the ABA induces stomatal closure and hence the cessation of carbon assimilation by photosynthesis. Since the treated plant is still exposed to light, these workers consider that H_2O_2 accumulates, resulting in oxidative damage that also contributes to weed phytotoxicity.

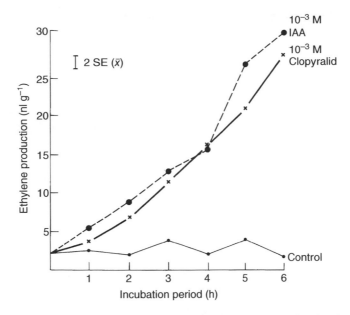

Figure 7.1   Ethylene evolution by scentless mayweed following application of 10^{-3} M clopyralid or indol-3-yl-acetic acid (IAA). Source: Thompson, L.M.L. and Cobb, A.H. (1987) The selectivity of clopyralid in sugar beet; studies on ethylene evolution. *British Crop Protection Conference, Weeds* **3**, 1097–1104.

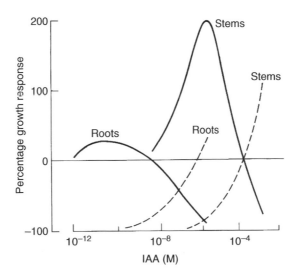

Figure 7.2   Effect of exogenous auxin (IAA) on growth (solid curves) and ethylene production (dashed curves) by roots and stems. Source: Goodwin, T.W. and Mercer, E.I. (1983) *Introduction to Plant Biochemistry*, 2nd edn. Oxford: Pergamon Press.

7.4 Biosynthesis and metabolism of auxins

The main pathway for auxin biosynthesis is thought to be a two-step process from L-tryptophan to indole pyruvic acid, catalysed by tryptophan aminotransferase (TAA). Its conversion to auxin, IAA, is catalysed by a group of enzymes, termed YUCCAs. These are flavin mono-oxygenase-type proteins, for which there are 11 genes in *Arabidopsis thaliana*. Both TAA and YUC genes have been identified in many species. Auxin concentration is also regulated by oxidation and conjugation, the main enzymes being Dioxygenase for Auxin Oxidation 1 (DAO1), which leads to the irreversible inactivation and degradation of IAA, and GH3 (Gretchen Hagen 3), a family of acyl amido synthetases which conjugate auxins to amino acids, and are then transferred to the vacuole, where they have a storage role (Figures 7.3 and 7.4). Westfall *et al.* (2017) have demonstrated that the GH3 protein can conjugate both auxins (IAA and phenylacetic acid) and benzoates (salicylic acid and benzoic acid). Consequently, amino acid conjugation not only contributes to auxin homeostasis, but also controls pathways involved in the pathogen response.

Auxin concentration *in vivo* is tightly controlled by the relative rates of biosynthesis and degradation, with a further layer of complexity evident when conjugation is taken into account (Figure 7.3). Auxin synthesis is complex and the pool size governed by oxidation and/or conjugation. It has been known since 1947 that plant tissues are capable of the oxidative degradation of IAA by a so-called IAA oxidase and that this enzyme activity is rapid and widespread in plant tissues. However, this activity is yet to be characterised and it remains to be convincingly demonstrated that it can be separated from plant peroxidases. In elongating tissues, low oxidase activity is thought to ensure a relatively high auxin concentration (approximately 10^{-6} M), and in roots lower auxin concentrations (approximately 10^{-10} M) result from measurably higher oxidase activity. Many *in vitro* studies with peroxidases, especially those isolated from horseradish, have suggested that the oxidation of auxin is under the control of many naturally occurring substances, including phenols and other growth regulators, but supporting data *in vivo* is lacking.

Auxin conjugation to glucose, amino acids and *myo*-inositol may serve as a storage form or auxin-reservoir, which may be hydrolysed to free auxin when necessary, especially following

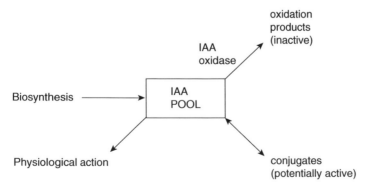

Figure 7.3 The control of auxin concentration *in vivo*. Source: Goodwin, T.W. and Mercer, E.I. (1983) *Introduction to Plant Biochemistry*, 2nd edn. Oxford: Pergamon Press.

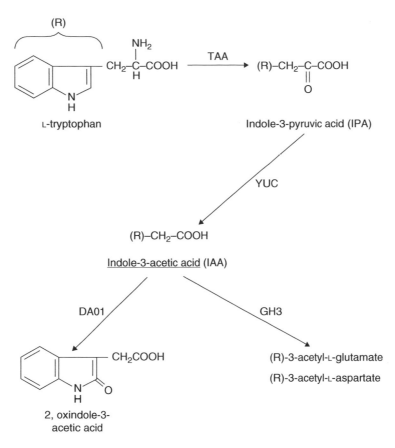

Figure 7.4   Auxin biosynthesis and metabolism (see text for details)

seed germination. This has recently assumed major physiological significance and importance with the finding that concentrations of conjugated auxins can be much higher *in vivo* than that of free IAA. The principal amino acid conjugate in vegetative tissues appears to be IAA–aspartate, which is formed by L-aspartate-*N*-acylase, an enzyme induced by all natural and synthetic auxins. Glucose esters are also common auxin conjugates and they are formed from pre-existing glucosyltransferases. Thus, although much rigorous work remains to be done, auxin synthesis, degradation and conjugation appear to interact, with the result that natural auxin concentrations appear to be tightly controlled *in vivo*.

In addition to IAA, plants synthesize three additional compounds that are also regarded as endogenous auxins, namely indole-3-butyric acid (IBA), 4-chloroindole-3-acetic acid and phenylacetic acid (PAA) (Simon and Petrasek, 2011; Figure 7.5). IBA comprises 25–40% of the auxin pool in *Arabidopsis* seedlings. It is formed from IAA by IBA synthase and can be converted back to IAA by *β*-oxidation enzymes in the peroxisome. It may therefore act as a stable storage form of IAA. Like IAA, IBA is actively transported in the plant and can induce responses independent of IAA. 4-Cl-IAA is found in developing seeds in legumes and its biosynthesis is via 4-Cl-tryptophan. There is little evidence for its involvement in auxin responses.

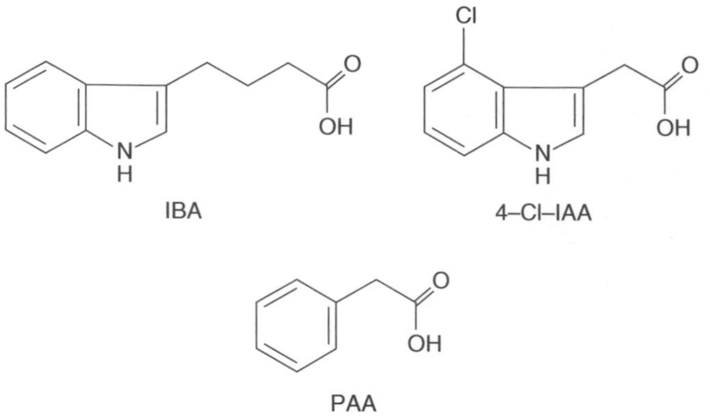

Figure 7.5 Structures of indole-3-butyric acid (IBA), 4-chloroindole-3-acetic acid (4-Cl-IAA) and phenyl acetic acid (PAA)

PAA shows weaker auxin responses than IAA, and is found in various plant species and tissues at a wide range of concentrations. It is thought that symbiotic bacteria in the root are able to synthesize PAA. PAA is not transported in a polar manner, but is able to bind to IAA receptors and regulate the expression of auxin-responsive genes. Aoi *et al.* (2020b), have shown that the arogenate dehydrogenase family (ADT) of enzymes in *Arabidopsis* catalyse the conversion of arogenate to phenylalanine, and that the overexpression of ADT4 or 5 leads to increased levels of PAA, which can partially restore auxin-deficient phenotypes. Also in 2020, the same team reported that the GH3 auxin-amido synthetases can alter the ratio of IAA and PAA, suggesting that PAA may play an important role in plant growth and development (Aoi *et al.*, 2020a).

7.5   Auxin receptors, gene expression and herbicides

Before an auxin can alter cell metabolism and tissue growth, it must first bind to a receptor and the signal be transmitted to the metabolic machinery of the cell. An auxin receptor may then be defined as a precise, cellular site of molecular recognition from which a series of reactions result in growth changes. Thus, the primary mechanism of auxin action is binding to an auxin receptor. According to Venis (1985) binding is predicted to be:

1   reversible, since reactions slow or stop when auxin is removed;
2   of high affinity, because endogenous concentrations of auxin are so low;
3   saturable, at concentrations similar to the saturation of physiological processes sensitive to auxin;
4   specific, to active auxins only;
5   confined, to a tissue sensitive to auxin; and
6   linked, to a biological response.

The search for plant hormone receptors has lagged far behind our understanding of animal and bacterial systems. Rapid advances have been made in recent years, although much complexity remains. The two main candidates as auxin receptors are the ABP1 (auxin binding protein) and the TIR1 (transport inhibitor response 1) proteins.

Current thinking suggests that auxins act by promoting rapid proton-flux at the plasma membrane and by regulating transcription via the action of three protein families,

- the TIR1/AFB F-Box proteins;
- auxin/IAA proteins;
- auxin response factor proteins.

7.5.1   *Auxin binding protein*

ABP1 was first detected in 1972 in crude membrane preparations of etiolated maize coleoptiles and purified in 1985. It has since been found in all green plants, including the bryophytes and pteridophytes, and in many tissues. The maize ABP1 cDNA encodes a 163-amino-acid protein and the mature protein has a molecular weight of 22 kDa containing a high-mannose oligosaccharide. Interestingly, it is localised at the endoplasmic reticulum (ER) and, like all proteins destined for delivery to the ER, carries a signal sequence 38 residues long at the C-terminus. To function as a receptor it is thought to associate with a membrane-bound 'docking' protein.

A comparison of ABP1 sequence data from a range of species indicated three highly conserved sequences, termed boxes A, B and C. Antibodies raised to box A were shown to have auxin-like activity, indicating an important role at the binding site. The exact roles of boxes B and C remain uncertain.

Edgerton *et al.* (1994) studied auxin binding to ABP1 in isolated maize microsomes and found the characteristics to be very similar to those predicted solely on biological data by Katekar (1979) (Figure 7.6).

More recently (2001), the protein has been crystallised and a role for a metal ion has been suggested as an ideal carboxylic acid coordination group. The metal ion, likely to be Zn^{2+}, is complexed to three histidine and one glutamic acid residue in box A.

All of the available data, reviewed by Napier *et al.* (2002), have shown ABP1 to be active to auxins at the surface of the plasma membrane, despite carrying the ER sequence. These include the early responses to auxin action, such as promoting ion fluxes.

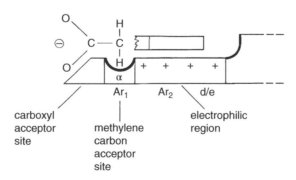

Figure 7.6   A topographic model of the auxin receptor viewed from the side. This model proposes that the auxin receptor possesses regions to accept a carboxyl group, the methylene carbon of IAA (α), the indole ring (Ar_1, Ar_2) and adjacent areas to the indole ring (d/e). Source: Katekar, G.F. (1979) Auxins: on the nature of the receptor site and molecular requirements for auxin activity. *Phytochemistry* **18**, 223–233. Reproduced with permission of Elsevier.

The suggestion that herbicides might bind to this site and alter plasma membrane function has been proposed by Hull and Cobb (1998). In their study, highly purified plasma membrane vesicles were isolated from the monocotyledonous weed black-grass (*Alopecurus myosuroides* Huds.) and the dicotyledonous crop sugar beet (*Beta vulgaris*) and H^+-efflux measured in the presence and absence of herbicides. They found that while auxin-type herbicides in general did not affect H^+-efflux, the aryloxyphenoxypropionate diclofop-methyl was highly inhibitory and 2,4-D gave a slight increase in activity. Auxins are known to antagonise the action of these 'fop' herbicides in the field when present in mixtures (Chapter 8). Since the fops are able to depolarise the plasma membrane potential by inhibiting ATP-ase, perhaps the auxin repolarises the potential by stimulating ATPase activity and so restores cytoplasmic homeostasis. It may also be speculated that an interaction of these herbicidal molecules could occur at ABP1. To speculate further, perhaps this interaction could also account for the differences in selectivity observed in the field between two herbicides sharing similar structures (Figure 7.7).

The question of how ABP1 transmits the auxin signal from the plasma membrane to the cytosol has at last been resolved by the discovery of a trans-membrane, ABP1-interacting molecule. This is a trans-membrane-like kinase, termed TMK1, which interacts with ABP1 in an auxin-dependent manner (Xu *et al.*, 2014). The ABP1–TMK1 complex activates a signalling pathway, such as the small GTPases of the Rho family (ROPs). Auxins can activate ROP2 and ROP6 within 30 s to promote growth of epidermal cells via changes in the actin cytoskeleton. ABP1 also influences auxin-regulated transcription. Mutants of ABP1 have impaired regulation of cell elongation and division, and display altered patterns of auxin-induced genes, such as those coding for the AUX/IAA family, SAUR and GH3. ROP-mediated signalling might also act via the TIR1. ABP1 also controls the regulation of cell wall-related genes, especially those controlling the xyloglucan side chains (Paque *et al.*, 2014). A connection has also been proposed between ABP1 and phytochromes A and B (Effendi *et al.*, 2013), such that ABP1 negatively regulates phytochrome B-dependent signalling, including hypocotyl elongation. These

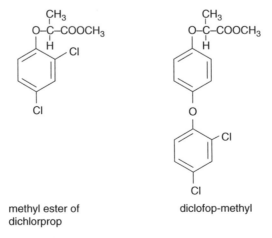

methyl ester of
dichlorprop

diclofop-methyl

Figure 7.7   Structures of the methyl ester of dichlorprop (which controls dicotyledonous weeds in monocotyledonous crops) and diclofop-methyl (which controls monocotyledonous weeds in dicotyledonous crops).

authors conclude that ABP1 directly or indirectly participates in auxin and light signalling. These observations suggest ABP1 involvement in auxin sensing at the cell membrane via the mediation of both non-transcriptional and transcriptional auxin responses. Further evidence is needed to confirm this linkage.

7.5.2   Transport inhibitor response 1 protein

Since 2005 the TIR1 protein has been closely associated with auxin binding and linked to auxin-induced gene expression. The TIR1 gene was first identified in 1997 in plants tolerant to the auxin transport inhibitor naphthylphthalamic acid (NPA) and named 'transport inhibitor response 1'. Since then, research has shown that auxin-mediated control of gene expression is achieved by the de-repression of genes owing to the ubiquitination and subsequent degradation of transcription factors. It is now believed that TIR1 is an auxin receptor in its own right, operating within the nucleus of the plant cell.

Sequence comparisons have found no similarities between TIR1 and ABP1, although binding affinities and pH optima indicate their different locations within the cell, that is pH 5 at the cell surface for ABP1 binding and pH 7 in the nucleus for auxin binding to TIR1. It is tempting to speculate that the rapid effects of auxins at the plasma membrane, such as H^+ efflux and ion movement, are linked to ABP1, while gene expression responses result from binding to TIR1 (Figure 7.8). Tan *et al.* in 2007 reported a structural model of the auxin receptor based on the crystallographic analysis of the TIR1 complex from *Arabidopsis*.

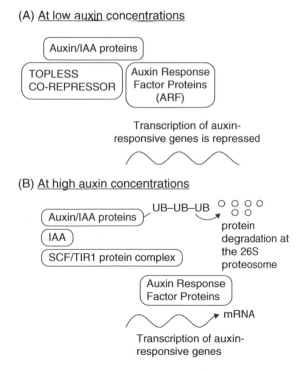

Figure 7.8   A simplified model describing how auxin concentration controls gene expression (see text for details).

According to these authors, auxin binds to the base of a site that can also bind 1-naphthalene acetic acid and 2,4-D. On top of this, the auxin/IAA polypeptide occupies the rest of the site and completely encloses the auxin binding site. They consider that IAA acts as a 'molecular glue' to enhance the TIR1-auxin/IAA protein interaction. This results in the ubiquitination of the auxin/IAA proteins, leading to their degradation at the proteosome. The loss of these repressor proteins then allows the auxin response factor proteins (ARFs) to activate the transcription of the auxin-response genes, according to the prevailing auxin concentration.

At low auxin concentrations, AUX/IAA transcriptional repressor proteins, together with TOPLESS co-repressor proteins, repress genes targeted by the ARF transcriptional activators. As auxin concentrations rise, auxin binds to TIR1, creating a high-affinity surface for the AUX/IAA co-receptor. Assembly of the complex leads to ubiquitination of the AUX/IAA proteins and their consequent degradation in the proteosome. This reduction in the concentration of the AUX/IAA proteins releases the ARFs, allowing transcription to commence. In the case of 2,4-D, for example, transcription of proteins leading to the biosynthesis of ethylene and abscisic acid follow, leading to the typical auxin-herbicide response.

In *Arabidopsis*, TIR1 is part of a family of six auxin receptors, with the other five referred to as Auxin F-Box proteins (i.e. AFB1–5). The family comprises three pairs, TIR1 and AFB5, AFB2 and AFB3, and AFB4 and AFB5. There is a large degree of sequence similarity and functional redundancy in this receptor family. AFB1, AFB2 and AFB3 show the closest homology to TIR1 with 67–72% amino acid similarity, and all bind IAA with different affinities. AFB5 differs from TIR1 in selectivity and is the primary site of action of the picolinate and pyridine carboxylic acid auxin-type herbicides. An *Arabidopsis* mutant line is sensitive to picloram, but remains sensitive to 2,4-D and IAA.

TIR1 is required for 2,4-D perception as it contains the binding pocket for both IAA and 2,4-D. This pocket contains two highly selective polar residues (arg 403 and ser 438), which, together with a less selective hydrophobic environment, form a fixed cavity. The co-receptor AUX/IAA proteins bind above the auxin once it is inside the TIR1 pocket to complete the co-receptor complex. It is therefore possible that mutations in these co-receptor proteins may contribute to altered selectivity and resistance to auxin-type herbicides (Busi *et al.*, 2018). Indeed, populations resistant to dicamba in *Kochia scoparia* have recently been observed which are also cross-resistant to 2,4-D and fluroxypyr (LeClere *et al.*, 2018). Transcriptome sequencing has identified a base-change in the resistant biotype from glycine to asparagine, in a highly conserved region of the AUX/IAA protein KsIAA16. This resistant allele confers a fitness penalty in glasshouse studies. This may have an important bearing on the control of this important weed, since dicamba has developed an increased importance in recent times because of its use in combination with glyphosate for the control of glyphosate-resistant weeds (Chapter 14).

It is now thought that the range of responses to auxins may, in part, be accounted for by the range of affinities measured for different co-receptor complexes, i.e. the range and structure–activity relationships of auxin-type herbicides (Lee *et al.*, 2014; Quareshy *et al.*, 2018a, b).

Through a combination of genetic, molecular and biochemical approaches we are at last beginning to unravel aspects of how auxins interact with gene expression. Auxin rapidly and transiently stimulates the transcription of three gene families, known as the primary auxin-responsive genes.

1   *Auxin/IAA proteins:* these short-lived nuclear proteins function as transcriptional regulators. They do not interact directly with DNA but exert regulation through another group of proteins known as the ARFs. The *Arabidopsis* genome contains 29 AUX/IAA

genes and 23 ARF genes. ARFs bind to conserved DNA sequences in the promoter regions of early auxin response genes, acting as transcriptional repressors. The AUX/IAA proteins turnover rapidly, over 10–80 min, and this degradation is controlled by a ubiquitin protein ligase activated by auxin binding.

2   *SAUR genes:* small Auxin Up/RNA transcripts accumulate rapidly after auxin exposure. They appear to have short half-lives, although their function remains unknown.

3   *GH3 genes:* auxin-induced GH3 gene transcription appears to encode IAA–amino acid conjugating enzymes. This may serve to dampen the auxin signal by inactivating the auxin itself by conjugation.

Kelley *et al.* (2006), have used auxin-induced gene expression as an indicator of auxin-type herbicide injury in soybean leaves. They found that GH3 expression was highly induced by dicamba and clopyralid, within 8 h after application, reaching a maximum at 1–3 days after treatment. GH3 expression was not affected by environmental stress or by viral attack, indicating its potential as a diagnostic assay of herbicide injury. It will be interesting to observe what use is made of such a test, since the crop inevitably recovers via the selective metabolism of these herbicides.

The reader is referred to the recent review by Quareshy *et al.* (2018a, b; Vernoux and Robert, 2017) for updates in this rapidly developing field. As already demonstrated with NPA, auxin-type herbicides can have an important role as probes in unravelling these complex but important aspects controlling plant growth and development. In addition, these newly discovered genes and gene products may become the targets for future generations of herbicides and plant growth regulators.

7.6   Signal transduction

Once bound to the ABP1 receptor, a signal must be transmitted to the rest of the cell to cause the specific changes in metabolism that result in altered growth. Brummell and Hall (1987) consider that for a relatively small number of auxin molecules binding to a relatively small number of receptors to produce a major effect, some sort of rapid amplification reaction is necessary. They suggest that both signal amplification and transduction are achieved through changes in intracellular concentrations of calcium ions with important roles for inositol triphosphate (IP_3) and diacylglycerol (DG) as additional secondary messengers.

The cytoplasmic concentration of free calcium ions in a 'resting' cell is kept low, in the region of 10^{-7}–10^{-8} M, owing to its continual active removal into the cell wall or into organelles, such as the ER and vacuole. However, when the cell is stimulated by auxin binding at receptors, a transient increase in cytoplasmic calcium ions (to 10^{-5}–10^{-6} M) results, which is then able to bind to a calcium-binding protein (calmodulin) and this complex may stimulate protein kinases. These in turn phosphorylate, and activate, key enzymes.

The link between auxin receptor occupancy and cytoplasmic calcium ion concentration is thought to involve the hydrolysis of phosphatidyl inositol bisphosphate (PIP_2) at the plasmalemma, which results in the production of IP_3 and DG. IP_3 can mobilise calcium ions from the ER, and is certainly metabolised in the presence of auxin (Ettlinger and Lehle, 1988), and DG may also activate calcium-dependent protein kinases.

A unifying, but still highly speculative, view of the consequences of binding to an auxin receptor in young tissues is presented in Figure 7.9. Auxin binding to the plasmalemmal

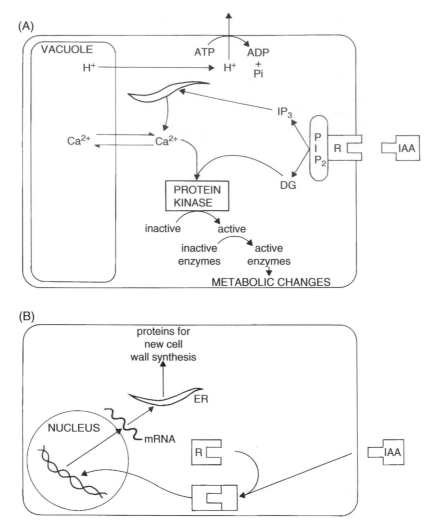

Figure 7.9   A speculative view of the immediate consequences of auxin binding to a plasmalemmal (A) or cytoplasmic (B) receptor. PIP$_2$, phosphatidyl inositol 4,5-bisphosphate; ER, endoplasmic reticulum; R, receptor; IP$_3$, inositol 1,4,5-triphosphate; DG, diacylglycerol. Source: based on Brummell, D.A. and Hall, J.L. (1987) Rapid cellular responses to auxin and the regulation of growth. *Plant, Cell and Environment* **10**, 523–543.

receptor causes hydrolysis of PIP$_2$ to IP$_3$ and DG. IP$_3$ causes enhanced mobilisation and cytoplasmic accumulation of calcium ions, which together with DG activates protein kinases, and the activation of other key enzymes follows. The elevated cytoplasmic concentration of calcium ions is reduced to resting state levels by transport into the vacuole or endoplasmic reticulum. Transport into the vacuole is in exchange for protons (H$^+$) which are then exported out of the cell by an H$^+$-ATPase.

We now know that auxin also binds to a soluble receptor in the nucleus where specific genomes become derepressed and specific mRNA sequences are synthesised.

These messengers are translated at the endoplasmic reticulum to products, for example, involved in new cell wall synthesis (Figure 7.9). This view is supported by the observation by many workers that specific mRNA synthesis can be detected within minutes after auxin application. These models are of real value for an appreciation of auxin-type herbicide action because they identify primary and measurable responses to auxin receptor occupancy, namely selective gene expression and enhanced H^+-efflux.

Useful information has been gained from studies of auxin-induced H^+-efflux. It is now well established that plant cells maintain large differences in electrical potential and pH across their intracellular compartments (Figure 7.10). Indeed, the electrogenic gradient generated by H^+ pumping provides the driving force for the transport of various solutes, including anions, cations, amino acids, sugars and auxins. These ATPases are associated with the plasmalemma and tonoplast and are the subjects of much current research. It has been known for some time that auxins are able to induce cell elongation in young tissues that is associated with H^+-efflux. Indeed, the traditional bioassays of measuring the elongation or curvature of intact tissues or tissue segments after a 24 h incubation was a cornerstone of auxin discovery. However, the relationship between auxin concentration and growth in these bioassays is poorly defined, such that a very large change in exogenous auxin is often necessary to cause a measurable difference in, for example, growth rate. Furthermore, most auxin bioassays have poor, or at least variable, sensitivity usually caused by poor or slow penetration of the exogenous auxin. Indeed, mecoprop was initially overlooked as a potential auxin herbicide because of its relative inactivity in the oat straight growth test (as cited in Kirby, 1980). On the other hand, the measurement of rates of H^+-efflux from sensitive tissues is rapid, linear with time and highly dependent on auxin concentration. In oat coleoptile tissues at least, it can be considered a primary auxin response in ion transport, which occurs as a direct consequence of auxin–receptor complex formation. By measuring H^+-efflux from oat segments, Fitzsimons (1989) and Fitzsimons *et al.* (1988) derived sensitivity parameters for a wide range of auxin-type herbicides (Figure 7.11). In this example, and with many other auxins tested, the maximum value of the

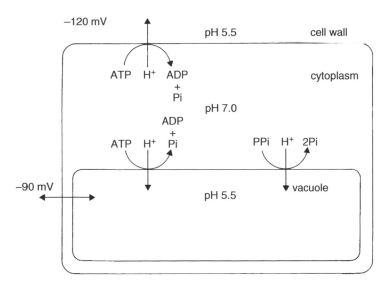

Figure 7.10 Approximate differences in electrical potential and pH in a plant cell.

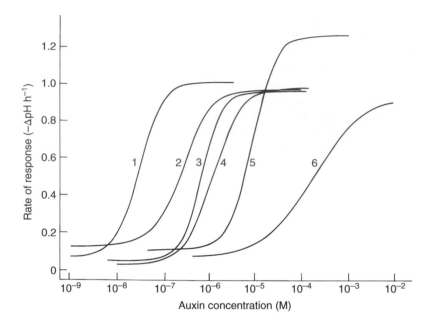

Auxin	*H_{50} (M)	Activity relative to IAA
1. IAA	2.8×10^{-8}	1
2. Mecoprop	2.4×10^{-7}	9
3. Fluroxypyr	6.5×10^{-7}	23
4. 2, 4-D	1.1×10^{-6}	39
5. Benazolin	8.3×10^{-6}	307
6. Clopyralid	1.7×10^{-4}	6071

*H_{50}, the auxin concentration giving half maximal response.

Figure 7.11 Dose–response curves for auxin-induced proton-efflux. Source: modified from Fitzsimons, P.J., Barnwell, B. and Cobb, A.H. (1988) A study of auxin-type herbicide action based on dose–response analysis of H+-efflux. In: *Proceedings of the European Weed Research Society Symposium, Factors Affecting Herbicidal Activity and Selectivity*. Wageningen: European Weed Research Society, pp. 63–68.

rate of efflux was always similar, with the unexplained exception of benazolin. Large-scale differences in the activity observed were attributed to the relative affinity of each herbicide for the auxin receptor in this species, which were computed by reference to a model of hormone–receptor interaction. These H_{50} values (the auxin concentration giving half maximal response) enable a more accurate and relevant comparison of auxin activity than data from growth bioassays, and may offer a means for a more detailed understanding of the auxin receptor in monocotyledonous species in the presence of herbicides.

7.7   Auxin transport

Auxin appears to regulate a wide range of plant developmental processes owing to its asymmetric distribution across adjacent cells and tissues during important stages of growth and development. Auxin is synthesised in meristematic tissues and is distributed throughout the plant either through the phloem or by a more controlled cell-to-cell, polar transport system. A now well-accepted model for polar auxin transport is known as the chemiosmotic hypothesis.

As shown in Figure 7.10, there is a pH gradient across the cell, from about pH 5.5 in the matrix of the cell wall to pH 7.0 in the cytoplasm. In the cell wall the IAA molecule exists primarily in the protonated form (IAAH) and enters the cell passively through the plasma membrane. It dissociates once inside the cell and the IAA$^-$ ion becomes trapped inside the cytoplasm. Specific auxin efflux carriers are required to transport the dissociated form out of the cell. These are predicted to be asymmetrically located in the cell to account for unidirectional auxin transport. We now know that an auxin carrier molecule also exists to further enable IAA$^-$ uptake (Figure 7.12).

The nature of the auxin-efflux carrier was identified from studies with the pin-1 mutant of *A. thaliana*. These mutants have needle-like stems that lack flowers. Molecular analysis of the *PIN1* gene revealed in 1998 that it encoded a transmembrane carrier protein. A further seven *PIN* genes have been found and all phenotypes related to aberrant auxin accumulation. All PIN proteins are localised in a polar fashion, according to the predictions of the chemiosmotic hypothesis.

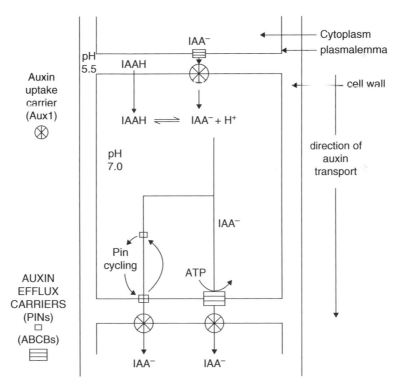

Figure 7.12   The chemiosmotic model for polar auxin (IAA) transport in xylem parenchyma cells. Sources: adapted from Vieten, A., Sauer, M., Brewer, P.B. and Friml, J. (2007) Molecular and cellular aspects of auxin-transport-mediated development. *Trends in Plant Science* **12**, 160–168; Busi, R., Goggin, D.E., Heap, I., Horak, M.J., Jugulam, M., Masters, R.A. *et al.* (2018) Weed resistance to synthetic auxin herbicides. *Pest Management Science* **74**, 2265–2276. See text for further details.

Arabidopsis mutants were also used to discover the auxin influx carrier in 1998. The AUX-1 mutant was identified as having resistance to movement of 2,4-D. The *AUX1* gene was cloned and found to encode a protein with similarity to amino acid permeases. The AUX1 protein is also localised asymmetrically in some cells.

An intriguing question is 'how do these auxin carriers end up at distinct sides of the cell?' Vieten *et al.* (2007) suggest that their movement and targeting result from the vesicle trafficking pathways in the cytoplasm. An accessory protein AXR4 is also thought to ensure that the protein is localised in the plasma membrane, although how this is achieved is not known.

In *Arabidopsis*, PIN proteins are thought to rapidly cycle between the plasma membrane and the endoplasmic reticulum via endosomes. This process is termed constitutive cycling and, according to Paciorek *et al.* (2005), is inhibited by auxin. This novel finding, shared by biologically active auxins, leads to increased auxin-efflux carriers at the cell surface where auxin concentration is least. Thus, auxins may promote their own efflux.

In plants, the ABCB-type auxin efflux transporters are involved in both cell-to-cell polar auxin transport and in facilitating long-distance transport of auxin through the phloem. These are members of the ABC-B subfamily. A number of studies have indicated that 2,4-D can bind to ABCB1, ABCB4 and ABC19, possibly suggesting that long-distance transport of auxin is affected. It is intriguing to note that the polar auxin-efflux inhibitor NPA (1-*N*-naphthylphthalamic acid) binds to ABCB-type auxin transporters disrupting their function. The individual roles of PIN- and ABCB-mediated auxin flows, and how they interact, are currently obscure

Inhibition of auxin transport has been often thought to be a potentially effective mechanism for herbicidal exploitation. Interestingly, one compound has emerged in recent years with these properties, diflufenzopyr (BAS 662H; Figure 7.13). When applied alone, diflufenzopyr stunts weed growth, but when combined with an auxin-type herbicide it appears to result in enhanced translocation of the auxin-herbicide to the weed apices to give more effective broadleaf weed control. The combination of dicamba with diflufenzopyr appears to be particularly effective for a broad spectrum of weed control and tolerance in maize (Bowe *et al.*, 1999), with relatively low dose rates of 100–300 g a.i. ha^{-1}. Diflufenzopyr acts by binding to a specific protein involved in transporting auxin away from the meristematic apices. It has a high affinity for this site, with an IC$_{50}$ of 19 nM diflufenzopyr. Could this be a PIN protein? Thus, both natural and synthetic auxins accumulate at these apices to induce an 'auxin-overdose' response. Interestingly, root geotropism is also inhibited by this treatment, with an IC$_{50}$ of 0.6 nM.

It is now known that the efficacy of auxin-type herbicides varies with time of day of application, although the underlying mechanisms explaining this observation are unclear.

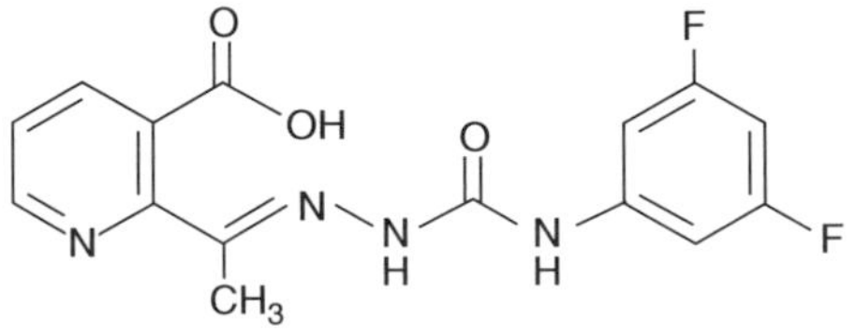

Figure 7.13   The structure of diflufenzopyr (BAS 662 H).

An increase in translocation has been observed with dawn applications of 2,4-D and dicamba, and the translocation of auxin-type herbicides has been implicated. Johnstone *et al.* (2020) studied the translocation of radiolabelled herbicides in *Amaranthus palmeri* and found an increase in the ATP-binding cassette subfamily at dawn compared with midday, and that increased ethylene evolution, related to phytotoxicity was also noted at dawn. They suggest that ethylene evolution is related to translocation, and that ethylene induces abscisic acid accumulation, resulting in the inhibition of growth. Perhaps an exploration of the relationship between gene expression, ethylene evolution and auxin-type herbicide translocation may explain the time of day response.

7.8   Resistance to auxin-type herbicides

The first cases of resistance to 2,4-D were reported in 1957 in climbing dayflower (*Commelina diffusa*) and wild carrot (*Daucus carota*). Since then 36 plants resistant to auxin-type herbicides have been reported, including 30 broad-leaf weeds and six grasses (see Figure 5 in Busi *et al.*, 2018, for a full list). Readers may also wish to consult I.M.Heap online for further details and regular updates (http://www.weedscience.org). Given the selection pressure for resistance to be observed owing to sustained use over 70 years, often in monocultures, the incidence of resistance is much lower when compared with the acetyl-CoA carboxylase (ACCase) and ALS inhibitors. This relatively low incidence of resistance may be due to (a) their proposed multiple sites of action, (b) resistance being conferred by recessive genes and (c) reduced phenotypes showing reduced fitness compared with the crop. Nonetheless, cross-resistance is widely observed and this can be a challenge to the development of effective control strategies, including the development of 2,4-D and dicamba-resistant crops in recent years using bacterial resistance genes. Regarding the mechanisms of resistance, however, our understanding is relatively limited and few resistant weed species have been fully investigated. The molecular understanding of auxin-type herbicide resistance mechanisms currently remains obscure. Understanding of a few weeds resistant to auxin-type herbicides is listed below.

1   Poppy *(Papaver rhoeas)* is a common weed in winter cereals in Europe and an increased resistance to auxin-type herbicides has been noticed in recent years. Torra *et al.* (2017) have studied resistance to 2,4-D in two resistant biotypes growing in Spain and conclude that enhanced metabolism of the herbicide to 2,3-D and 2,5-D was evident in resistant populations, conferred by cytochrome P450-linked enzymes. Reversal from resistant to susceptible was noted when the plants were pre-treated with a P450 inhibitor. It is also thought that reduced translocation may also contribute to herbicide resistance in this species.

2   Kochia (*Kochia scoparia*) is a summer annual weed that infests the plains of North America and has become resistant to atrazine, dicamba, glyphosate and several ALS inhibitors, although it remains sensitive to fluroxypyr. Reduced translocation in resistant plants is a possible cause.

3   Prickly lettuce (*Lactuca serriola*) is an annual or biennial weed found in the Pacific Northwest of the USA. It is resistant to glyphosate and 2,4-D. Inheritance of 2,4-D resistance in this species is governed by a single co-dominant gene, which remains to be identified

(Riar *et al.*, 2011). After application of radiolabelled 2,4-D to resistant and susceptible biotypes for 96 h, the resistant biotype absorbed less and retained more 2,4-D in the treated leaf, when compared with the susceptible biotype. Less translocation of 2,4-D appears to be an important factor in resistance in this species.

4   Wild radish (*Raphanus raphanistrum*) is an important dicot weed in Southern Australia, resulting in yield losses and increased weed control costs. Goggin *et al.* (2016) applied radiolabelled 2,4-D to leaves and found that phloem loading and transport was impaired in resistant plants. The auxin transporting ATP-binding cassette (ABC transporters) of the B *sub*family facilitate polar auxin transport in plants and are often associated with the transport regulating proteins (TWD1) and auxin efflux proteins (PINs) to ensure effective auxin transport complexes at the plasma membrane (Figure 7.12). It is thought that mutations in these proteins will impair the transport of auxin-type herbicides. Note that in this species, there is no evidence of differences in detoxification rates via cytochrome P450 enzymes. It may follow that sequence analyses from ABCB genes and gene products may resolve the basis of resistance in this and other resistance species (Schulz and Segobye, 2016).

Finally, since climatic projections indicate an increase in global mean temperature and carbon dioxide concentrations in the atmosphere by the end of the twenty-first century, it is important to investigate and predict how climate change might influence herbicide resistance in decades to come. In an important publication, Refatti *et al.* (2019) examined the effects of increasing temperature and carbon dioxide on the absorption, translocation and efficacy of cyhalofop-butyl on multiple resistant and susceptible *Echinchloa colonum* under simulated climate change conditions. The multiple-resistant biotype has extreme resistance to the auxin-type herbicide quinclorac and also propanil, which involves non-target site resistance mechanisms. Although plants grown at higher temperatures and carbon dioxide concentrations were taller than ambient control plants, herbicide efficacy was broadly the same. Herbicide efficacy, however, was reduced in multiple-resistant plants by about 50% when compared with plants grown in ambient conditions. One suggestion for this is an increased activity of non-target site resistance mechanisms in herbicide metabolism, resulting in detoxification or reduced activity. Indeed, evidence exists for increased activity of GST enzymes in resistant grasses (e.g. Milner *et al.*, 2001). These findings imply that a changing climate might increase selection pressure in favour of resistance, which is a warning call to weed management practitioners everywhere.

7.9   An 'auxin overdose'

It is evident from the preceding sections that natural auxins are present in very low, but controlled, concentrations in plant tissues, and that they can have a profound influence on plant growth and development. Studies since the early 1970s using combined gas chromatography–mass spectrometry (GC-MS) have demonstrated a range of natural auxin concentrations in plant tissues from 1 to 100 µg kg^{-1} fresh weight. The lower values are commonly found in fleshy tissues and the higher amounts are reported in seeds. It follows that if a broadleaf weed seedling weighs 10 g, then the approximate amount of auxin present in the plant will be in the range 10–100 ng. Given that a field rate of 2,4-D is 0.2–2.0 kg a.i. ha^{-1}, we can safely assume that a broadleaf weed will intercept at least 100 µg of this synthetic auxin, which is at least 1000 times more auxin than is already present in the plant. Since auxins are rapidly absorbed into the leaf and translocated throughout the plant, an imbalance of growth regulator is clearly evident, the control systems are overloaded and growth is drastically altered by supra-optimal auxin concentrations.

7.10 How treated plants die

Weed death occurs as a consequence of an auxin-overdose and is due to uncontrolled growth. The exact sequence of events depends on the age and physiological state of the tissues affected and varies considerably between species. Nonetheless, three phases of symptom development are commonly observed.

1 *Phase 1, the first day*. Profound changes in membrane permeability to cations can be discerned within minutes of auxin application. For example, rapid and sustained proton efflux results in measurable cell elongation within an hour, and an enhanced accumulation of potassium ions in guard cells causes increased stomatal apertures and a transient stimulation of photosynthesis (Figure 7.14). Furthermore, a rapid mobilisation of cellular carbohydrate and protein reserves is commonly observed as large rises in soluble reducing sugar and amino acid concentrations. This coincides with enhanced mRNA synthesis and large increases in rates of protein synthesis. In addition, ethylene evolution is typically detected from treated plants (see Figure 7.1). Research into dicamba-regulated gene expression with *Arabidopsis* mutants has given an insight into how auxin

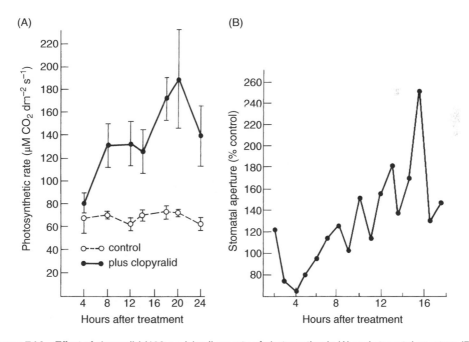

Figure 7.14 Effect of clopyralid (100 g a.i. ha⁻¹) on rate of photosynthesis (A) and stomatal aperture (B) in leaves of scentless mayweed (modified from Thompson, 1989).

herbicides act in the sensitive plant cell. In a comprehensive study, Gleason *et al.* (2011) identified over 550 genes that were upregulated following a 10 h exposure to dicamba, leading to lasting effects on gene expression. Their work supports the existing dogma that auxin-type herbicides activate the TIR1 auxin receptors and related receptor proteins to give a permanent upregulation of auxin-induced genes, leading to increased ethylene and ABA synthesis that eventually results in plant death. Three ACCases were induced by dicamba with the overproduction of ABA. They also noted the overexpression of stress proteins, glutathione and GH3 enzymes, which would represent an in-vain attempt to regulate auxin concentrations.

2   *Phase 2, within a week.* In the days following herbicide treatment major growth changes become apparent as visible symptoms initiated by new genome expression and powered by reserve mobilisation in Phase 1. For example, increased cell division and differentiation in the cambium lead to adventitious root formation at stem nodes in some species (Sanders and Pallett, 1987), and general tissue swelling caused by the division and elongation of the cortical parenchyma is typically observed in other tissues, such as the petiole. Classic symptoms of stem, petiole and leaf epinasty are now observed in young tissues in response to ethylene evolution, and abnormal apical growth is sometimes observed (Figure 7.15). Lateral buds may also be released from apical dominance and all other meristems increase in activity. Until recently, it was thought that the classic epinastic symptoms observed following application of auxin-type herbicides were due to ethylene production alone. Studies by Rodriguez *et al.* (2014), however, have demonstrated that active oxygen species, (hydrogen peroxide, superoxide free radicals) and nitrous oxide are involved in 2,4-D-induced epinasty in young tissues, while the overproduction of active oxygen species triggers senescence in older tissues. The overproduction of active oxygen species is generated in the peroxisomes via the actions of xanthine oxidase,

Figure 7.15   Typical symptom development of auxin-type herbicides. Plants of *Tripleurospermum maritimum* subsp. *Inodora* (scentless mayweed) were treated with 0, 50, 100, 200 and 400 g a.i. ha^{-1} clopyralid, and foliar symptoms were photographed after 7 days (from Thompson, 1989).

dehydrogenase and acetyl-CoA-oxidase. Initially, the main system that is changed as a result of increased active oxygen species is the cell wall, leading to a re-organisation and changes in cell wall structure. They also observed that 2,4-D affects the cytoskeleton by inducing oxidative and *S*-nitrosylation post-translational modifications on actin and expansins, leading to structural modifications. The authors consider that these structural changes are responsible for epinasty.

3   *Phase 3, within 10 days.* Ultrastructural studies reveal progressive disruption of intracellular membranes culminating in the disruption of the plasmalemma, organelle breakdown and tissue collapse. Root disintegration, leaf chlorosis and senescence are rapidly followed by plant death.

7.11   Selectivity and metabolism

The success of the auxin-type herbicides is principally due to their highly selective action. However, since these herbicides behave differently in many species, it is thought that many factors interact and contribute to selectivity in various ways. Several examples of this variation are cited by Pillmoor and Gaunt (1981) in their extensive review of phenoxyacetic acid herbicides. At an extreme level, a few micrograms may kill a susceptible species, but a tolerant crop may withstand milligram doses, although increasing the dose will eventually induce phytotoxicity in the crop. Furthermore, species sensitivity clearly varies with plant and tissue morphology and age. For example, young seedlings of cucumber (*Cucumis sativus*) and wild carrot (*D. carota*) are sensitive to 2,4-D but develop tolerance as tissues mature.

Generally, herbicide selectivity is achieved either by differences in herbicide concentration reaching an active site or by differences in sensitivity at an active site. The former involves a consideration of herbicide uptake, movement and metabolism, since the amount of herbicide in a sensitive tissue is determined by its import and transport from the site of application and its metabolic fate in the largest tissue. A full balance sheet is therefore ideally needed for all of these factors in both resistant and susceptible species to satisfactorily account for selectivity. However, such detailed information is invariably lacking with auxin-type herbicides. In general, differences in uptake and movement have been reported in many species, but no correlation between uptake, movement, and selectivity has been convincingly demonstrated. On the contrary, uptake and movement are sometimes faster in tolerant species! Studies with radiolabelled herbicides show that foliar uptake is typically rapid, and that active ingredients accumulate at major growth regions, especially the apical meristem, as a result of phloem transport (Figure 7.16). Slow but significant rates of root excretion of auxin-type herbicides have also been reported in some species, but how this is achieved, and to what extent it contributes to selectivity, remains unclear.

The pattern and extent of metabolism of auxin-type herbicides are also highly variable. Conjugation, hydroxylation and side-chain cleavage are the principal routes for the metabolism of phenoxyalkanoic acids, the eventual products depending on the sequence of these processes.

Thus, the 2-methyl group of MCPA is highly susceptible to oxidation, and the hydroxyl group so formed is rapidly used in glycoside formation (Figure 7.17A). Direct conjugation of phenoxyacetic acids to form glucose and aspartate esters has also been widely reported, as has side-chain degradation to a corresponding phenol. This oxidation also yields glycolic acid, which is subsequently metabolised in photorespiration to carbon dioxide.

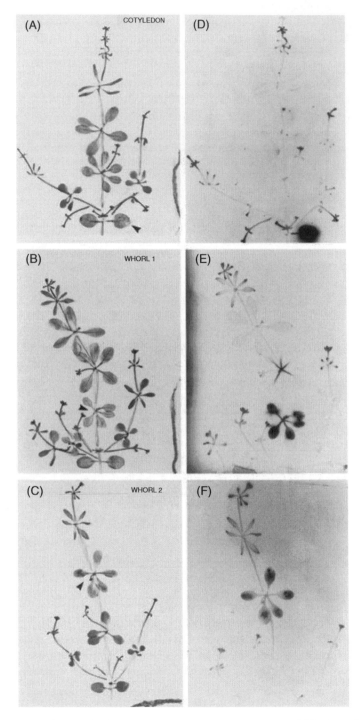

Figure 7.16   The use of autoradiography to study the translocation patterns of ^{14}C-clopyralid in *G. aparine* (cleavers). (A–C) Photographs of treated plants showing position of ^{14}C-clopyralid application (arrowed). (D–F) Autoradiographs of plants A–C, respectively, showing patterns of ^{14}C-translocation from sites of application to regions of active growth (from Thompson, 1989).

In addition, a reaction unique to the metabolism of 2,4-D, known as the NIH shift, is commonly observed in many species. Here, the migration of a chloride atom, usually from the 4- to the 5-carbon position, is probably evidence for an epoxide intermediate in ring hydroxylation, and the product is a substrate for further glucosylation (Figure 7.17B).

Further sugar conjugates that are more polar than either the glucose esters or glycosides mentioned earlier can also be formed. Monocotyledonous species in particular appear to be able to form glycosides with two or more sugar residues, although their possible contribution to selectivity is uncertain (Pillmoor and Gaunt, 1981). In addition, studies using radiolabelled

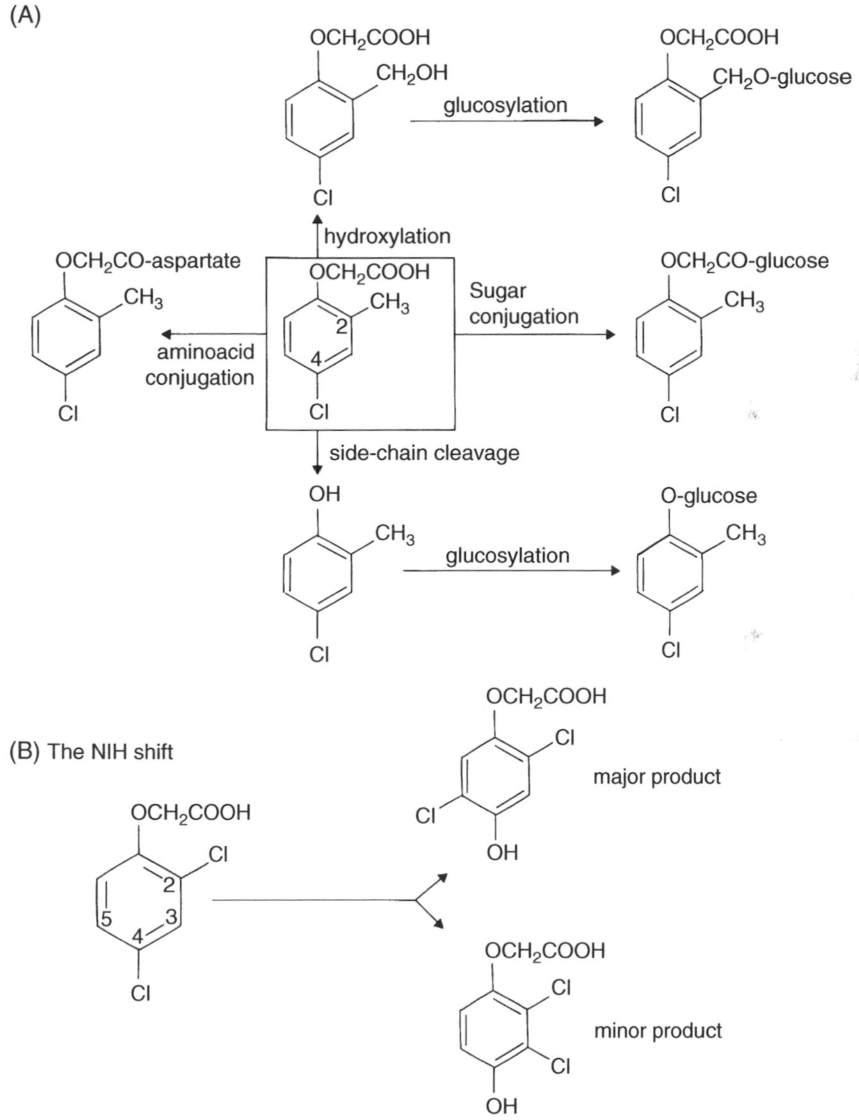

Figure 7.17 Pathways of metabolism of (A) MCPA and (B) 2,4-D.

herbicides have shown that some phenoxyacetic acids and their metabolites can also become bound to insoluble fractions in monocotyledons. Structural polymers in the cell wall are often implicated and lignin, pectin and cellulose have all been suggested to bind auxin metabolites. Further studies are now needed to identify the ligands and characterise these binding phenomena. Such information is clearly needed since these bound residues are seldom found in dicotyledonous plants. Indeed, resistant monocots generally contain very low levels of free auxin-type herbicide in contrast to susceptible dicots, and this may be an important feature of selectivity to these herbicides. In most instances the products of metabolism are more hydrophilic, non-phytotoxic and polar than the parent herbicide, and can be stored or sequestered in the vacuole or become bound to structural polymers. Each factor will contribute to the lowering of the cytoplasmic pool of free herbicide, reducing the level of auxin-receptor occupancy in sensitive tissues.

These observations on metabolism are not confined to the phenoxyalkanoic acids. In a study of triclopyr selectivity in wheat, barley and chickweed, Lewer and Owen (1990) were able to correlate rates of metabolism with species selectivity. They found that resistant wheat plants rapidly metabolised triclopyr to a glucose ester within 12 h, but susceptible chickweed slowly converted the herbicide to triclopyr-aspartate over a 48 h period. In addition, levels of free herbicide remained higher in the weed than in the crop plants.

Metabolism forms the basis of selectivity of the phenoxybutyric acid herbicides, MCPB and 2,4-DB. MCPA and 2,4-D cannot be used in legume crops because they kill both legumes and weeds, but their butyric acid derivatives are selective in these crops. Selectivity is achieved by the conversion of the inactive phenoxybutyric acid derivative to an active phenoxyacetic acid *only* in broadleaf weeds by the process of β-oxidation, which successfully removes two CH_2 residues from the side-chain so that an active auxin-herbicide is only produced when an odd number of CH_2 residues is originally present (Figures 7.18 and 7.19).

In this way the phenoxycaproic acids ($n = 5$) may also have theoretical uses as selective herbicides in legume crops, although only the phenoxybutyric acid derivatives ($n = 3$) have been commercially developed.

The possibility that selectivity is due to differential receptor sensitivity has already been raised in Section 7.4. It is tempting to suggest that the auxin receptor in dicots is more accessible to auxin-type herbicides than the receptor in monocots. However, no direct supporting evidence is currently available. Alternatively, it may be argued that monocots are less sensitive to auxin-type herbicides because their leaves intercept and retain less herbicide, and that since their mature vascular tissues lack a layer of cambium (cells capable of cell division), they may not possess sensitive cells capable of auxin reception. In conclusion, the selectivity

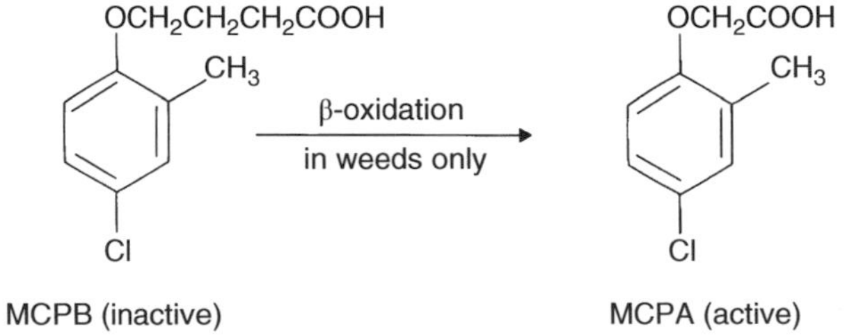

Figure 7.18   Bioactivation of MCPB in susceptible broadleaf weeds.

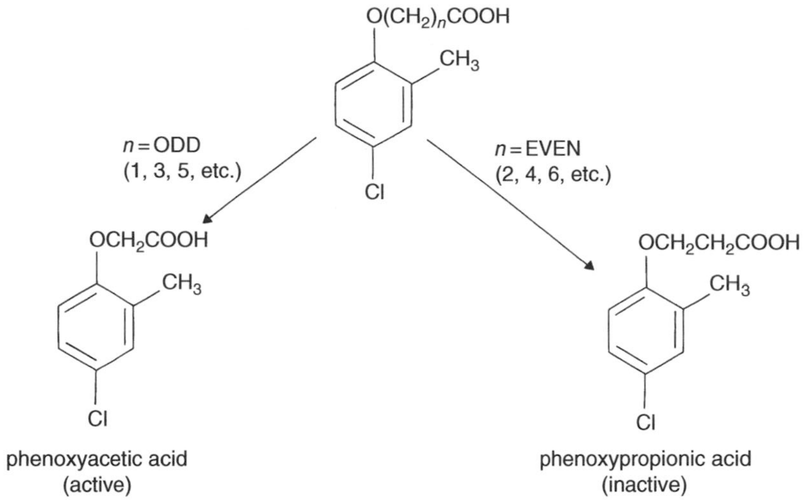

$O(CH_2)_n COOH$

CH_3

Cl

$n = ODD$
(1, 3, 5, etc.)

$n = EVEN$
(2, 4, 6, etc.)

$OCH_2 COOH$

CH_3

Cl

phenoxyacetic acid
(active)

$OCH_2 CH_2 COOH$

CH_3

Cl

phenoxypropionic acid
(inactive)

Figure 7.19 The effect of β-oxidation on phenoxyacids containing an odd number and an even number of CH_2 groups. Only the former gives rise to active auxin-herbicides.

of the auxin-type herbicides is clearly a complex topic, dependent on many interacting aspects of herbicide behaviour, metabolism and plant physiology.

Finally, Herud-Sikimic *et al.* (2021) report the development of a genetically encoded biosensor for the quantitative *in vivo* visualisation of auxins in plants. Their sensor is based on the *Escherichia coli* tryptophan repressor, the binding product of which is engineered to be specific to auxin. The authors have coupled the auxin-binding moeity with fluorescent proteins, to enable the energy transfer signal to be used. This exciting discovery enables the direct visualisation and real-time monitoring of auxin movement in individual plant cells and compartments. This technique should offer new perspectives of auxin and hence herbicide action.

References

Aoi, Y., Tanaka, K., Cook, S.D., Hayashi, K.-I. and Kasahara, H. (2020a) GH3 auxin-amido synthases alter the ratio of indole-3-acetic acid and phenylacetic acid in Arabidopsis. *Plant Cell Physiology* **61**, 596–605.

Aoi, Y., Oikawa, A., Sasaki, R., Huang, J., Hayashi, K.-I. and Kasdahara, H. (2020b) Arogenate dehydratases can modulate levels of phenylacetic acid in *Arabidopsis*. *Biochemical and Biophysical Research Communication* **524**, 83–88.

Atwood, D. and Paisley-Jones, C. (2017) Pesticides industry sales and usage 2008–2012 estimates. Office of Pesticide Programs. US Environmental Protection Agency.

Bowe, S., Landes, M., Best, J., Schmitz, G. and Graben, M. (1999) BAS 662 H: an innovative herbicide for weed control in corn. *Proceedings of the Brighton Crop Protection Conference, Weeds* **1**, 35–40.

Brinkworth, L.A, Egerton, S.A, Bailey, A.D. and Bernhard, U. (2005) Aminopyralid, a new active substance for long-term control of annual and perennial broad-leaved weeds in grassland. Proceedings of the BCPC International Congress. *Crop Science and Technology* **1**, 43–48.

Brummell, D.A. and Hall, J.L. (1987) Rapid cellular responses to auxin and the regulation of growth. *Plant, Cell and Environment* **10**, 523–543.

Burnside, O.C. (ed.) (1996) Biologic and economic assessment of benefits from use of phenoxy herbicides in the United States. USDA National Agricultural Pesticide Impact Assessment Program (NAPIAP) Report number 1-PA-96.

Busi, R., Goggin, D.E., Heap, I., Horak, M.J., Jugulam, M., Masters, R.A. *et al.* (2018) Weed resistance to synthetic auxin herbicides. *Pest Management Science* **74**, 2265–2276.

Edgerton, M.D., Tropsha, A. and Jones, A.M. (1994) Modelling the auxin-binding site of auxin binding protein 1 of maize. *Phytochemistry* **35**, 111–1123.

Effendi F., Jones, A.M. and Scherer, G.F.E. (2013) Auxin-Binding-Protein 1 in phytochrome B-controlled responses. *Journal of Experimental Botany* **64**, 5065–5074.

Ettlinger, C. and Lehle, L. (1988) Auxin induces rapid changes in phosphatidyl inoitol metabolites. *Nature* **331**, 176–178.

Fitzsimons, P.J. (1989) The determination of sensitivity parameters for auxin-induced H^+-efflux from *Avena* coleoptile segments. *Plant, Cell and Environment* **12**, 737–746.

Fitzsimons, P.J., Barnwell, B. and Cobb, A.H. (1988) A study of auxin-type herbicide action based on dose–response analysis of H^+-efflux. In: *Proceedings of the European Weed Research Society Symposium, Factors Affecting Herbicidal Activity and Selectivity*. Wageningen: European Weed Research Society, pp. 63–68.

Gleason, C., Foley, R.C. and Singh, K.B. (2011) Mutant analysis in *Arabidopsis* provides insight into the molecular mode of action of the auxinic herbicide dicamba. *PLoS ONE* 6, e17245.

Goggin, D.E., Cawthray, G.R. and Poweles S.B. (2016) 2,4-D resistance in wild radish: reduced herbicide translocation via inhibition of cellular transport. *Journal of Experimental Botany* **67**, 3223–3235.

Goodwin, T.W. and Mercer, E.I. (1983) *Introduction to Plant Biochemistry*, 2nd edn. Oxford: Pergamon Press.

Grones, P. and Friml, J. (2015) Auxin transporters and binding proteins at a glance. *Journal of Cell Science* **128**, 1–7.

Grossmann, K., Kwiatkowski, A. and Tresch, S. (2001) Auxin herbicides induce H_2O_2 overproduction and tissue damage in cleavers (*Galium aparine* L.). *Journal of Experimental Botany* **52**, 1811–1816.

Herud-Sikimic, O., Stiel, A.C., Shanmurgaratnam, S., Berendzen, K.W., Feldhaus, C., Hocker, B. and Jurgens, G. (2021) A biosensor for the direct visualisation of auxin. *Nature* **592**, 768; doi. org/10.1038/s41586-021-03425-2.

Hull, M.R. and Cobb, A.H. (1998) An investigation of herbicide interaction with the H^+-ATPase activity of plant plasma membranes. *Pesticide Science* **53**, 155–164.

Johnstone, C.R., Malladi, A., Vencill, W.K., Grey, T.L., Culpepper, S., Henry, G. *et al.* (2020) Investigation of physiological and molecular mechanisms conferring diurnal variation in auxinic herbicide efficacy. *PLoS ONE* **15**, e0238144; doi: 10.1371/journal.pone.0238144

Katekar, G.F. (1979) Auxins: on the nature of the receptor site and molecular requirements for auxin activity. *Phytochemistry* **18**, 223–233.

Kelley, K.B., Zhang, Q., Lambert, C.N. and Riechers, D.E. (2006) Evaluation of auxin-responsive genes in soybean for detection of off-target plant growth regulator herbicides. *Weed Science* **54**, 220–229.

Kirby, C. (1980) *The Hormone Weedkillers*. Farnham: British Crop Protection Council.

LeClere, S., Wu, C., Westra, P. and Sammons, R.D (2018) Cross resistance to dicamba, 2,4-D and fluroxypyr in *Kochia scoparia* is endowed by a mutation in an AUX/IAA gene. *Proceedings of the National Academy of Sciences* **115**, E2911–2920.

Lee, S., Sundaram, S, Armitage, L., Evans, J.P., Hawkes, T., Kepinski, S., *et al.* (2014) Defining binding efficiency and specificity of auxins for SCF/TIR1/AFB-Aux/IAA co-receptor complex formation. *ACS Chemical Biology* **9**, 673–682 (2014)

Lewer, P. and Owen, W.J. (1990) Selective action of the herbicide triclopyr. *Pesticide Biochemistry and Physiology* **36**, 187–200.

Lewis, D., Roten, R, Everman, W. and Yelverton, F. (2013) Absorption, translocation and metabolism of Aminocyclopyrachlor in tall fescue *(Lolium arudinaceum) Weed Science* **61**, 348–352.

Martin, T.J. (1987) Broad versus narrow-spectrum herbicides and the future of mixtures. *Pesticide Science* **20**, 289–299.

Milner, L.J., J.P.H. Reade and A.H. Cobb (2001) Developmental changes in glutathione S-transferase activity in herbicide-resistant populations of *Alopecurus myosuroides* Huds. (black-grass) in the field. *Pest Management Science* **57**, 1100–1107.

Napier, R.M., David, K.M. and Perrot-Rechenmann, C. (2002) A short history of auxin-binding proteins. *Plant Molecular Biology* **49**, 339–348.

Paciorek, T., Zazimalova, E., Ruthardt, N., Patrasek, J., Sticrhof, Y.D., Kleine-Vehn, J. *et al.* (2005) Auxin inhibits endocytosis and promotes its own efflux from cells. *Nature* **435**, 1251–1256.

Paque, S., Mouille, G., Grandont, L., Alabadi, D., Gaertner, C., Goyallon, A. *et al.* (2014) ABP1 links cell wall remodelling, auxin signalling and cell expansion in Arabidopsis. *The Plant Cell* **26**, 280–295.

Parry, G., Calderon-Villalobos, L.I., Prigge, M., Peret, B., Dharmasiri, S., Itoh, H. *et al.* (2009) Complex regulation of the TIR1/AFB family of auxin receptors. *Proceedings of the National Academy of Science* **106**, 22540–22545.

Peterson, M.A., McMaster, S.A., Riechers, D.E., Skelton, J. and Stahlman, P.W. (2016) 2,4-D past, present, and future: A review. *Weed Technology* **30**, 303–345.

Pillmoor, J.B. and Gaunt, J.K. (1981) The behaviour and mode of action of the phenoxyacetic acids in plants. In: Hutson, D.H. and Roberts, T.R. (eds) *Progress in Pesticide Biochemistry*, vol. 1. Chichester: Wiley, pp. 147–218.

Quareshy, M., Prusinska, J., Li, J. and Napier, R. (2018a) A cheminformatics review of auxins as herbicides. *Journal of Experimental Botany* **69**, 265–2775.

Quareshy, M., Prisinska, J., Kieffer, M., Fukui, K., Pardal, A.J., Lehman, S. *et al.*(2018b) The tetrazole analogue of the auxin indole-3-acetic acid binds preferentially to TIR1 and not AFB5. *ACS Chemical Biology* **13**, 2585–2594.

Refatti, J.P., de Avila, L.A., Camargo, E.R., Ziska, L.H., Salas-Pctrcz, R. and Roma-Burgos, N. (2019) High carbon dioxide concentrations and temperature increase resistance to cyhalofop-butyl in multiple-resistant *Echinochloa colona. Frontiers in Plant Science* **10**, 1–11.

Riar, D.S., Burke, I.C., Yenish, J.P., Bell, J. and Gill, K. (2011) Inheritance and physiological basis for 2,4-D resistance in prickly lettuce (*Lactuca serriola* L.) *Journal of Agriculture and Food Chemistry* **59**, 9417–9423.

Robatscher, P., Eisenstecken, D., Innerebner, G., Roschatt, C., Rafer, B., Rohregger, H. *et al.* (2019) 3-chloro-5-trifluoromethylpyrimidine-2-carboxylic acid, a metabolite of the fungicide fluopyram, causes growth disorder in *Vitis vinifera. Journal of Agriculture and Food Chemistry* **67**, 7223–7231.

Rodriguez-Serrano, M., Sparkes, I., Rochetti, A., Hawes, C., Romero-Puertas, M.C. and Sandalio, L.M. (2014) 2,4-D promotes S-nitrosylation and oxidation of actin affecting cytoskeleton and peroxisomal dynamics. *Journal of Experimental Botany* **65**, 4783–4793.

Sanders, G.E. and Pallett, K.E. (1987) Physiological and ultrastructural changes in *Stellaria media* following treatment with fluroxypyr. *Annals of Applied Biology* **111**, 385–398.

Schmitzer, P.R., Balko, T.W., Daeuble, J.F., Epp, J.B., Satchivi, N.M., Siddall, T.L. *et al.* (2015) Discovery and SAR of halauxifen-methyl: a novel auxin-herbicide. *ACS Symposium Series*, **1204**, Chapter 18, pp 247–260.

Schulz, B. and Segobye, K. (2016) 2,4-D transport and herbicide resitance in weeds. *Journal of Experimental Botany* **67**, 3177–3179.

Simon, S. and Petrasek, J. (2011) Why plants need more than one type of auxin. *Plant Science* **180**, 454–460.

Sperry, B.P., Dias, J.-L.C.S., Prince, C.M., Ferrell, J.A. and Sellers, B.A. (2020) Relative activity comparison of aminocycloprachlor to pyridine carboxylic acids. *Weed Technology* **34**, 402–407; doi: org10/1017/wet.2019.129

Tan, X., Calderon-Villalobos, L.I.A, Sharon, M., Zheng, C., Robinson, C.V., Estelle, M. and Zhing, N (2007) Mechanism of auxin perception by the TIR1 ubiquitin ligase. *Nature* **446**, 640–645.

Thompson, L.M.L. (1989) An investigation into the mode of action and selecticity of 3,6-dichloropicolinic acid. PhD Thesis, Nottingham Polytechnic, UK.

Thompson, L.M.L. and Cobb, A.H. (1987) The selectivity of clopyralid in sugar beet; studies on ethylene evolution. *British Crop Protection Conference, Weeds* **3**, 1097–1104.

Torra, J., Rojano-Delgado, A.M., Rey-Caballero, J., Salas, M. and De Prado, R (2017) Enhanced 2,4-D metabolism in two resistant populations of *Papaver rhoeas* from Spain. *Frontiers of Plant Science*, **8**, 1584.

Venis, M. (1985) *Hormone Binding Sites in Plants*. Harlow: Longman.

Vernouz, T. and Robert, S. (2017) Auxin 2016: A burst of auxin in the warm south of China. *Development* **144**, 533–540.

Vieten, A., Sauer, M., Brewer, P.B. and Friml, J. (2007) Molecular and cellular aspects of auxin – transport-mediated development. *Trends in Plant Science* **12**, 160–168.

Ware, G.W. (1983) *Pesticides. Theory and Application*. San Francisco, CA: W.H. Freeman.

Westfall, C., Sherp, A.M., Zubieta, C., Alvarez, S., Scraft, E., Marcellin, R. *et al.* (2017) GH3.5 acyl acid amido synthetase mediates cross-talk in amino auxin and salicylic acid homeostasis. *Proceedings of the National Academy of Sciences* **113**, 13917–13922.

Xu, T., Di, N., Chen, J., Nagawa, S., Cao, M., Li, H. *et al.* (2014) Cell surface ABP-1-TMK auxin-sensing complex activates ROP GTPase signalling. *Science* **343**, 1025–1028.

Chapter 8

Inhibitors of Lipid Biosynthesis

No man can be a statesman who is ignorant of wheat.

Socrates (469–399 BC)

8.1　Introduction

Current agricultural practices of reduced cultivation, cereal crop monoculture and the widespread use of broadleaf weedkillers has resulted in an increased spread of grass weeds. The scale of the problem became evident in the 1970s, and by 1981 it was estimated that, for example, 55% of the 209,000 ha of sugar beet in the UK was infested with wild oats (*Avena* spp.) and black-grass (*Alopecurus myosuroides*), and approximately 10% with couch grass (*Elytrigia repens*) (Siddall and Cousins, 1982). Wild oats and black-grass are highly competitive and invasive weeds, especially in cereals, and various models predict significant yield losses when as few as 10 (*Avena* spp.) or 50 (*A. myosuroides*) weed plants are present per square metre. Furthermore, a frequently reported problem is the abundance of volunteer wheat and barley populations in succeeding broadleaf crops, such as sugar beet, potatoes, or oilseed rape. Consequently, considerable effort has been directed towards the development of selective grass weedkillers, or graminicides, in both narrowleaf and broadleaf crops.

The selective control of wild oats in cereals first became possible in the late 1950s with the introduction of the chlorophenylcarbamate, barban and the thiocarbamate diallate. However, barban has proven to be phytotoxic to certain barley cultivars, and the activity of the soil-applied diallate varies with soil type, condition and placement. Triallate has superseded diallate in cereals, since wheat and barley are more tolerant to this compound. These volatile herbicides must be rapidly incorporated into the top 2.5 cm of soil for maximum efficacy. Grasses which germinate from seed grow through this treated layer and are killed, although surface germinating seeds or those at greater depth are not controlled. No soil incorporation is required for the pre-emergent chloroacetamides, since these herbicides have a residual action in the soil, and so can kill annual broadleaf weeds and annual grasses for up to 8 weeks after application.

The post-emergent alaninopropionates were introduced between 1969 and 1972 by Shell for the control of wild oats already present in cereals. The control of a broad spectrum of annual

Herbicides and Plant Physiology, Third Edition. Andrew H. Cobb.
© 2022 John Wiley & Sons Ltd. Published 2022 by John Wiley & Sons Ltd.

and perennial grasses in a wider range of crops was made possible by the development of the aryloxyphenoxypropionates (AOPPs) in the 1970s (also termed the 'fops', such as diclofop, fluazifop and quizalofop) and the cyclohexanediones (CHDs; also termed the 'dims', such as sethoxydim, tralkoxydim and clethodim) in the 1980s. These active ingredients offer widespread and successful post-emergence grass-weed control at doses as low as 100–200 g ha^{-1}. The commercialisation of pinoxaden in 2006, a phenylpyrazolin that acts at the same enzyme target site as fops and dims, but is chemically distinct from these groups, has resulted in a third class of acetyl-CoA carboxylase (ACCase)-inhibiting graminicides which are termed the 'dens'.

These modern graminicides have been used extensively, with an estimated target area of 120 million ha/year in dicot crops, and have been especially valuable in controlling glyphosate-resistant grasses in South America (Takano and others, 2019). In addition to these herbicides, which inhibit lipid synthesis at the ACCase step, a number of other lipid biosynthesis inhibitors are available that prevent plants from synthesising very-long-chain fatty acids (VLCFAs), which result in disruption to plasma membranes and the development of the epicuticular wax that covers aerial plant surfaces. The earlier pre-emergence graminicides have made something of a comeback in the last decade in the control of herbicide-resistant grasses, as their use is considered to slow the development of herbicide resistance.

8.2   Structures and uses of graminicides

Graminicides may be categorised into seven chemical families, namely the thiocarbamates, chloroacetamides, alaninopropionates, AOPPs, CHDs, phenylpyrazolines and benzylethers (Table 8.1).

In the cases of the alaninopropionates and the AOPPs, the C2 of the propanoic acid group is a chiral centre so that both (R) and (S) enantiomers exist. Interestingly, only the (R) enantiomer is biologically active, so the removal of the inactive (S) half of an isomeric mix represents a doubling of activity. Thus, (R) isomers are commercially produced and used at half the dose rate of the racemic mixture.

The fops, dims and den all have a molecular weight between 327 and 400 g mol^{-1}, a pK_a of 3.5-4.1 as weak acids and a logK_{ow} of 3.6–4.2. They have limited residual activity in the soil.

8.3   Inhibition of lipid biosynthesis

The first hint of a specific target process for graminicide inhibition came from the studies of Hoppe (1980). He found that although diclofop-methyl did not interfere with photosynthesis, respiration, protein synthesis or nucleic acid synthesis, the inhibition of acetate incorporation into fatty acids could be demonstrated in susceptible species. Lichtenthaler and Meier (1984), later reported that the CHDs also disrupted *de novo* lipid biosynthesis in developing barley seedlings, and in 1987–88 four different research teams independently reported that the enzyme ACCase was the target of inhibition by both graminicide classes (see Lichtenthaler *et al.*, 1989 and Secor *et al.*, 1989 for further details).

Plant membranes contain unique fatty acids that have crucial structural and biochemical roles. For example, at least 70% of the total leaf fatty acids consist of the unsaturated α-linolenic acid (18:3), which itself makes up between 40 and 80% of the lipid fraction in the chloroplast. Indeed, the unique functioning of the thylakoid membrane to permit the movement of protons, electrons and their carriers is considered by many workers to result from the property of membrane fluidity

Table 8.1 Structures of graminicides.

(A) Thiocarbamates

$$C_3H_7 \diagdown N - \underset{\underset{O}{\|}}{C} - S - CH_2 - R$$
$$C_3H_7 \diagup$$

R	Common name
$-\underset{\underset{H}{}}{\underset{\|}{C}} = \underset{\underset{}{}}{\underset{\|}{C}}$ (Cl, Cl)	Diallate
$-\underset{}{\underset{\|}{C}} = \underset{\underset{Cl}{}}{\underset{\|}{C}}$ (Cl, Cl)	Triallate
$-CH_3$	EPTC
ring-N(H)—CO—O—$CH_2C \equiv C - CH_2Cl$ (Cl-phenyl)	Barban

(B) Chloroacetamides

Aromatic ring with substituents R_1, R_2 and N bearing R_3 and $CO - CH_2Cl$

R_1	R_2	R_3	Common name
CH_3	C_2H_5	$CH_2OC_2H_5$	Acetochlor
C_2H_5	C_2H_5	CH_2OCH_3	Alachlor
CH_3	CH_3	$CH_2 - N - N$ (pyrazole ring)	Metazachlor
CH_3	C_2H_5	$\underset{\underset{CH_3}{\|}}{\overset{\overset{H}{\|}}{C}} - CH_2OCH_3$	Metolachlor
H	H	$\underset{\underset{CH_3}{\|}}{\overset{\overset{H}{\|}}{C}} - CH_3$	Propachlor

(Continued)

Table 8.1 (*Continued*)

(C) Alaninopropionates

R_1		R_2	R_3	R_4	Common name
Cl		Cl	H	C_2H_3	Benzoylprop-ethyl
H		F	Cl	C_3H_7	Flamprop-isopropyl
H		F	Cl	CH_3	Flamprop-methyl

(D) Aryloxyphenoxypropionates

R_1	R_2	Common name
	CH_3	Diclofop-methyl
	C_4H_9	Fluazifop-butyl
	CH_3	Haloxyfop-methyl
	C_2H_5	Quizalofop-ethyl
	C_2H_5	Fenoxaprop-ethyl
	C_2H_5	Fenthiaprop-ethyl
	C_4H_9	Cyhalofop-butyl

Table 8.1 (*Continued*)

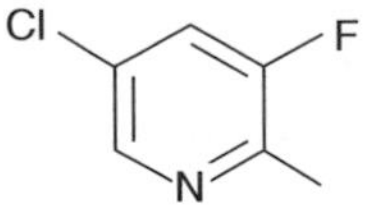	propargyl	Clodinofop-propargyl
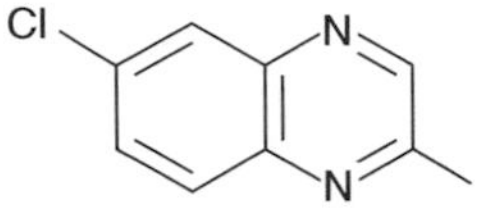	$(CH_3)_2C{=}NOCH_2CH_2$	Propaquizafop

(E) Cyclohexanediones

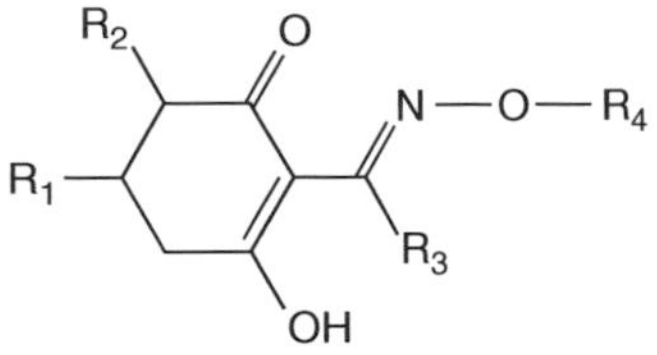

R_1	R_2	R_3	R_4	Common name
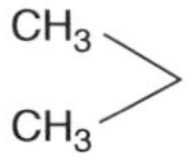	$-C(=O)CH_3$	C_3H_7	$CH_2CH{=}CH_2$	Alloxydim
$C_2H_5S-C(CH_3)(H)-CH_2-$	H_2	C_2H_5	$CH_2CH{=}CHCl$	Clethodim
$C_2H_5S-C(CH_3)(H)-CH_2-$	H_2	C_3H_7	C_2H_5	Sethoxydim
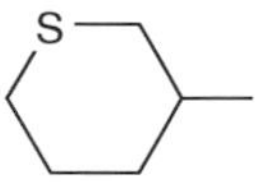	H_2	C_3H_7	C_2H_5	Cycloxydim
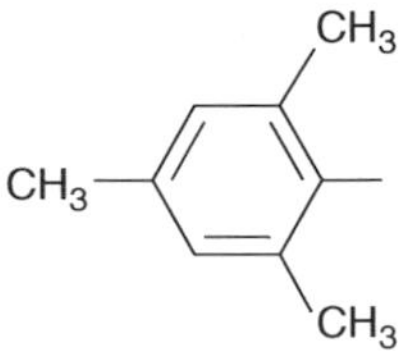	H_2	C_2H_5	C_2H_5	Tralkoxydim
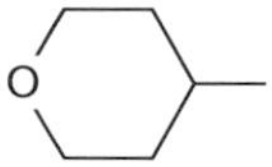	H_2	C_2H_5	$CH_2CH{=}CHCl$	Tepraloxydim
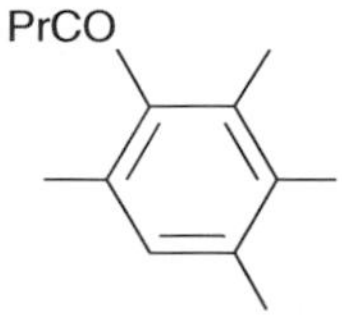	H_2	C_2H_5	C_2H_5	Butroxydim

(*Continued*)

Table 8.1 (*Continued*)

conferred by this unsaturated fatty acid. Trans-Δ_3-hexadecanoic acid and linoleic acid (18:2) are additional examples of important thylakoid fatty acids.

The synthesis of fatty acids in plants involves two major enzymes: ACCase and fatty acid synthetase. Fatty acids are synthesised both in the chloroplast stroma and the cytoplasm (Figure 8.1). Essentially, malonyl-CoA is formed from acetyl-CoA and converted to the saturated palmitate (16:0) by the action of a soluble stromal enzyme complex, termed fatty acid synthetase. This complex contains seven enzymes covalently bound to an acyl carrier protein, which transfers intermediates between the seven enzymes. Thus, seven enzyme cycles are needed for the condensation of seven additional C_2 units into one palmitate. Two metabolic routes are now possible from palmitate. On the one hand, a soluble condensing enzyme and elongases bound to the endoplasmic reticulum are able to add further C_2 units in the cytoplasm to yield the very-long-chain saturated fatty acids found in suberin and the epicuticular waxes on plant surfaces, and desaturases are present in the chloroplast to form the unsaturated fatty acids mentioned earlier (Harwood, 1988).

Acetyl-CoA carboxylase (acetyl-coenzyme A: bicarbonate ligase [ATP], E.C. 6.4.1.2.), or ACCase, is the first committed step for fatty acid biosynthesis in plants, and catalyses the formation of malonyl-CoA (Harwood, 1989). ACCase is a high-molecular-weight, multifunctional protein with three distinct functional regions (two enzymic regions and one carrier protein region), which involves biotin as an essential cofactor that functions as a CO_2 carrier (Figure 8.2A). Initially, a carboxyl group is donated from a bicarbonate anion, and ATP hydrolysis is used to allow the formation of a carboxybiotin intermediate by biotin carboxylase (BC). Carboxybiotin is attached to an ε-amino group of a lysine residue on the biotin carboxyl carrier protein (BCCP). Carboxybiotin then functions as a CO_2 donor in

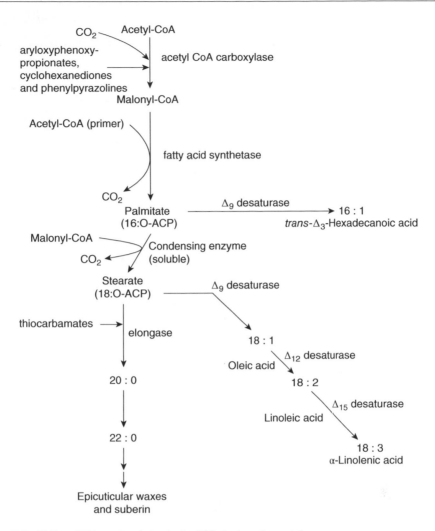

Figure 8.1   Fatty acid biosynthesis in plants. ACP, Acyl carrier protein.

malonyl-CoA formation (Figure 8.2A). Malonyl-CoA is a substrate for fatty acid synthetase and also for fatty acid elongation (and the subsequent production of a number of important secondary metabolites, including flavonoids and phytoalexins). It is therefore likely that inhibition of ACCase affects a number of malonyl-CoA-requiring metabolic pathways. ACCase activity can be regulated at the transcription and post-transcription level, but is also regulated by a number of metabolic factors. ACCase is most active when a plant is in light. During photosynthesis the stromal pH rises from 7 to 8 and Mg^{2+} concentration rises from approximately 1 to 3 mM. Plastid ACCase activity reaches a maximum at pH 8 and at Mg^{2+} concentrations of 2–5 mM. This ensures that ACCase is most active in the light, when photosynthesis is producing ATP, reductant and photosynthate, all of which are necessary for lipid biosynthesis. Laboratory studies by Kozaki and Sasaki (1999) have demonstrated that reducing agents will also activate ACCase, specifically the carboxytransferase (CT) domain of the enzyme. This appears to be due to the formation of a disulphide bridge between two cysteine residues located in the α- and β-CT regions of ACCase (Kozaki *et al.*, 2001). During photosynthesis concentration of reductant rises in the plastid and this represents a further

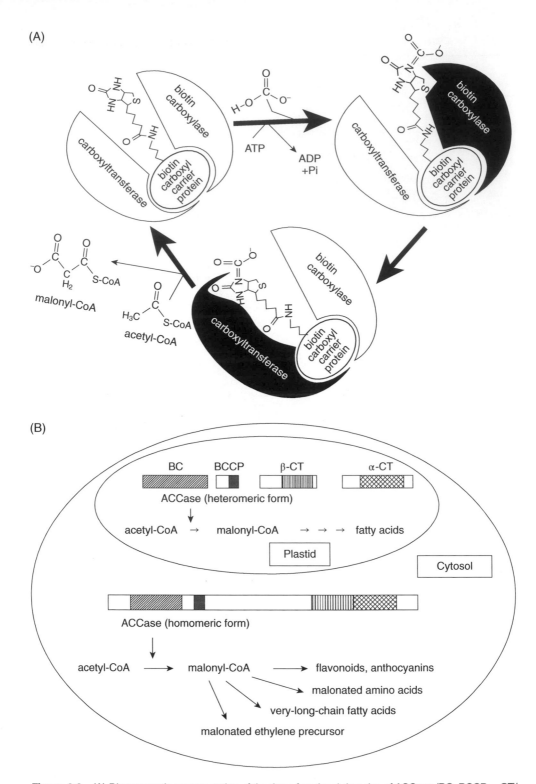

Figure 8.2 (A) Diagrammatic representation of the three functional domains of ACCase (BC, BCCP, α-CT/ β-CT). Source: Ohlrogge, J. and Browse, J. (1995) Lipid biosynthesis. *Plant Cell* **7**, 957–970. Reproduced with permission of Oxford University Press. (B) Composition and compartmentalisation of the two forms of ACCase found in higher plants. See text for abbreviations. Source: Sasaki, Y. and Nagano, Y. (2004) Plant acetyl-CoA carboxylase: structure, biosynthesis, regulation, and gene manipulation for plant breeding. *Bioscience, Biotechnology and Biochemstry* **68**, 1175–1184. Reproduced with permission of Oxford University Press.

light-mediated regulation of ACCase activity. It has also been postulated that ACCase may undergo phosphorylation, and this may also regulate its activity, although the mechanisms involved have yet to be fully elucidated (Savage and Ohlrogge, 1999).

Rendina *et al.* (1990) have characterised the kinetics of ACCase inhibition by graminicides. AOPPs and CHDs appear to be non-competitive inhibitors with respect to Mg^{2+} ATP, HCO_3 and acetyl-CoA. It therefore seems likely that this class of herbicides inhibits the carboxytransferase rather than the carboxylation step of ACCase activity. It also appears that both AOPPs and CHDs compete for the same site on the enzyme and that the inhibition is reversible. The inhibition of ACCase is both rapid and concentration dependent so that, for example, about 1 μm haloxyfop or tralkoxydim can inhibit the enzyme *in vitro* by 50% within 20 min. Furthermore, haloxyfop-acid is more than 100-fold more potent than the methyl-ester and only the *R*(+) enantiomer is herbicidally active (Secor *et al.*, 1989). Recent observations that amino acids in the CT region are responsible for ACCase inhibition may lead to further information on the mechanism of herbicide binding to the enzyme. Indeed, the observation that Ile1781→Leu results in resistance to all fops and most dims, but that Ile2041→Asn only results in resistance to fops, clearly indicates the importance of Ile2041 in fop but not dim binding (Délye *et al.*, 2003). Additionally, a homozygous herbicide-resistant black-grass biotype that contains Gly at position 2078 shows decreased fitness in the absence of herbicide, highlighting the importance of position 2078 in ACCase activity (Menchari *et al.*, 2008).

Two forms of ACCase are found in higher plants (Figure 8.2B) and this plays an important role in the selectivity demonstrated by AOPPs and CHDs. In dicotyledons a heteromeric form (termed the prokaryotic form) of ACCase, located in plastids, is insensitive to AOPPs and CHDs. The heteromeric form is composed of BCCP, BC, α-CT and β-CT polypeptides. It has been postulated that this form of ACCase is $BCCP_4$ BC_2 α-CT_2 β-CT_2, similar to bacterial ACCase (Choi-Rhee and Cronan, 2003). This form of ACCase is absent from the monocotyledon grasses and evidence suggests that this is due to the absence of the *accD* gene that encodes the β-CT polypeptide (Konishi and Sasaki, 1994). A homomeric form of ACCase, located in plastids, is found in grasses and is sensitive to AOPPs and CHDs. This homomeric form is also found in the cytosol of both dicotyledons and grasses. It is a large polypeptide (~250 kDa) which contains BCCP, BC, α-CT and β-CT regions. It appears to be active as a dimer. These forms of ACCase are summarised in Table 8.2 and Figure 8.2B. This explains the selectivity of ACCase inhibitors between dicotyledons and grasses, as the presence of the herbicide-insensitive heterodimeric ACCase in dicotyledons allows the synthesis of fatty acids in the presence of these herbicides.

Table 8.2 Forms of ACCase found in higher plants.

	Prokaryotic form	Eukaryotic form
Structure	Heteromeric (separate BCC, BC and CT subunits)	Homomer; single multifunctional polypeptide
Grasses	Absent	Plastids and cytosol
Dicotyledonous species	Plastids	Cytosol
Sensitivity to 'fops' and 'dims'	Insensitive	Sensitive (plastidic) Insensitive (cytosolic)

BCC, biotin carboxyl carrier; BC, biotin carboxylase; CT, carboxytransferase.
Source: Sasaki, Y., Konishi, T. and Nagano Y. (1995) The compartmentation of acetyl-coenzyme A carboxylase in plants. *Plant Physiology* **108**, 445–449.

In the monocot maize two isoforms of ACCase are reported. The plastid form (ACCase I) predominates and is inhibited by AOPPs and CHDs. A cytosolic form (ACCase II) is 2000-fold less sensitive to these herbicides. Further studies have found a similar situation in other grasses, and it appears that some naturally occurring resistant biotypes of *Lolium multiflorum* possess an altered form of the plastid ACCase that is less sensitive to herbicides (Evenson *et al.*, 1997; Table 8.3).

Fops, dims and den all share two common anchoring points on the carboxyl transferase, at ile 1735 and ala 1627, and form a hydrogen bond for binding at ser 698.

ACCase is not the only site of graminicide action. Weisshaar *et al.* (1988) have demonstrated that micromolar concentrations of the chloroacetamides also inhibit fatty acid biosynthesis by preventing the elongation of palmitate and the desaturation of oleate in the green microalga *Scenedesmus acutus*. It has subsequently been confirmed that both the elongation and desaturation steps of VLCFA biosynthesis are targets for a number of herbicides.

Very-long-chain fatty acids are important components of the plant cell plasma membrane (plasmalemma) and are enriched in the leaf epicuticular waxes. Here they are embedded in a matrix and ensure the hydrophobicity of the leaf surface (Chapter 3.2). The cuticle is the main barrier against invasion by external agents and microorganisms, while preventing the loss of water and solutes from the leaves. Disrupting the plasma membrane will lead to a loss of permeability, transport and hormone receptor functions. It therefore follows that inhibition of VLCFA biosynthesis is a valuable target for herbicide action and development.

Members of several herbicide groups are now known to act as specific inhibitors of VLCFA biosynthesis, with alkyl chains longer than C_{18}. They strongly inhibit elongase activity in reactions taking place outside the chloroplast, at the endoplasmic reticulum and the Golgi complex. Chemical groups include the chloroacetamides (e.g. alachlor), oxyacetamides (e.g. mefenacet), carbomylated five-membered nitrogen heterocycles (e.g. cafenstrole), oxiranes (e.g. indanofan) and thiocarbamates and others (such as ethofumesate). Some examples of herbicide structures acting as inhibitors of VLCFA biosynthesis are shown in Figure 8.3.

The VLCFAs have proved to be effective herbicides. For example, Palmer amaranth (*Amaranthus palmeri*) and Common waterhemp (*Amaranthus rudis*) are difficult to control and can reduce yields in many crops. Hay *et al.* (2018) have noted effective control when VLCFA inhibitors are applied pre-emergence as residual herbicides.

The chloracetamides have been successfully used in maize, soybean and rice for several decades, and remain an important means of weed control today, especially in maize. They are persistent herbicides, taken up from the soil, and several safeners have been developed for use in mixtures to extend their range of use. While germination is unaffected, early

Table 8.3 Sensitivity of ACCase I and II from *Lolium multiflorum* biotypes to diclofop.

Biotype	Isoform	Source	Diclofop concentration (µM)	Inhibition (%)
Susceptible	ACCase I	Plastid	0.2	50
	ACCase II	Cytosol	125	42
Resistance	ACCase I	Plastid	7	50
	ACCase II	Cytosol	127	31

Source: Evenson, K.J., Gronwald, J.W. and Wyse, D.L. (1997) Isoforms of acetyl-coenzyme A carboxylase in *Lolium multiflorum*. *Plant Physiology and Biochemistry* **35**, 265–272.

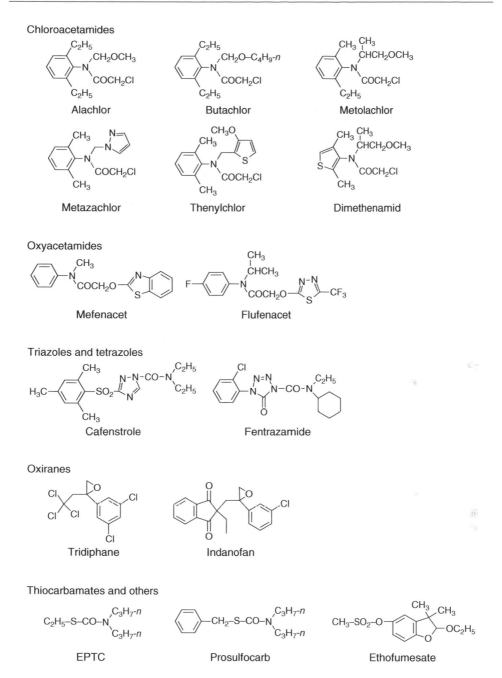

Figure 8.3 Structures of herbicides acting as very long-chain fatty acid biosynthesis inhibitors;

seedling growth is typically inhibited and the seedlings do not emerge from the soil or are severely stunted. Cell division and expansion are both inhibited.

The enzymes of acyl elongation are membrane-bound and thought to be associated with the endoplasmic reticulum and the Golgi apparatus in the cytoplasm, and perhaps with the plasma membrane itself. There are 21 genes encoding VLCFA elongases in *Arabidopsis thaliana*. Trentkamp *et al.* (2004) investigated the expression and activity of six gene products in the presence of VLCFA biosynthesis inhibiting herbicides and found wide substrate specificity. They suggest that such complex patterns of substrate specificity may explain why resistance to these herbicides is rare.

Cinmethylin was first reported by Shell in 1981 and commercialised in 1989 for grass weed control in rice. The development of several new and successful herbicide groups in the 1980s, however, such as the ACCase, ALS and VLCFA inhibitors, relegated cinmethylin to a minor herbicide. In more recent decades the emergence of herbicide-resistant weeds has resulted in the re-evaluation of herbicides with different mechanisms of action, and cinmethylin was found by BASF to have good activity against resistant grasses, such as *Alopecurus myosuroides* and *Lolium* sp., with good selectivity in wheat and barley. In 2019, the Australian Pesticides and Veterinary Medicines Authority approved cinmethylin as the active ingredient of Luximax, trade name Luximo.

Campe *et al.* (2018) have demonstrated that cinmethylin inhibits the release of fatty acids from the chloroplasts by binding to fatty acid thioesterases (FAT) with high affinity. Normally, the C16 and C18 fatty acids are re-esterified in the cytoplasm and then used for the synthesis of glycerolipids, triacylglycerol and VLCFAs at the endoplasmic reticulum (Figure 8.4).

The substituted pyridazinones have been found to inhibit lipid biosynthesis in addition to their known inhibition of carotenoid biosynthesis (see Section 6.3). Norflurazon can

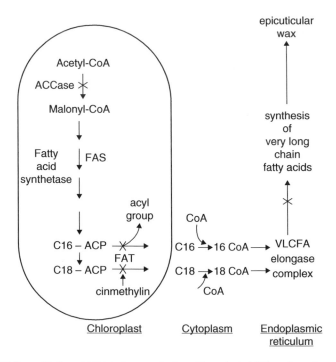

Figure 8.4 Inhibition of fatty acid thioesterases in the chloroplast. ACP, acyl carrier protein; FAS, fatty acid synthetase; FAT, fatty acid thioesterase.

inhibit the Δ_{15}-desaturase to prevent α-linolenic acid synthesis, and metflurazon may additionally prevent the formation of *trans*-Δ_3-hexadecanoate by the inhibition of the Δ_9-desaturase. These herbicides are classed as carotenoid biosynthesis inhibitors and their effects on lipid biosynthesis probably represent a secondary mode of action. That they have two points of action should not be surprising as they inhibit a desaturase enzyme in both metabolic pathways.

8.4   Activity of graminicides in mixtures

Although most authors accept the selective inhibition of ACCase as a primary target of the AOPPs, CHDs and the DEN, it has also been proposed that these and other graminicides can also act as anti-auxins. Diclofop-methyl, for example, has no auxin activity alone, but can inhibit several auxin-mediated processes such as coleoptile elongation and proton-efflux (Figure 8.5), and alter cell membrane potentials, possibly by acting as a protonophore (Wright and Shimabukuro, 1987). Disruption of membrane function in the form of rapid depolarisation of the plasma membrane electrogenic potential is reported after ACCase inhibitor treatment of susceptible species. This phenomenon is not observed when the same herbicides are applied to ACCase-resistant biotypes or species. However, this alternative hypothesis of graminicide action has been rejected by many workers as a secondary

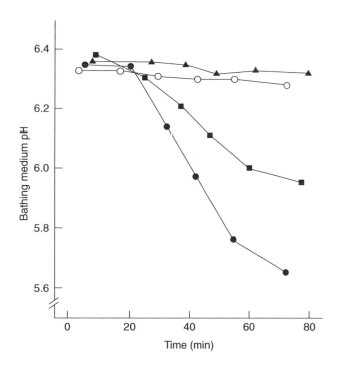

Figure 8.5   Inhibition of auxin-induced proton-efflux by diclofop-methyl (DM). ●, Control; ▲, 50 μM DM; ●, 10 μM MCPA; ■, 10 μM MCPA; +, 50 μM DM. Source: Cobb, A.H. and Barnwell, P. (1989) Anti-auxin activity of graminicides. *Brighton Crop Protection Conference, Weeds* **1**, 183–190.

consequence of the inhibition of lipid biosynthesis which would, it is argued, alter membrane function. Furthermore, for this hypothesis to gain acceptance it would be necessary for this anti-auxin activity to satisfy the following four criteria:

1 shows at least equal sensitivity to the micromolar concentrations known to inhibit ACCase;
2 is observed within minutes;
3 shows stereospecificity; and
4 is selective in grasses.

Several studies have provided supportive or circumstantial evidence to fulfil each of the above criteria, and more detailed and convincing data are slowly emerging. For example, micromolar concentrations of several AOPPs can inhibit auxin-induced proton-efflux from sensitive coleoptile segments within minutes (Figure 8.5; Cobb and Barnwell, 1989), and stereospecificity has been demonstrated (Andreev and Amrhein, 1976). Evidence is also accumulating for a selective anti-auxin activity being confined to grasses. Young, rapidly elongating monocotyledonous tissues are especially sensitive to graminicides, and a rapid retardation of grass internode elongation is commonly observed, with the result that sensitive grasses show significant stunting when compared with control plants and cereal crops. The need for a rapid rate of tissue extension has recently been identified as a prerequisite for optimal graminicide activity, since decreased extension rates associated with water stress, for example, severely reduce diclofop efficacy (Andrews *et al.*, 1989). Finally, it is surely pertinent that a foliar-applied graminicide must first come into contact with a cell membrane before its subsequent intracellular metabolism and translocation to the grass meristem. The opportunity therefore exists *in vivo* for a graminicide to demonstrate anti-auxin activity in addition to an inhibition of ACCase.

This additional activity of graminicides has commercial significance when herbicide mixtures are taken into account. Since an arable field will contain a mixture of both monocotyledonous and dicotyledonous weeds, it would be of obvious advantage to the farmer to mix a broadleaf weed, auxin-type herbicide with a graminicide. This single treatment could ensure a broad spectrum of weed control with additional savings in time, labour and fuel costs. However, it is well known that such tank mixtures commonly result in reduced efficacy and crop damage is often observed. Antagonism of ACCase inhibitor action by a number of herbicides including 2,4-D, dicamba, ALS inhibitors, bentazone and acifluorfen has been reported. This antagonism does not appear to be at the sites of action of the herbicide and may be due to the decreased uptake and movement of the ACCase inhibitors. Indeed, broadleaf weed herbicides such as bentazone are able to reduce the uptake of some graminicides, and so account for a reduction in graminicide efficacy.

The literature continues to provide reports of graminicide antagonism by auxin-type herbicides. A recent example is the antagonism of clethodim and quizaolofop-*P*-tefuril efficacy when mixed with 2,4-D in *Digitaria insularis* (sourgrass), a major grass weed in soybean crops in South America (Gomes *et al.*, 2020). These authors noted that antagonism was not observed when there was a 9- to 12-day window after 2,4-D application. Perhaps the auxin-type herbicide may cause the expression of cytochrome p450 and other enzymes that result in the detoxification of the graminicides, leading to reduced grass weed control. Or perhaps the auxin-type herbicide reduces phloem loading and unloading (Perkins *et al.*, 2021).

On the other hand, if the broadleaf weedkiller is an auxin-type herbicide, it may be argued that the anti-auxin activity of the graminicide is overcome by the addition of 'extra auxin'. Hence, less stunting is observed in treated grasses and plants may recover to set seed. The real significance of these observations remains to be established, although further understanding of this interaction may eventually lead to the design of more compatible graminicide/broadleaf weed-killer mixtures in the future.

Curiouser and curiouser! Windmillgrass (*Chloris verticillata* Nutt.) is a summer perennial grass weed native to temperate regions of the USA and is becoming problematic in turfgrass and residential lawns. In Australia *Chloris truncata* R.Br. can be used for summer grazing by livestock. It can be controlled by ACCase inhibitors and fenoxaprop-*p*-ethyl is particularly effective (Smeda and Xi Xiong, 2019). Combinations of the HPPD inhibitors mesotrione and topramezone have also proved effective, but only when tank-mixed with the auxin-type herbicide triclopyr-ester. How triclopyr is able to synergise the activity of these herbicides is unclear. It may alter (increase) the uptake and movement of mesotrione in smooth crabgrass (Yu and McCullough, 2016).

Others have observed that ACCase inhibitors can control Bermuda grass better in turfgrass in the presence of triclopyr, while overcoming injury to zoysia grass, also implying a safener effect (Lewis *et al.*, 2010). In a more recent study, however, Webster *et al.* (2019) studied the efficacy of quizalofop-*p*-ethyl when mixed with different auxin-type herbicides and ACCase inhibitors for barnyard grass and weedy rice control in rice. They observed that cyhalofop, fenoxaprop, 2,4-D, quinclorac and triclopyr all antagonised quizalofop for barnyard grass control, 14 days after treatment, but gave a neutral response for the control of red rice. Underwood *et al.* (2016) described field trials to evaluate whether there was antagonism when dicamba was added to quizalofop-*p*-ethyl or clethodim for the control of volunteer glyphosate-resistant maize. They found that there was greater antagonism when a high rate of dicamba was tank-mixed with a lower rate of graminicide. Finally, Harre *et al.* (2019) studied the control of glyphosate-resistant maize with mixtures of 2,4 D or dicamba in combination with glyphosate and clethodim. The auxin-type herbicides did not affect clethodim efficacy, but the addition of glyphosate caused a reduction in clethodim efficiency, which could be overcome by the inclusion of an activator adjuvant, crop oil concentrate. Can the findings simply be explained by differences in uptake and movement over time? Again, this author is of the opinion that the application of one herbicide may also cause the *de novo* expression of genes coding for detoxification enzymes, leading to the metabolic breakdown of other herbicides in the mixture. Clearly, more work is needed to test this hypothesis and better understand these interactions.

Why are such interactions important? It is possible that antagonism plus herbicide resistance may eventually lead to a failure in weed control, and so accelerate the evolution of herbicide resistance. It follows that preventing antagonism may aid the prevention of herbicide resistance.

8.5 How treated plants die

The foliar uptake of the AOPPs is very rapid, and studies using radiolabelled herbicides have shown that almost all of the applied dose remains at the site of application, and so contact damage is commonly observed at the treated leaf (Figure 8.6; Carr *et al.*, 1986). De-esterification occurs in the leaf tissues, and the phytotoxic acid accumulates at the apical meristem, which becomes necrotic. Active plant growth and warm temperatures encourage active transport

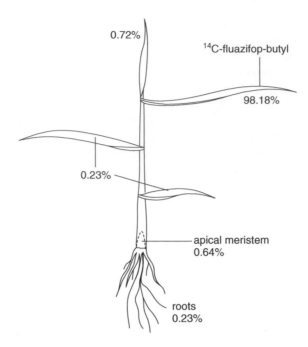

Figure 8.6 Percentage ^{14}C-activity recovered from plants of *Setaria viridis* 72 h after treatment with radiolabelled fluazifop-butyl. Source: modified from Carr, J.E., Davies, L.G., Cobb, A.H. and Pallett, K.E. (1986) Uptake, translocation and metabolism of fluazifop-butyl in *Setaria viridis*. *Annals of Applied Biology* **108**, 115–123.

within the phloem and passive movement in the xylem, so that the acid accumulates in all meristematic zones. However, most herbicidal injury is noted in the apical meristem owing to the limited translocation of these herbicides, with far fewer symptoms noted in the root meristem. Interference with lipid biosynthesis causes an irreversible disruption in membrane synthesis and membrane integrity, causing metabolite leakage and cell death, so that normal plastid development is not observed and metabolism is drastically altered. Leaf elongation and growth ceases within 2 days and plastid disruption is most marked in young leaves, which appear chlorotic. The main site of action is the apical meristem where *de novo* fatty acid biosynthesis is taking place to support growth. Within 2–4 days new tissue at the meristem can be easily detached from the rest of the plant. Chlorosis spreads throughout the aerial parts of the plant and grass death follows within 2–3 weeks after application. Grasses treated with the alaninopropionates and the CHDs show similar symptoms to those described above, although the CHDs have slower rates of penetration into the treated leaf.

Pre-planting or pre-emergent treatments with the thiocarbamates and chloroacetamides effectively provide a chemical barrier for the growth of grass seedlings. These compounds penetrate the mesocotyl of germinating weeds to inhibit lipid biosynthesis in these young tissues. Consequently, any tissues that emerge are chlorotic and short-lived. Indeed, seedling death will occur when seed reserves of fatty acids are exhausted.

8.6   Plant oxylipins: lipids with key roles in plant defence and development

While ACCase, FAT and VLCFA have been exploited as target sites for herbicides, a novel area of lipid biochemistry may show promise for new target sites. It has been known for some time that plants are able to respond to abiotic and biotic stresses, including the effects of grazing insects, bacteria, fungi and other pathogens by a process termed induced systemic resistance. In this process, a signal is transmitted from, for example, a damaged leaf to the rest of the plant conferring defence responses leading to resistance. Among the mobile signals jasmonic acid (JA) and ethylene are thought to be involved.

The last decade has seen a dramatic increase in the number of publications related to this phenomenon and much attention has been given to the involvement of a large and diverse family of lipids, termed the oxylipins, defined as more than 200 metabolites derived from the oxidation of polyunsaturated fatty acids. Oxygenation can arise either by enzymatic and non-enzymatic reactions, generating a complex array of molecules, including hydroperoxides, alcohols, aldehydes, ketones, acids and hydrocarbon gasses (Griffiths, 2015).

8.6.1   The jasmonate pathway

Herbivory, pathogen infections and mechanical wounding induce the synthesis of jasmonates, which activate defence responses and inhibit plant growth. Several intracellular signalling mechanisms have been proposed following the disruption of the plant cell wall and cell membranes, including an increased calcium ion concentration in the cytoplasm and active oxygen species in the apoplast leading to MAPK cascades and CDPKs that lead to transcriptional reprogramming in the nucleus (Mielke and Gasperini, 2019). In the chloroplast, lipases release α-linolenic acid from the thylakoids which are subsequently oxidised by a lipoxygenase (p450 sub-family CYP94) to generate hydroperoxides and the intermediate cyclic-12-oxo-phytodienoic acid (OPDA) which moves to adjacent peroxisomes for subsequent reduction and β-oxidation to produce JA. Jasmonic acid is conjugated to *iso*-leucine in the cytoplasm to JA-ile which migrates to the nucleus where it initiates new gene expression. JA-ile is perceived by a co-receptor complex consisting of JASMONATE ZIM-DOMAIN repressors (JAZ) and the substrate receptor for an E3 ubiquitin ligase CORONATINE-INSENSITIVE1 (COI-1). When JA-ile is present the JAZ repressors are ubiquitinylated and degraded, liberating JE-dependent transcription factors that activate JA-dependent responses (Figures 8.7 and 8.8). In this example, new gene expression is for cell wall synthesis and repair.

OPDA and KODA (α-ketol of octadecadienoic acid) are xylem mobile oxylipins that are thought to regulate induced systemic resistance in maize, indicating that these oxylipins are long-distance signals, linking roots to leaves. While allene oxidase synthase leads to non-volatile jasmonate signalling, hydroperoxide lyases generate pathogen and pest defensive volatiles (including hexanals, hexenols, hexanyl-acetates and green leaf volatiles) and divinyl ether synthases and peroxygenases produce strong anti-microbial compounds (such as divinyl ether fatty acids, keto-fatty acids, hydroxy fatty acids and fatty acid hydroperoxides

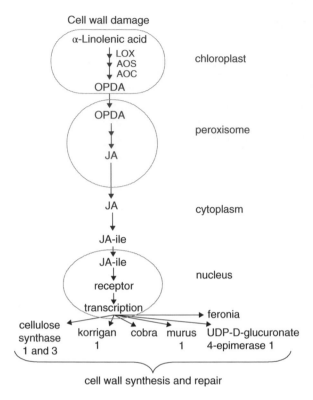

Cell wall damage

α-Linolenic acid
LOX
AOS
AOC
OPDA   chloroplast

OPDA   peroxisome
JA

JA   cytoplasm

JA-ile
JA-ile
receptor   nucleus
transcription
                                    feronia
cellulose
synthase   korrigan   cobra   murus   UDP-D-glucuronate
1 and 3   1   1   4-epimerase 1

cell wall synthesis and repair

Key

   LOX: lipoxygenase
   AOS: allene oxide synthase
   AOC: allene oxide cyclase
  OPDA: cyclic 12-oxo-phytodienoic acid
     JA: jasmonic acid
 JA-ile: jasmonyl-L-isoleucine

Figure 8.7   Biosynthesis of jasmonyl-L-isolencine (JA-ile) and the jasmonic acid (JA) pathway (see text for details)

(Genva *et al.*, 2018; Farmer and Goossens, 2019). Prost *et al.* (2005) performed *in vitro* growth inhibition bioassays to investigate the direct anti-microbial activities of 43 natural oxylipins against 13 pathogenic micro-organisms, and found that 41 showed activity against all tested bacteria and fungi. The volatile, esterified oxylipins have also been shown to modify chloroplast function, induce stomatal opening and plant senescence, and inhibit root growth.

Deboever *et al.* (2020) suggest that the plant oxylipins may be regarded as natural biocontrol agents that act across the biological kingdoms. The reader is referred to the references cited above to obtain a more detailed review of this complex biochemistry which is now being unravelled. It follows that the enzymes involved could become new targets for future herbicide development.

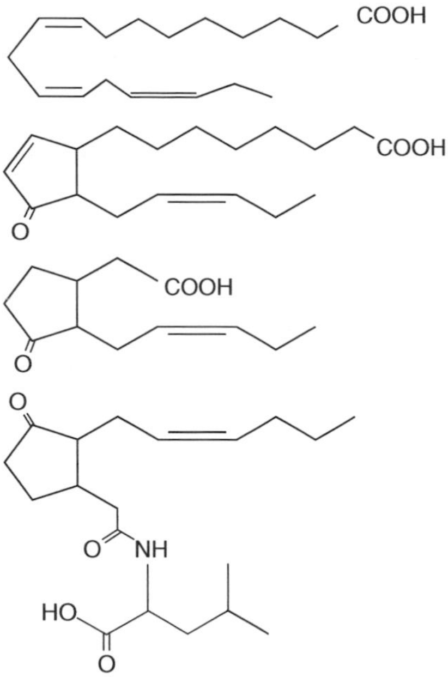

α-linolenic acid

12-oxo-phytodienoic acid (OPDA)

(+)-7-iso-jasmonic acid (JA)

jasmonyl-isoleucine (JA-ile)

Figure 8.8 Structures of some molecules involved in Jasmonic acid biosynthesis

8.7 Selectivity

Crop selectivity to the thiocarbamates is achieved through depth protection and metabolism. These herbicides are volatile and so need to be rapidly incorporated into the soil to be effective. EPTC, the most volatile example of this class, must be incorporated within 15 min of application! Such instability ensures that EPTC can be used to clear a soil of germinating couch grass (*E. repens*) or wild oats (*Avena* spp.) and that it mostly disappears before a crop is planted in the following week or so. Triallate is incorporated into the top 2.5 cm of soil and also controls grasses at germination. Cereal seeds are then drilled at a minimum 4 cm depth and their seedlings grow through this zone, the sensitive meristem being protected by the coleoptile and primordial leaves. On the other hand, germinating wild oat seedlings have an elongating mesocotyl (first internode) that extends the unprotected wild oat meristem into the phytotoxic chemical barrier. However, wild oats germinating at the soil surface or at greater depths than 4 cm may not be controlled by this treatment, and so mixtures with more persistent soil-applied herbicides, such as atrazine, are often used in practice. In addition, thiocarbamates appear to undergo bio-activation in susceptible species by the process of sulphoxidation. The sulphoxides so formed create an electrophilic centre in the molecule which seems to be linked to an increase in phytotoxicity. Tolerant crops such as cereals, and especially maize, appear able to detoxify the sulphoxides by conjugation with glutathione.

The chloroacetamides, such as alachlor and metolachlor, are absorbed onto soil colloids and may remain active for two or three months without major soil leaching. Selectivity appears to be achieved by rapid metabolism to inactive glutathione conjugates in barley, sorghum, maize and sugarcane seedlings.

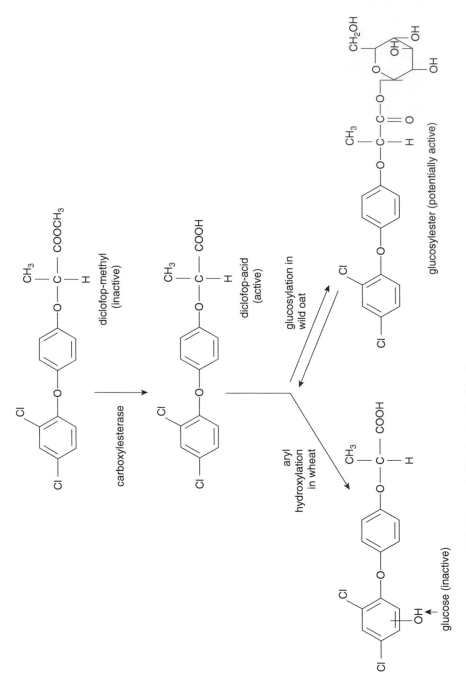

Figure 8.9 Metabolism of diclofop-methyl in wheat and wild oat.

Ester hydrolysis appears central to the selectivity of the alaninopropionates in the control of wild oats in cereals. The inactive esters of benzoylprop and flamprop, for example, are rapidly hydrolysed in susceptible wild oats to their respective phytotoxic acids. In contrast, de-esterification is far slower in wheat, and any acid formed is rapidly inactivated by glycosylation. The carboxylesterase responsible for benzoylprop-ethyl hydrolysis in *Avena fatua* has been studied in some detail by Hill *et al.* (1978), and has also been implicated in the selective metabolism of the AOPPs. In this case Shimabukuro *et al.* (1979) found that diclofop-methyl was rapidly de-esterified in both wild oat and wheat. However, aryl hydroxylation rapidly inactivated the phytotoxic acid in the crop, and an ester glucoside was formed in the weed from which the toxic species could be easily and rapidly regenerated (Figure 8.9).

Metabolism also appears to form the basis of selectivity of the CHDs. Sulphoxidation, aryl hydroxylation and molecular rearrangement have been observed with cycloxydim in tolerant species, and rapid conjugation of these groups leads to more polar and inactive by-products.

The striking selectivity of the ACCase inhibitors haloxyfop and tralkoxydim has been investigated in some detail by Secor *et al.* (1989). In this study, both susceptible and tolerant plants were sprayed with a range of herbicide concentrations to establish the dose at which growth was inhibited by 50% (ED_{50}) and values compared with the herbicide concentrations needed to inhibit *in vitro* ACCase activity by 50% (IC_{50}). Their results, presented in Table 8.4, indicate that plant tolerance to these herbicides is clearly related to the insensitivity of ACCase. The identification of two forms of ACCase with very different sensitivities to ACCase inhibitors has confirmed that this is the major basis of selectivity between monocotyledonous and dicotyledonous crops (see Section 8.3 and Table 8.4). Of the five species examined, soybean was the most tolerant at both the whole plant and the enzyme levels, although the opposite was the case in maize. Interestingly, wheat proved to be tolerant to tralkoxydim even though the isolated ACCase was sensitive to inhibition, which may imply metabolic inactivation in this crop. Further differences in grass sensitivity to the ACCase inhibitors have been characterised by Lichtenthaler *et al.* (1989). It is therefore clear that grass tolerance to these graminicides is a consequence of both metabolism and lower sensitivity of ACCase, and that the apparent resistance of many dicotyledonous plants (e.g. soybean in Table 8.4) is due to insensitivity of the target enzyme itself.

Table 8.4   Effect of haloxyfop and tralkoxydim on plant growth (ED_{50}) and inhibition of ACCase *in vitro* (IC_{50}).

Species	ED_{50} (µM)		IC_{50} (µM)	
	Haloxyfop	Tralkoxydim	Haloxyfop	Tralkoxydim
Maize	19 (S)	18 (S)	0.50	0.52
Wheat	83 (S)	>760 (T)	1.22	0.91
Tall fescue	133 (S)	225 (S)	0.94	0.40
Red fescue	1250 (T)	>6000 (T)	23.32	13.83
Soybean	>6000 (T)	>6000 (T)	138.50	516.72

S, Susceptible; T, tolerant.
Source: Secor, J., Csèke, C. and Owen, J. (1989) The discovery of the selective inhibition of acetyl- coenzyme A carboxylase activity by two classes of graminicides. *Brighton Crop Protection Conference, Weeds* 1, 145–154.

References

Andreev, G.K. and Amrhein, N. (1976) Mechanism of action of the herbicide 2-chloro-3-(4-chlorophenyl) propionate and its methyl ester: interaction with cell responses mediated by auxin. *Physiologia Plantarum* **37**, 175–182.

Andrews, M., Dickson, R.L., Foreman, M.H., Dastgheib, F. and Field, R.J. (1989) The effects of different external nitrate concentrations on growth of *Avena sativa* cv. Amuri treated with diclofop-methyl. *Annals of Applied Biology* **114**, 339–348.

Campe, R., Hollenbach, E., Kammerter, L., Hendricks, J., Hoffken, H.-W., Kraus, H. *et al.* (2018) A new herbicidal site of action: cinmethylin binds to acyl-ACP thioesterase and inhibits plant fatty acid biosynthesis. *Pesticide Biochemistry and Physiology* **148**, 116–125.

Carr, J.E., Davies, L.G., Cobb, A.H. and Pallett, K.E. (1986) Uptake, translocation and metabolism of fluazifop-butyl in *Setaria viridis*. *Annals of Applied Biology* **108**, 115–123.

Choi-Rhee, E. and Cronan, J.E. (2003) The biotin carboxylase–biotin carboxyl carrier protein complex of *Escherichia coli* acetyl CoA carboxylase. *Journal of Biological Chemistry* **278**, 30806–30812.

Cobb, A.H. and Barnwell, P. (1989) Anti-auxin activity of graminicides. *Brighton Crop Protection Conference, Weeds* **1**, 183–190.

Deboever, E., Deleu, M., Mongrand, S., Lins, L. and Fauconnier, M.-L. (2020) Plant–pathogen interactions: underestimated roles of phyto-oxylipins. *Trends in Plant Science* **25**; doi: org/10.1016/j.tplants.2019.09.009

Délye, C., Zhang, X.Q., Chalopin, C., Michel, S. and Powles, S.B. (2003) An isoleucine residue within the carboxyl-transferase domain of multidomain acetyl-coenzyme A carboxylase is a major determinant of sensitivity to aryloxyphenoxypropionate but not to cyclohexanedione inhibitors. *Plant Physiology* **132**, 1716–1723.

Evenson, K.J., Gronwald, J.W. and Wyse, D.L. (1997) Isoforms of acetyl-coenzyme A carboxylase in *Lolium multiflorum*. *Plant Physiology and Biochemistry* **35**, 265–272.

Farmer, E.E. and Goossens, A. (2019) Jasmonates: what allene oxide synthase does for plants. *Journal of Experimental Botany* **70**, 3373–3378; doi: org/10.1093/jxb/erz254

Genva, M., Akong, F.O, Andresson, M.X., Deleu, M., Lins, L. and Fauconnier, M.-L. (2018) New insights into the biosynthesis of esterified oxylipins and their involvement in plant defence and developmental mechanisms. *Phytochemistry Reviews*; doi: org/10.1007/s11101-018-9595-8

Gomes, H.L.L., Sambatti, V.C. and Dalazen, G. (2020) Sourgrass control in response to the association of 2,4-D to ACCase inhibitor herbicides. *Bioscience Journal Uberlandia* **36**, 1126–1136; doi: org/10.14393/BJ-v36n4a2020-47895

Griffiths, G. (2015) Biosynthesis and analysis of plant oxylipins. *Free Radical Research* **49**, 565–582; doi: org/10.3109/10715762.2014.1000318

Harre, N.T., Young, J.M. and Young, B.G. (2019) Influence of 2,4-D, dicamba and glyphosate on clethodim efficacy of volunteer glyphosate-resistant corn. *Weed Technology*; doi: org/10.1017/wet.2019.124

Harwood, J.L. (1988) Fatty acid metabolism. *Annual Review of Plant Physiology* **39**, 101–38.

Harwood, J.L. (1989) The properties and importance of acetyl-coenzyme A carboxylase in plants. *Brighton Crop Protection Conference, Weeds* **1**, 155–162.

Hay, M.M., Shoup, D.E. and Peterson, D.E. (2018) Palmer amaranth and common waterhemp control with very-long-chain-fatty-acid-inhibiting herbicides. *Crop Forage and Turfgrass Management* **4**, 1–9.

Hill, B.D., Stobbe, E.H. and Jones, B.L. (1978) Hydrolysis of the herbicide benzoylprop-ethyl by wild oat esterase. *Weed Research* **18**, 149–154.

Hoppe, H.H. (1980) Veränderungen der membranpermeabilitdt, des kohlenhydragehaltes, des lipid-zusammensetzung in keimwurzel von *Zea mays* L. nach behandlung mit diclofop-methyl. *Zeitschrift für Pflanzenphysiologie* **100**, 415–426.

Konishi, T. and Sasaki, Y. (1994) Compartmentation of two forms of acetyl CoA carboxylase in plants and the origin of their tolerance towards herbicides. *Proceedings of the National Academy of Sciences, USA* **91**, 3598–3601.

Kozaki, A. and Sasaki, Y. (1999) Light-dependent changes in redox status of the plastidic acetyl-CoA carboxylase and its regulatory component. *Biochemical Journal* **339**, 541–546.

Kozaki, A., Mayumi, K. and Sasaki, Y. (2001) Thiol–disulfide exchange between nuclear-encoded and chloroplast-encoded subunits of pea acetyl-CoA carboxylase. *Journal of Biological Chemistry* **276**, 39919–39925.

Lewis, D.F., Mcelroy, J.S., Dorochan, J.C. and Mueller, T.C. (2010) Efficacy and safening of aryloxyphenoxyproprionate herbicides when tank-mixed with triclopyr for bermudagrass control on zoysiagrass turf. *Weed Technology* **24**, 489–494.

Lichtenthaler, H.K. and Meier, D. (1984) Inhibition by sethoxydim of chloroplast biogenesis, development and replication in barley seedlings. *Zeitschrift für Naturforschung* **39c**, 115–122.

Lichtenthaler, H.K., Kobek, K. and Focke, M. (1989) Differences in sensitivity and tolerance of monocotyledonous and dicotyledonous plants towards inhibitors of acetyl-coenzyme A carboxylase. *Brighton Crop Protection Conference, Weeds* **1**, 173–182.

Menchari, Y., Chauvel, B., Darmency, H. and Delye, C. (2008) Fitness costs associated with three mutant acetyl-coenzyme A carboxylase alleles endowing herbicide resistance in black-grass *Alopecurus myosuroides*. *Journal of Applied Ecology* **45**(3), 939–947.

Mielke, S. and Gasperini, D. (2019) Interplay between plant cell walls and jasmonate production. *Plant Cell Physiology* **60**, 2629–2637; doi: org/10.1093/pcp/pcz119

Ohlrogge, J. and Browse, J. (1995) Lipid biosynthesis. *Plant Cell* **7**, 957–970.

Perkins, C.M, Muller, T.C. and Strekel, L.E. (2021) Junglerice control with glyphodate and clethodim as influenced by dicamba and 2,4-D mixtures. *Weed Technology* **35**, 419–425; doi: org/10.1017/wet.2021.5

Prost, I., Dhondt, S., Rothe, G., Vicente, J., Rodriguez, M.-J., Kift, N. *et al.* (2005) Evaluation of the antimicrobial activities of plant oxylipins supports their involvement in defence against pathogens. *Plant Physiology* **139**,1902–1913.

Rendina, A.R., Craig-Kennard, A.C., Beaudoin, J.D. and Breen, M.K. (1990) Inhibition of acetyl-coenzyme A carboxylase by two classes of grass-selective herbicides. *Journal of Agricultural and Food Chemistry* **38**, 1282–1287.

Sasaki, Y. and Nagano, Y. (2004) Plant acetyl-CoA carboxylase: structure, biosynthesis, regulation, and gene manipulation for plant breeding. *Bioscience, Biotechnology and Biochemstry* **68**, 1175–1184.

Sasaki, Y., Konishi, T. and Nagano Y. (1995) The compartmentation of acetyl-coenzyme A carboxylase in plants. *Plant Physiology* **108**, 445–449.

Savage, L.J. and Ohlrogge, J.B. (1999) Phosphorylation of pea chloroplast acetyl-CoA carboxylase. *Plant Journal* **18**, 521–527.

Secor, J., Csèke, C. and Owen, J. (1989) The discovery of the selective inhibition of acetyl-coenzyme A carboxylase activity by two classes of graminicides. *Brighton Crop Protection Conference, Weeds* **1**, 145–154.

Shimabukuro, R.H., Walsh, W.C. and Hoerauf, R.A. (1979) Metabolism and selectivity of diclofop-methyl in wild oat and wheat. *Journal of Agricultural and Food Chemistry* **27**, 615–623.

Siddall, C.J. and Cousins, S.F.B. (1982) Annual and perennial grass weed control in sugar beet following sequential and tank-mix application of fluazifop-butyl and broad-leaf herbicides. *British Crop Protection Conference, Weeds* **2**, 827–833.

Smeda, R.J. and Xi Xiong (2019) Post-emergence control of windmillgrass using selected herbicides. *American Journal of Plant Sciences* **10**, 1300–1312.

Takano, H.K., Lopez-Ovejero, R.F., Gross-Belchoir, G. and Leal-Maymone, G.P. (2019) ACCase-inhibiting herbicides: mechanisms of action, resistance evolution and stewardship. *Scientia Agricola* **78**.doi.org/10.1590/1678-992X-2019-0102.

Trentkamp S., Martin, W. and Tietjen, K. (2004) Specific and differential inhibition of very long-chain fatty acid elongases from *Arabidopsis thaliana* by different herbicides. *Proceedings of the National Academy of Sciences, USA* **101**, 11903–11908.

Underwood, M.G., Soltani, N., Hooker, D.C., Robinson, D.E., Vink, J.C. *et al.* (2016) The addition of dicamba to POST applications of quizalofop-*p*-methyl or clethodim antagonises volunteer glyphosate-resistant corn control in dicamba-resistant soybean. *Weed Technology* **30**, 639–647; doi: org/10.1614/WT-D-16-00016.1

Webster, E.P., Rustom, S.Y., McKnight, B.M., Blouin, D.C. and Telo, G.M. (2019) Quizalofop-*p*-ethyl mixed with synthetic auxin and ACCase-inhibiting herbicides for weed management in rice production. *International Journal of Agronomy*; doi: org/10.1155/2019/6137318

Weisshaar, H., Retzlaff, G. and Böger, P. (1988) Chloroacetamide inhibition of fatty acid biosynthesis. *Pesticide Biochemistry and Physiology* **32**, 212–216.

Wright, J.P. and Shimabukuro, R.H. (1987) Effects of diclofop and diclofop-methyl on the membrane potentials of wheat and oat coleoptiles. *Plant Physiology* **85**, 188–193.

Yu, J. and McCullough, P.E. (2016) Triclopyr reduces foliar bleaching from mesotrione and enhances efficacy for smooth crabgrass control by altering uptake and translocation. *Weed Technology* **30**, 516–523.

Chapter 9
The Inhibition of Amino Acid Biosynthesis

All life on Earth emanates from the green of the plant.

Jay Kordich (1923–2017)

9.1 Introduction

Since the early 1970s a new generation of herbicides of great agronomic and commercial importance has been developed that specifically inhibit the biosynthesis of amino acids. Inhibition of plant amino acid biosynthesis has become a major target of herbicide development as only plants and microorganisms can synthesise all of their amino acids. Consequently, the chances of mammalian toxicity are slight because animals cannot synthesise histidine, isoleucine, leucine, lysine, methionine, threonine, tryptophan, phenylalanine and valine, which are termed essential amino acids and which have considerable dietary significance.

Microorganisms have proved key to our understanding of the herbicidal inhibition of amino acid biosynthesis, since our knowledge of the biochemistry and genetics of microorganisms has been far greater than our understanding of plant metabolism. In particular, mutations in several bacteria and yeasts have been raised with relative ease, and the study of their biochemistry and genetics has been directly applied to green plants. Furthermore, the developed technologies to isolate genes from microorganisms responsible for herbicide resistance and to transfer them to crop plants have provided exciting opportunities in crop protection, as discussed in Chapter 14. Indeed, the major success of glyphosate, glufosinate and the development of the very low dosage acetolactate synthase inhibitors has led to an increased interest in this area of plant metabolism.

9.2 Overview of amino acid biosynthesis in plants

Our understanding of the biosynthesis of amino acids in plants is fragmentary. Although the general pathways are known with some certainty, their cellular location and details of regulation often remain obscure. Similarly, information of amino acid metabolism during plant development and in different tissues is also lacking, as are details of the structure and

Herbicides and Plant Physiology, Third Edition. Andrew H. Cobb.
© 2022 John Wiley & Sons Ltd. Published 2022 by John Wiley & Sons Ltd.

function of many key enzymes. In fact, herbicides have proved immensely powerful and useful tools to probe this aspect of plant metabolism. One certainty, however, is the central role played by the chloroplast in amino acid biosynthesis. Although most of the enzymes involved are nuclear-encoded, they are synthesised in the cytoplasm and imported into the chloroplast where carbon skeletons from photosynthesis provide the starting materials for amino acid biosynthesis. All amino acids derive their nitrogen from glutamate by the combined action of glutamine synthase and glutamate synthase in the chloroplast stroma. These enzymes are also central to the recycling of nitrogen within the plant from the storage amides, glutamine and asparagine. In addition, the photosynthetic carbon oxidation cycle generates phosphoglycollate, and the ammonia released from photorespiration is efficiently re-assimilated by glutamine synthase. Arginine and proline are synthesised from glutamate, and aspartate, produced by a glutamate–oxaloacetate aminotransferase, is the starting point for the synthesis of methionine, threonine, lysine and the branched-chain amino acids leucine, isoleucine and valine. Pyruvate plays a central role in the synthesis of alanine and the branched-chain amino acids. On the other hand, the aromatic amino acids tryptophan, phenylalanine and tyrosine are derived from erythrose 4-phosphate in the photosynthetic carbon reduction cycle, and histidine is similarly formed via ribose 5-phosphate. These major reactions and interconversions are summarised in Figure 9.1. Individual pathways susceptible to herbicidal inhibition will be considered in more detail in the following sections. For further details the reader is referred to Maeda and Dudareva (2012), Hildebrandt *et al.* (2015) and Galili *et al.* (2016).

9.3   Inhibition of glutamine synthase

The ammonium salt of glufosinate (DL-homoalanin-4-yl [methyl] phosphonic acid) is a major non-selective, post-emergence herbicide introduced in 1981 for the rapid control of vegetation. It is a potent inhibitor of glutamine synthase and, in common with methionine sulfoximine and tabtoxinine-β-lactam, is a structural analogue of glutamic acid (Figure 9.2).

The discovery and development of glufosinate is of particular interest since this herbicide may be regarded as having a natural origin. Glufosinate, also termed phosphinothricin, and a tripeptide containing glufosinate bound to two molecules of L-alanine (phosphinothricyl-alanyl-alanine) were first described in 1972 as products of the soil bacterium *Streptomyces viridichromogenes*. Independent studies in Japan in 1973 also discovered this tripeptide in culture filtrates of another soil bacterium, *Streptomyces hygroscopicus*, and this compound was introduced commercially as bialaphos in the 1980s. Bialaphos itself is inactive, but it is rapidly hydrolysed in plants to form the phytotoxic glufosinate (Wild and Ziegler, 1989). An additional tripeptide, phosalacine, also produced by a *Streptomyces* sp. has been reported by Omura *et al.* (1984). This tripeptide differs from bialaphos in that one alanine moiety is replaced by leucine. Evidence for the inhibition of plant glutamine synthase by glufosinate was first provided by Leason *et al.* (1982) who demonstrated a K_i value of 0.073 mM, and this potent inhibition has since been confirmed in a wide variety of plants and algae.

Glutamine synthase (E.C. 6.3.1.2) catalyses the conversion of L-glutamate to L-glutamine in the presence of ATP and ammonia (Figure 9.3). This enzyme has a molecular weight of about 400 kDa and has eight subunits, each with an active site. It exists as separate isozymes

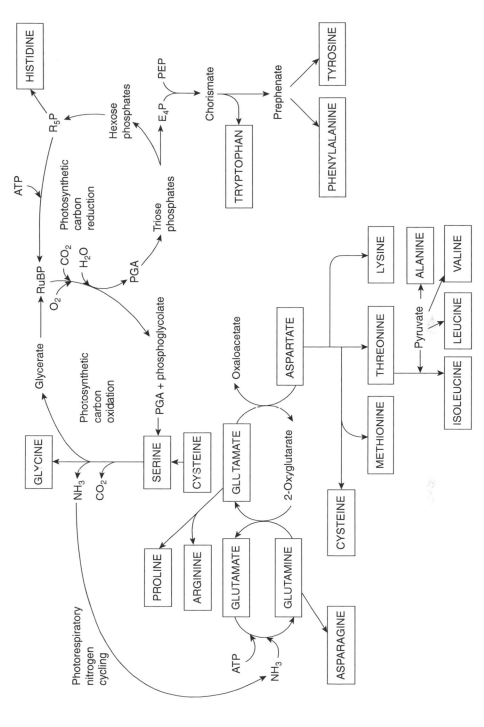

Figure 9.1 Overview of amino acid biosynthesis. R_5P, Ribulose 5-phosphate; PEP, phosphoenoylpyruvate; E_4P, erythrose 4-phosphate; RuBP, ribulose 1,5-bisphosphate; PGA, 3-phosphoglyceric acid.

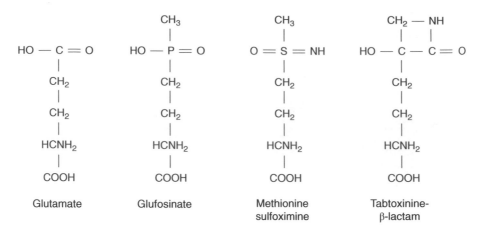

Figure 9.2   The structures of glutamate and the glutamine synthase inhibitors glufosinate, methionine sulfoximine and tabtoxinine-β-lactam.

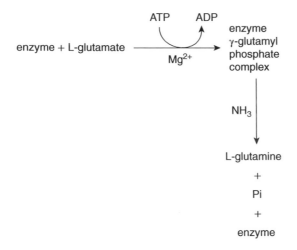

Figure 9.3   The two-step process in the catalysis of glutamine synthesis by glutamine synthase.

in the leaf cytoplasm, chloroplast and roots, and in legumes as an isoform specific to root nodules. The isoform found in the chloroplast is expressed in greater amounts in the presence of light and at high sucrose concentrations, an indicator of high levels of photosynthetic activity.

Glutamine synthase forms glutamine in a two-step process. Normally, γ-glutamyl phosphate is first produced from ATP and L-glutamate, and then ammonia reacts with this complex to release inorganic phosphate (Pi) and L-glutamine. However, glufosinate, as an L-glutamate analogue, can also be phosphorylated to produce an enzyme–glufosinate–phosphate complex to which ammonia cannot bind and the enzyme is irreversibly inhibited (Manderscheid and Wild, 1986). Glufosinate can inhibit bacterial, plant and mammalian glutamine synthase *in vitro*, but is non-toxic to mammals, apparently because of its inability to cross the blood–brain barrier and its rapid clearance by the kidneys.

As with other herbicides that inhibit amino acid biosynthesis, the sequence of events leading from glufosinate application to plant death has been open to much debate. Phytotoxicity is certainly rapid, since leaf chlorosis, desiccation and necrosis may be observed within 2 days after treatment, and plant death results 3 days later. Since glutamine synthase is potently inhibited, the rapid accumulation of ammonia was the presumed toxic species which was thought to uncouple electron flow from proton transport in the thylakoid, so that photosynthesis was rapidly inhibited. This view was supported experimentally by the observation that phytotoxic symptoms only develop in plants exposed to light, especially when photorespiration is favoured. Thus, the ammonia generated in photorespiration is not re-assimilated and may directly uncouple proton gradients.

Further and more recent observations clearly imply an additional and very rapid action of glufosinate on membrane transport processes that precede visual symptom development. Treated plants accumulate ammonia and show increased rates of cell leakage of potassium ions, and the uptake of nitrate and phosphate is adversely affected. Ullrich *et al.* (1990) have convincingly demonstrated that glufosinate, like glutamate, can directly cause an electrical depolarisation of the plasmalemma, although unlike glutamate, recovery is often incomplete and a secondary depolarisation is evident. These decreased membrane potentials inhibit or alter transport processes so that, for example, the co-transport of glutamate/ H^+ was irreversibly inhibited, but potassium flux was increased. In this way, the accumulation of ammonia may be seen to uncouple or interfere with membrane function and transport at the plasmalemma as well as at the thylakoid.

Increases in ammonia concentrations in treated tissues have been reported as 10 times higher than in untreated tissues 4 h after treatment and as high as 100 times within 1 day. Free ammonia is toxic to biological systems and it is for this reason that a number of processes are found in biological systems that aim to maintain free ammonia at very low concentrations.

An associated decrease in the concentration of a number of amino acids is also noted where treated plants remain in the light. A decrease in photosynthetic carbon fixation is also noted under the same conditions. These symptoms develop far more slowly in the dark and this is mirrored in field observations that glufosinate activity is observed far quicker under conditions of full sunlight.

Free ammonia has a severe effect on the pH gradient across biological membranes, collapsing membrane potentials. Glufosinate does not have a direct inhibitory effect on photosynthetic carbon fixation, but was assumed to inhibit this process via the reduction in ATP production by photophosphorylation, owing to the effect of ammonia on membrane potentials. However, Wild *et al.* (1987) reported that under conditions in which photorespiration was not taking place, photosynthesis was not inhibited by the increased ammonia concentrations resulting from glufosinate treatment.

Photorespiration is an alternative metabolic pathway where ribulose 1,5-bisphosphate carboxylase-oxygenase (RuBisCo), the enzyme responsible for carbon fixation in the Calvin cycle using the products of the light stage of photosynthesis, uses oxygen in place of carbon dioxide as a substrate to react with ribulose 1,5-bisphosphate. Photorespiration takes place in conditions of elevated oxygen and, although it does produce glyceraldehyde 3-phosphate for the carbon reduction cycle, it does this at an energetically far less economical rate compared with the carboxylase activity of RuBisCo. In addition, photorespiration produces free ammonia that must be detoxified by re-assimilation into organic molecules. Glufosinate will inhibit this re-assimilation and will therefore reduce or inhibit photorespiration, owing to reduced concentrations of glutamate as an amino donor for glyoxylate. This will result in a reduction in photosynthetic carbon fixation, owing to increased glyoxylate concentrations, and may also result in an increase in active oxygen species from triplet state chlorophyll in the light stage of photosynthesis. The result would be lipid peroxidation as reported for a number of other herbicide classes (Chapter 5). Indeed, Takano *et al.* (2019) have clearly demonstrated that the rapid desiccation of foliage induced by glufosinate is due to a massive production of active oxygen species that causes lipid peroxidation of cell membranes, resulting in rapid cell death.

It therefore appears that glufosinate can affect a number of metabolic pathways in plants by its indirect effect on membrane polarisation, reduced peptide, protein and nucleotide concentration, increased protein degradation (to release free amino acids) and the inhibition of photosynthetic carbon assimilation via the inhibition of photorespiration.

More recent studies, as reviewed by Takano and Dayan (2020), confirm that the herbicidal activity of glufosinate is due to the accumulation of active oxygen species followed by lipid peroxidation. A further paper by Takano *et al.* (2020) considers how the active oxygen species are formed, and how the presence of glufosinate affects the delicate balance of the generations versus the scavenging of active oxygen species. Thus, the accumulation of ammonia and intermediates in photorespiration in the light leads to the accumulation of hydrogen peroxide and glycolate which overwhelms the enzymes that scavenge reactive oxygen species.

Introduction of genes encoding a specific glufosinate-metabolising enzyme has allowed the successful development of glufosinate-resistant, genetically-modified crops. These are discussed in detail in Chapter 14.

9.4   Inhibition of aromatic amino acid biosynthesis

9.4.1   Inhibition of EPSP synthase

The herbicide glyphosate (*N*-phosphonomethyl glycine; Figure 9.4) is a major non-selective, post-emergence herbicide used in circumstances where the total control of vegetation is required. Its success lies in very low soil residual activity, broad spectrum of activity, low non-target organism toxicity and great systemicity in plants, so that even the most troublesome rhizomatous weeds can be controlled. Monsanto received a patent for use of phosphoric acid derivatives, including glyphosate, as non-selective herbicides in 1974. The Stauffer Chemical Company, which had patented a number of phosphonic and phosphinic acids as industrial cleaning agents in 1964, subsequently released sulfosate for development in 1980.

The search for the target site of glyphosate action began with the observation by Jaworski (1972) that the control of duckweed (*Lemna* sp.) by glyphosate could be overcome by the addition of aromatic amino acids to the growth medium. However, it was not until 1980 that Steinrücken and Amrhein identified the enzyme 5-enoyl-pyruvyl shikimic acid 3-phosphate (EPSP) synthase (E.C. 2.5.1.19) as being particularly sensitive to glyphosate. This enzyme is involved in the biosynthesis of the aromatic amino acids tryptophan,

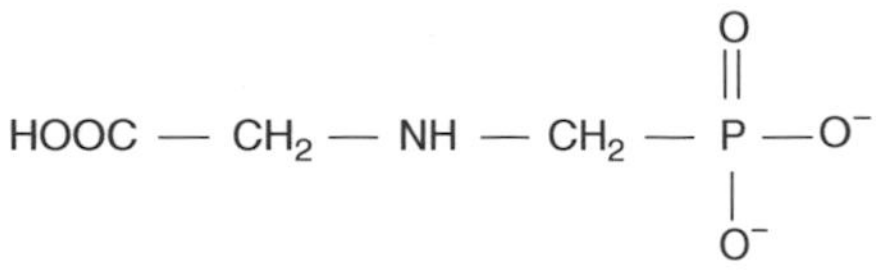

$$HOOC - CH_2 - NH - CH_2 - \overset{\overset{\displaystyle O}{\|}}{\underset{\underset{\displaystyle O^-}{|}}{P}} - O^-$$

Figure 9.4 The structure of glyphosate.

phenylalanine and tyrosine, and also leads to the synthesis of numerous secondary plant products (Figure 9.5). Approximately 30% of carbon fixed by green plants is routed through the shikimate pathway with an impressive number of significant end products, including vitamins, lignins, alkaloids and a wide array of phenolic compounds such as flavonoids. A common pathway exists from erythrose 4-phosphate, provided from photosynthetic carbon reduction in the chloroplast stroma, to chorismic acid, so inhibition at EPSP synthase by glyphosate is at a particularly strategic location. Additionally, the enzyme is not found in animals and so the chance of non-target organism toxicity is reduced.

The higher plant enzyme has a molecular weight of 45–50 kDa, found mainly in the chloroplast, which can be reversibly inhibited by glyphosate with an IC_{50} of about 10–20 µM. EPSP synthase is encoded in the nuclear genome, but the shikimate pathway is located in the chloroplast. The peptide sequence is 444 amino acids long, with an additional 72 amino acid-long transit peptide to enable access to the chloroplast. For glyphosate to be active it must enter into the cell and then the chloroplast in sufficient concentration to inhibit the enzyme. Glyphosate is one of only a handful of herbicides for which a carrier protein has been identified that aids in the crossing of plasma membranes. In the case of glyphosate, it is a phosphate carrier that is involved (Denis and Delrot, 1993).

Studies indicate that glyphosate acts as a competitive inhibitor with respect to phospho-enolpyruvate (PEP, K_i; 0.1–10 µM), but as a non-competitive inhibitor with respect to shikimic acid 3-phosphate (S3P). Mechanistically, S3P forms a complex with the enzyme to which glyphosate binds before PEP addition. Glyphosate appears to bind away from the active site of EPSP synthase, at a putative allosteric site, and also has very little affinity for the free enzyme. However, it is 'trapped' on the enzyme in the presence of EPSP. The binding of glyphosate appears to prevent the binding of PEP to the enzyme. Indeed, the dissociation rate for glyphosate is 2300 times slower than that for PEP, so that the enzyme is effectively inactivated by glyphosate. Observations that even minor changes in the structure of glyphosate result in loss of binding and subsequent herbicidal activity suggest a very specific enzyme–herbicide interaction.

A number of studies have attempted to use X-ray crystallography to resolve the order and binding characteristics of the substrates and glyphosate to EPSP synthase (Stilling et al., 1991; Franz et al., 1997; Schöenbrunn et al., 2001). Observations suggest that EPSP has two distinct hemispherical domains that come together in a 'screw-like' movement that causes the active site to be revealed. It appears that the binding of S3P is responsible for causing this conformational change. The active site is postulated to be near the 'hinge' where the two domains of the EPSP synthase enzyme meet. A number of important, conserved, amino acids have been identified that appear to play important roles in active site binding and activity. The cleft formed between the two domains is largely electropositive and this plays a role in the attraction of the anionic ligands that are EPSP substrates (Figure 9.6).

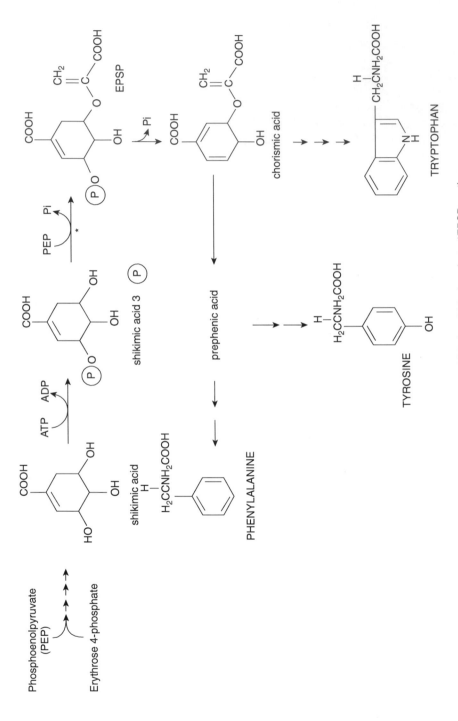

Figure 9.5 Biosynthesis of aromatic amino acids. EPSP, 5-Enoylpyruvate shikimic acid 3-phosphate; *EPSP synthase.

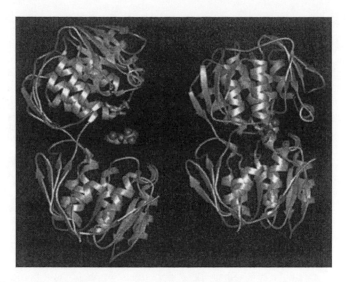

Figure 9.6   The X-ray crystal structure of *Escherichia coli* EPSP synthase. On the left is the open form. On the right is the closed configuration. A modelled glyphosate molecule in the left form is shown, leading to the closed formation on the right. Source: CaJacob, C.A., Feng, P.C.C., Heck, G.R., Alibhai, M.F., Sammons, R.D. and Padgette, S.R. (2004) Engineering resistance to herbicides. In: Christou, P. and Klee, H. (eds) *Handbook of Plant Biotechnology Vol. 1*. Chichester: John Wiley and Sons, pp. 353–372.

Since the enzyme is more active above pH 7, as may be expected in the chloroplast stroma in the light, it is the ionised form of glyphosate which is most likely to be the inhibitory species. Thus, S3P and shikimate accumulate, the inhibition cannot be reversed by PEP and the pool of aromatic amino acids declines.

Additional proof that EPSP synthase is the target enzyme for glyphosate has come from the elegant studies of Comai *et al.* (1985). These workers isolated a mutant form of EPSP synthase from *Salmonella typhimurium* which was resistant to glyphosate because of a single amino acid substitution, from proline to serine. Furthermore, they isolated the *aroA* gene encoding the resistant enzyme and successfully transferred it to tobacco, making this crop glyphosate resistant. Other studies have cloned the *aroA* gene and an overproduction of EPSP synthase in plant cell cultures has also generated tolerance to glyphosate. Subsequent isolation of glyphosate-resistant EPSP synthase and incorporation into a number of crop species is discussed in detail in Chapter 14.

9.4.2   *How glyphosate-treated plants die*

Glyphosate treatment causes growth inhibition, chlorosis, necrosis and subsequent plant death (Table 9.1). Plants treated with glyphosate, however, may not show symptoms for 7–10 days. This slow action in the field probably reflects the time taken for the depletion of the available carbon pool in general and the aromatic amino acid pool in particular, causing decreasing rates of protein synthesis. Therefore, treated weeds that are young and small may show phytotoxicity in a week or two, while treated shrubs may take a year for control.

Table 9.1 Typical symptoms following glyphosate treatment.

(A) Physiological	
Within minutes	Shikimate and shikimate 3-phosphate accumulate.
Within hours	Inhibition of aromatic amino acid biosynthesis.
	Stomatal closure and lower photosynthetic rate.
Within days	Reduced starch synthesis and translocation of sucrose.
	Inhibition of growth and protein synthesis in meristems.
	Reduced synthesis of chlorophylls and protective carotenoids.
Within weeks	Chlorosis followed by necrosis.
(B) Morphological	
Within days	Cessation of leaf growth and stem elongation.
	Chlorosis in new leaves and apices.
	Tissue death.
Within weeks	Root uptake of water is reduced, leading to loss of turgor.
	Leaf abscission. Desiccation of shoots and roots.
	Necrosis.

It is now known that glyphosate has additional effects on phenol and pigment metabolism. Some studies have suggested that glyphosate induces a transient increase in the activity of phenylalanine ammonia lyase, an important enzyme in phenylpropanoid metabolism, so that enhanced concentrations of natural growth inhibitory phenols accumulate. This enzyme activity declines when phenylalanine pools become limiting. Since many phenols are regarded as natural inhibitors of auxin oxidation, then greater auxin metabolism occurs in glyphosate-treated plants with the result that apical dominance is overcome. Consequently, lateral growth of dicots and increased tillering in monocots are often observed in the field.

Relating EPSP synthase inhibition to plant death is not easy owing to the multiple effects this has on a number of metabolic processes. Inhibition of the enzyme results in reduced production of chorisimic acid and a subsequent increase in shikimate and shikimate-3-phosphate owing to the blocking of the shikimic acid pathway. In glyphosate-treated tissue, shikimate and shikimate-3-phosphate have been reported as making up 16% of plant dry matter (Schuktz *et al.*, 1990). This causes a reduction in protein synthesis owing to depletion of amino acid pools. This alone would result in growth retardation and subsequent plant death. However, blocking of the shikimic acid pathway and subsequent reduction in chorisimate causes an increase in flow of carbon compounds out of the photosynthetic carbon reduction cycle. This results in reduction in both photosynthesis and starch production. Reduced carbon compound translocation and, ironically, glyphosate translocation from treated tissue are subsequently observed. In addition, blockage of the shikimic acid pathway results in reduction in the production of auxin growth regulators, lignin, plant defence compounds, UV protectants and photosynthetic pigments, either directly or via reduced synthesis of plastoquinone.

A further metabolic consequence of glyphosate treatment is the development of chlorotic areas on leaves. This may be due to an inhibition of δ-aminolaevulinic acid (δ-ALA) synthetase, an early reaction in the biosynthesis of all porphyrin containing molecules, including the chlorophylls and cytochromes (Chapter 6, Section 6.2). Since these compounds are central to plant metabolism, it would appear that many important biochemical pathways are therefore subject to interference by glyphosate treatment. Indeed, the observation that only 10 µM glyphosate will inhibit EPSP synthase *in vitro* from field rates of greater than 10,000 μM suggests the involvement of many secondary sites of action which produce an overall herbicidal effect.

Since glyphosate is a non-selective total herbicide, little is known about its metabolism in plants, which is presumed to be either non-existent or ineffectual. It is, however, rapidly biodegraded by soil microorganisms which are able to utilise glyphosate as a sole phosphate source. Specific enzymes operate in a *Pseudomonas* sp. to cleave the phosphate group, and the phosphonomethyl C–N bond is broken to release glycine. The main intermediate of glyphosate metabolism is aminomethylphosphonic acid, which is degraded further by soil microorganisms to phosphoric acid and eventually to inorganic phosphate and carbon dioxide.

9.4.3   Glyphosate and hormesis

It has often been observed that the application of sub-lethal doses of herbicides may stimulate a plant response. Such 'hormetic' effects (from the Greek 'hormo', to excite) are usually observed within a narrow dose range, with increasing doses becoming phytotoxic or less than an appropriate control. The literature suggests that hormesis is found in all organisms and is induced by chemical or physical stressors, including glyphosate, although the effect was not observed in glyphosate-resistant plants (Velini *et al.*, 2008). Reports of consistent effects in the field, however, remain to be demonstrated. Brito *et al.* (2018) have reviewed the hermetic effect of glyphosate. They noted previous work in the literature on 25 species and the dose range causing the hermetic effect was typically between 1.8 and 25 g of acid equivalent of glyphosate, although doses causing hormesis and phytotoxicity are invariably close. In most studies, the effect was transient and few yield increases were reported under field conditions. The authors also noted physiological effects of glyphosate in the literature, but concluded that how glyphosate causes them remains unknown. Sublethal doses of many herbicides, such as 2,4-D and paraquat, in addition to those reported for glyphosate, are commonly reported in the literature (Velini *et al.*, 2010; Silva *et al.*, 2019). In a recent review, Jalal *et al.* (2021) consider that the use of glyphosate as a hormesis promotor shows promise, i.e. that reducing lignin biosynthesis may improve crop growth and productivity. They also propose that hormesis may improve plant tolerance to adverse conditions, enabling survival.

9.4.4   Inhibition of dehydroquinate synthase

Brilisaur *et al.* (2019) reported the discovery of 7-deoxy-sedoheptulose (7dSH, Figure 9.7) from culture supernatants of the cyanobacterium *Synechoccus elongans*, and found that it inhibited the growth of various phototrophs, including species of cyanobacteria, yeasts and *Arabidopsis*, at micromolar concentrations. They found that all treated organisms accumulated 3-deoxy-D-arabino-heptulosonate-7-phosphate, indicating a target site of 3-dehydroquinate synthase (DQS), a key enzyme in the shikimate pathway, before EPSPS (Figure 9.8). Symptoms of inhibition at DQS following 7dSH application are similar to those caused by glyphosate. Initial studies suggest that DQS shows promise as a new target site for the post-emergence control of broad-leaf weeds.

The pathway of tyrosine metabolism in plants is a road-map for the synthesis of a great diversity of natural, i.e. secondary, metabolites, which include tocopherols, plastoquinone, ubiquinone, betalains, salidroside and several families of alkaloids (Figure 9.9). Many of the enzymes and intermediates have been elucidated in recent years (Xu *et al.*, 2019).

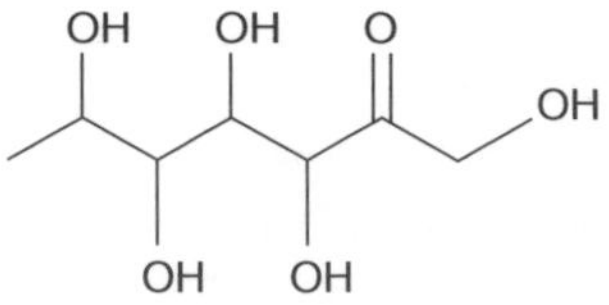

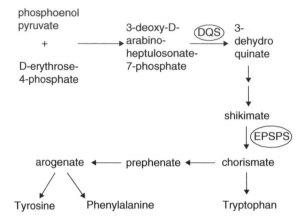

Figure 9.7   Structure of 7-deoxy-sedoheptulose.

Figure 9.8   Shikimate pathway highlighting DQS and EPSPS as target enzymes for the inhibition of aromatic amino acids.

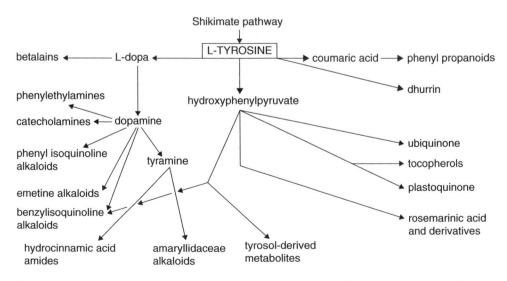

Figure 9.9   A simplified version of tyrosine pathways in plants (modified from Xu *et al.*, 2019)

9.5 Inhibition of branch-chain amino acid biosynthesis

9.5.1 Inhibition of acetolactate synthase

Since the 1980s five new herbicidal classes have emerged that all share the same site of action. These have proved to be potent, selective, broad-spectrum inhibitors of plant growth at field rates measured in grams rather than kilograms per hectare. The sulphonylureas (SUs), imidazolinones (IMs), triazolopyrimidines (TPs), sulphonylaminocarbonyl-triazolines (often classed as SUs) and pyrimidinyl-oxy-benzoates (pyrimidinyl-carboxy herbicides; PCs) are chemically different (Table 9.2), yet all share the same site of action, namely acetolactate synthase (ALS, E.C. 4.1.3.18, also known as acetohydroxyacid synthase, AHAS), a key

Table 9.2 Structures of a selection of amino acid biosynthesis inhibitors at acetolactate synthase (ALS).

(A) Sulphonylureas

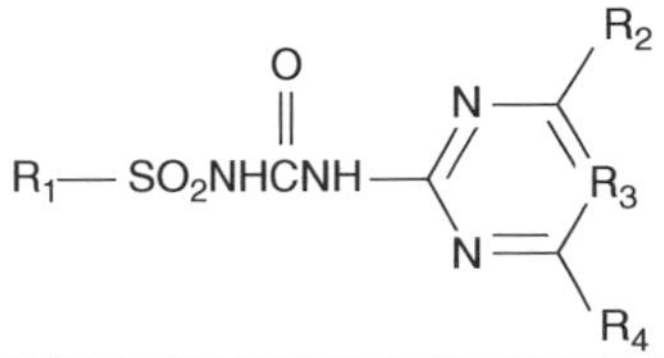

Core structure: R_1—SO_2NHCNH— (with C=O) attached to a triazine/pyrimidine ring bearing R_2, R_3, R_4 substituents and two ring N atoms.

R_1	R_2	R_3	R_4	Common name (crop/dosage, g ha^{-1})
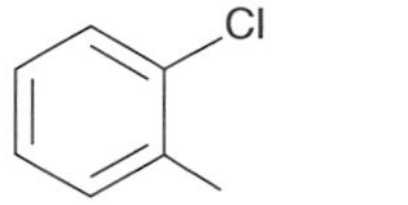 Phenyl with Cl (ortho)	$-CH_3$	N	$-OCH_3$	Chlorsulfuron (cereals/4–26)
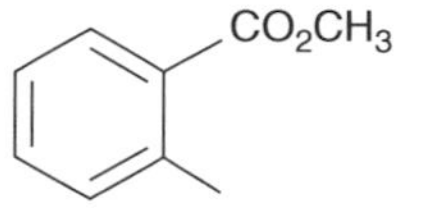 Phenyl with CO_2CH_3 (ortho)	$-CH_3$	N	$-OCH_3$	Metsulfuron-methyl (cereals/2–8)
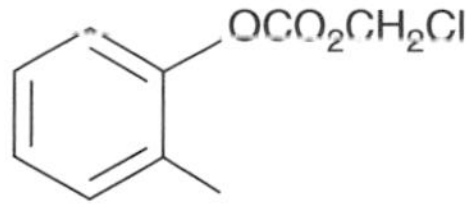 Phenyl with OCO_2CH_2Cl (ortho)	$-CH_3$	N	$-OCH_3$	Triasulfuron (cereals/10–40)
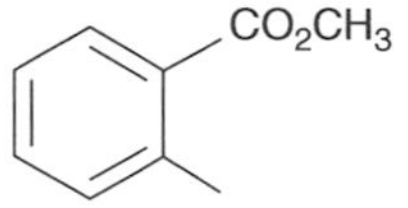 Phenyl with CO_2CH_3 (ortho)	$-OCO_3$	CH	$-OCH_3$	Bensulfuron-methyl (rice/20–75)
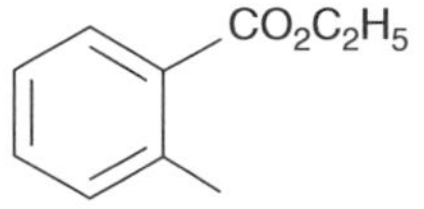 Phenyl with $CO_2C_2H_5$ (ortho)	$-Cl$	CH	$-OCH_3$	Chlorimuron-ethyl (soybean/8–13)
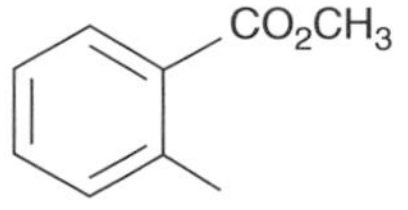 Phenyl with CO_2CH_3 (ortho)	$-CH_3$	CH	$-CH_3$	Sulfometuron-methyl (non-crop/70–840)
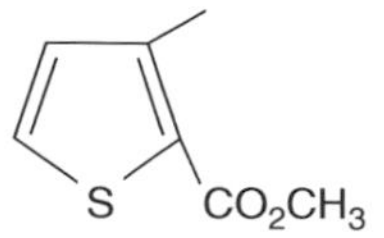 Thiophene with CO_2CH_3	CH_3	N	OCH_3	Thifensulfuron (cereals, soybean/17–35)

(Continued)

Table 9.2 (*Continued*)

R_1	R_2	R_3	R_4	Common name (crop/dosage, g ha^{-1})
Benzene ring with CO_2CH_3 and CH_3 substituents	CH_3	N	OCH_3	Tribenuron (cereals/5–30)
Benzene ring with $C(=O)OCH_3$ and CH_3 substituents	$OHCF_2$	CH	$OCHF_2$	Primisulfuron (maize/20–40)
Pyridine ring with $CON(CH_3)_2$ and CH_3 substituents	OCH_3	CH	OCH_3	Nicosulfuron (maize/40–60)
Benzene ring with CO_2CH_3 and CH_3 substituents	OCH_2CH_3	N	$NHCH_3$	Ethametsulfuron (oilseed rape/10–120)
Pyridine ring with $SO_2CH_2CH_3$ and CH_3 substituents	OCH_3	CH	OCH_3	Rimsulfuron (potato/5–15)
Benzene ring with CO_2CH_3, CH_3 and CH_3 substituents	$N(CH_3)_2$	N	OCH_2CF_3	Triflusulfuron (sugarbeet/10–25)
Benzene ring with CH_3 and $OCH_2CH_2OCH_3$ substituents	OCH_3	N	OCH_5	Cinosulfuron (oilseed rape/10–40)
Benzene ring with $CH_2CH_2CF_3$ and CH_3 substituents	CH_3	N	OCH_3	Prosulfuron (cereals/12–40)
Pyrazole ring with Cl, CO_2CH_3, CH_3 and N–CH_3 substituents	OCH_3	N	OCH_3	Halosulfuron (maize/18–140)
Imidazo[1,2-a]pyridine ring with CH_3 and $SO_2CH_2CH_3$ substituents	OCH_3	N	OCH_3	Sulfosulfuron (wheat/10–30)

Table 9.2   (*Continued*)

(B) Imidazolinones	
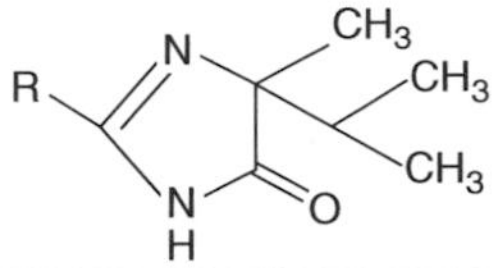	

R	Common name (crop/dosage g ha^{-1})
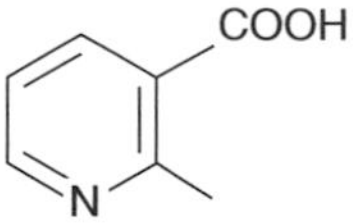	Imazapyr (non-crop/500–2000)
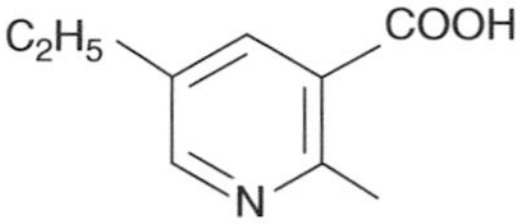	Imazethapyr (legumes/30–150)
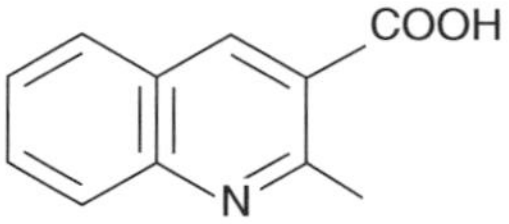	Imazaquin (soybean/140–280)
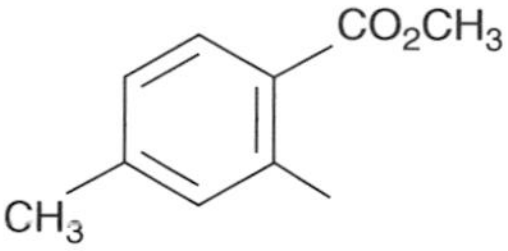	Imazamethabenz (cereals/400–750)
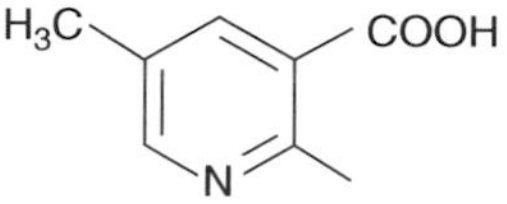	Imazapic (non-crop/50–105)
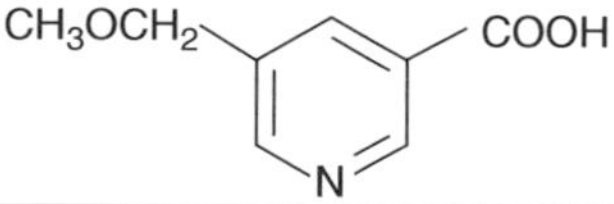	Imazamox (soybean/35–45)

(C) Triazolopyrimidines	
	Common name (crop/dosage, g ha^{-1})
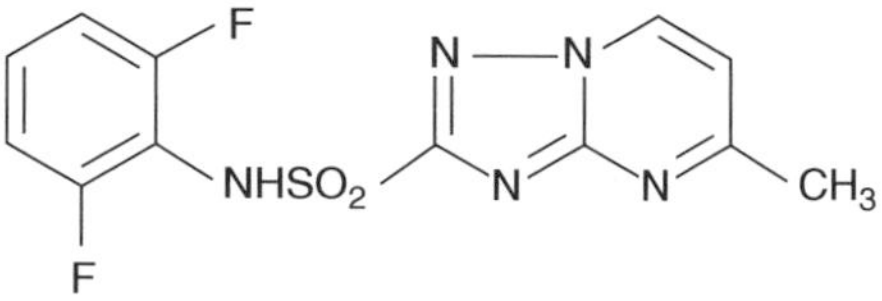	Flumetsulam (soybean/17–70)
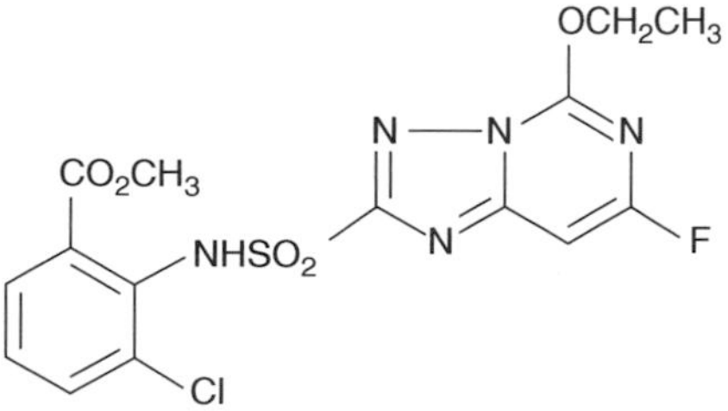	Cloransulam-methyl (soybean/35–44)

(*Continued*)

Table 9.2 (*Continued*)

	Diclosulam (soybean/26–35)

(D) Sulphonylaminocarbonyltriazoline

	Common name (crop/dosage, g ha^{-1})
	Flucarbazone-sodium (cereals/30)

(E) Pyrimidinyloxybenzoates

	Common name (crop/dosage, g ha^{-1})
	Pyrithiobac (cotton/30–100)

enzyme in the biosynthesis of the branched-chain amino acids leucine, isoleucine and valine. In each case, growth inhibition may be overcome by the addition of these amino acids. The efficacy and potency of the ALS inhibitors has ensured the continued success of these herbicides, which have rapidly challenged, and in some instances replaced, traditional products, especially in cereals and soybeans. SUs and TPs are active at field rates of 10–100 g ha^{-1} whereas application of IMs is required at 100–1000 g ha^{-1} to give a similar degree of weed control. Currently, ALS inhibitors represent the second biggest class of herbicidal active ingredients. A major research effort has progressed to understand their mode(s) of action and to develop new products to inhibit the synthesis of branched-chain amino acids. Further information on this class of herbicide can be found in reviews by Babczinski and Zelinski (1991) and Duggleby and Pang (2000).

The highest activity obtained with the SUs is when the aryl group has an *ortho* substituent. Thiophene, furan, pyrimidine and naphthalene groups are also active herbicides when replacing the aryl group, but the *ortho* substitution is still essential with respect to the sulphonylurea bridge. The heterocycle configurations for optimal activity appear to be a symmetrical pyridine or a symmetrical triazine with low alkyl or alkoxy substituents (Beyer *et al.*, 1987). The sulphonylaminocarbonyl moiety also results in the herbicidally active SUs procarbazone and flucarbazone (Amann *et al.*, 2000; Müller *et al.*, 1992). In the case of the imidazolinones, the highest biological activity has been observed when an imidazolinone ring, ideally substituted with methyl and isopropyl groups, is attached to an aromatic ring containing a carboxyl group in an *ortho* position (Los, 1986). The more recently discovered PCs demonstrate activity at similar field rates to SUs and TPs. However, the inhibitory mechanism and cross-resistance patterns appear to be a hybrid between the SUs and the IMs. Elegant studies by Shimizu *et al.* (2002), based upon the synthesis of novel analogues based upon phenoxyphenoxypyrimidine, suggest that PCs require both esteric bonding and an appropriate substituted pyrimidine ring in order to demonstrate ALS-inhibiting activity. The highest inhibition was observed when a COOMe group was at the *ortho* position to the pyrimidinyloxy group. A pyrimidine ring imparted greater inhibiting activity than structures containing other *N*-heterocyclics. Interestingly, the replacement of the O-bridge with an S-bridge reduced ALS-inhibiting activity but did increase crop tolerance in some cases. The presence of an S-bridge also increased herbicide mobility, both via root uptake and by translocation. The base S-containing compound used in these synthesis studies was pyrithiobac-sodium. Studies of this type aid not only in elucidation of structure–function relationships, but also in the discovery of new ALS-inhibiting herbicides.

All five classes of ALS inhibitors possess remarkable herbicidal properties. They are able to control a very wide spectrum of troublesome annual and perennial grass and broad-leaf weeds at very low doses. Furthermore, formulations have proved to be both foliar- and soil-active with very low mammalian toxicity. Chlorsulfuron, metsulfuron-methyl, and imazamethabenz give selective weed control in cereals, chlorimuron-ethyl and imazaquin are selective in soybean, imazethapyr is selective in other legumes as well as soybean, bensulfuron-methyl is effective in rice and sulfometuron-methyl and imazapyr have found industrial and non-crop uses for total vegetation control. Imazapyr is effective in forest management by controlling deciduous trees in conifers, and a coformulation of imazapyr and imazethapyr is being developed as a growth retardant in grassland and turf areas, with an additional control of broadleaf weeds.

ALS, like EPSP synthase, is a nuclear-encoded, chloroplast-localised enzyme in higher plants, and also occupies a strategic location in the biosynthetic pathway of essential amino acids. This pathway has been well studied in microorganisms and is becoming increasingly understood in higher plants. Essentially, synthesis occurs in the stroma from threonine and pyruvate in a common series of reactions (Figure 9.10; Ray, 1989). In isoleucine synthesis, threonine is first deaminated to 2-oxobutyrate by threonine dehydratase, which is controlled by feedback regulation by valine and isoleucine. ALS catalyses the first common step of branched-chain amino acid biosynthesis to yield acetohydroxy acids which undergo oxidation and isomerisation to yield derivatives of valeric acid. Dehydration and transamination then yield isoleucine and valine. 2-Oxoisovalerate reacts with acetyl-CoA to form α-isopropylmaleate which is then isomerised, reduced and transaminated to yield leucine. ALS demonstrates feedback inhibition to leucine, valine and isoleucine.

ALS has been extensively studied in microorganisms and is subject to increasing scrutiny in plants, although plant ALS is labile and constitutes less than 0.01% of total plant protein. As many as six ALS isozymes have been reported in bacteria to allow carbon flux through this pathway at varying concentrations of pyruvate. However, isozymes are not required for ALS in the chloroplast stroma where more reliable concentrations of substrates are assumed. Study of ALS extracted from pea seedlings identified a 320 kDa ALS that

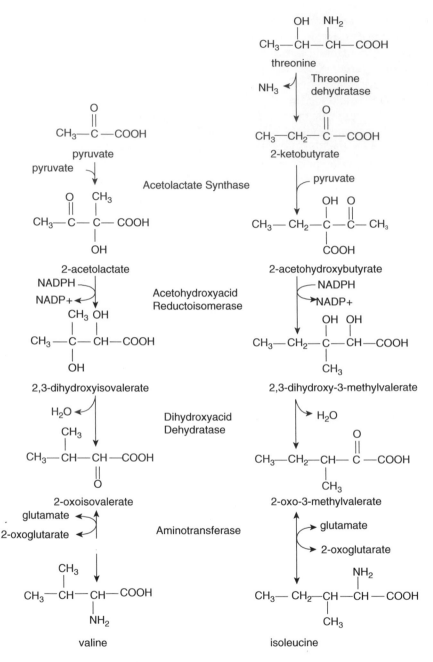

Figure 9.10 Biosynthesis of branched-chain amino acids. Source: Ray, T.B. 1989, Herbicides as inhibitors of amino acid biosynthesis. In: Böger, P. and Sandmann, G. (eds) *Target Sites of Herbicide Action*. Boca Raton, FL: CRC Press, pp. 105–125.

dissociated to a 120 kDa ALS in the absence of flavin adenine dinucleotide (FAD). The larger ALS demonstrated feedback inhibition from valine, leucine and isoleucine whereas the smaller ALS, although still demonstrating enzyme activity, did not exhibit feedback inhibition. This suggests that there are separate, regulatory, subunits that require FAD in order to remain attached to the catalytic subunits of ALS. Current understanding is that the ALS enzyme in pea seedlings consists of at least four catalytic and two regulatory subunits (Shimizi *et al.*, 2002). Regulation of ALS activity is carried out by leucine, valine and iso-leucine (feedback inhibition). This inhibition is also observed for leucine/isoleucine and leucine/valine mixtures but for valine/isoleucine mixtures an antagonism of feedback was observed. This suggests that there are two regulatory sites on the enzyme, one for leucine and one for valine/isoleucine. Two regulatory sites are also suggested by regulatory pro-moter studies (Hershey *et al.*, 1999). Analysis of the ALS gene from a number of plant species has revealed that it is highly conserved, with very little difference between the rice, maize and barley ALS genes. The isozyme II from *S. typhimurium* requires FAD, thiamine pyrophosphate and Mg^{2+} for complete activation, and the reaction proceeds in a biphasic manner. First, a pyruvate molecule binds to thiamine pyrophosphate at the active site and is decarboxylated to yield an enzyme–substrate complex plus CO_2. A second pyruvate then reacts with this complex and acetolactate is released. A number of ALS herbicides bind slowly, but tightly, to the enzyme–substrate complex to prevent the addition of the second pyruvate molecule (LaRossa and Schloss, 1984). However, imidazolinones are uncompeti-tive inhibitors with respect to pyruvate (Shaner *et al.*, 1984), and in the case of sulphonylu-reas and triazolopyrimides both competitive and non-competitive inhibition is exhibited (mixed-type inhibition) (Subramanian and Gerwick, 1989; Durner *et al.*, 1991). This indi-cates that the herbicide-binding site for ALS is distinct from the enzyme's active site. Other studies have shown that the herbicide does not bind at the allosteric site either.

The absolute requirement for FAD where no oxidation or reduction reactions are involved has puzzled many workers. Schloss *et al.* (1988) demonstrated that ALS shows considerable sequence homology with pyruvate oxidase, suggesting that both enzymes may have evolu-tionary similarities. Indeed, ALS does demonstrate an oxygen-consuming side-reaction when its activity is inhibited (Durner *et al.*, 1994). Durner *et al.* discovered that pyruvate oxidase binds both FAD and a quinone for redox reactions, and that the binding of pyruvate oxidase from *E. coli* with ubiquinone-40 is tightest in the presence of pyruvate. Since the binding of ALS with herbicides is also tightest in the presence of pyruvate, these authors have concluded that the herbicide binding site is derived from an evolutionary vestige of a quinone-binding cofactor site that is no longer functional in ALS. Hence, the herbicide site is extraneous to, or outside of, the ALS active site. McCourt *et al.* (2006) have reported that SUs and IMs block a channel in ALS through which substrates access the active site of the enzyme. These studies further reported that SUs approach to within 5 Å of the catalytic centre, whereas IMs bind at least 7 Å from it. This confirms earlier observations of different binding sites for these two classes of herbicide. It seems likely that TPs and PCs will bind to ALS in a similar manner to SUs. Ten amino acids have been identified that are involved in the binding of both SUs and IMs, six that are only involved in SU binding and two that are only involved in IM binding. These observations may lead to the development of novel ALS inhibitors that are unaffected by mutations in ALS genes that currently result in target site-mediated herbicide resistance. Zhou *et al.* (2007) have recently reviewed current understanding of the mechanisms of action of ALS-inhibiting herbicides.

9.5.2 Inhibition of threonine dehydratase

Threonine dehydratase (TD), EC 4.2.1.16, is the first enzyme in isoleucine biosynthesis. It deaminates and dehydrates L-threonine to produce 2-ketobutyrate and ammonia (Figure 9.10). Its activity is feedback-inhibited by isoleucine. Szamosi *et al.* (1994) noted that 2-cyclohexene-D-glycine (Figure 9.11) was an inhibitor of bacterial TD and demonstrated that it also inhibited plant TD and the growth of *Arabidopsis*. The inhibition was competitive and could be reversed by the addition of isoleucine. The authors concluded that TD could become a potential new target for herbicide screening, but a review of the literature does not record that a TD inhibitor has ever reached the herbicide market.

9.5.3 Inhibition of dihydroxy-acid dehydratase

Since the ALS inhibitors have had such spectacular commercial success as herbicides, it is not surprising that acetohydroxyacid reductoisomerase (KARI) and dihydroxyacid dehydratase (DHAD) have also been investigated as potential target sites for the development of new herbicides. Several inhibitors of KARI are known, but none have been commercialised as herbicides. On the other hand, DHAD (Figure 9.10) shows promise. Shimada *et al.* (2002) used *Arabidopsis thaliana* to screen for regulators of reproductive development, using fungal metabolites. Their bioassay identified aspterric acid (Figure 9.12), isolated from *Aspergillus terreus* as a potent inhibitor of pollen development at micromolar concentrations. More recently, Yan *et al.* (2018), demonstrated that aspterric acid also targets DHAD. While aspterric acid is only weakly phytotoxic, its herbicidal properties might be enhanced by chemical modification and the demonstration of activity in the field.

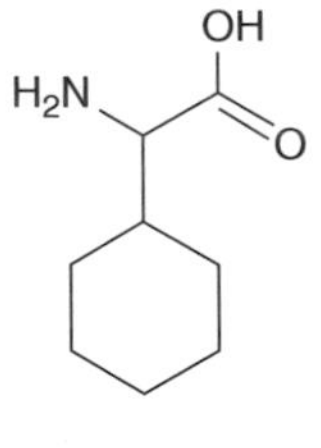

Figure 9.11 Structure of 2-cyclohexene-D-glycine, an inhibitor of threonine dehydratase. Source: based on Szamosi, I.T., Shaner, D.L. and Singh, B.K. (1994). Inhibition of threonine dehydratase is herbicidal. Plant Physiology **106**, 1257–1260.

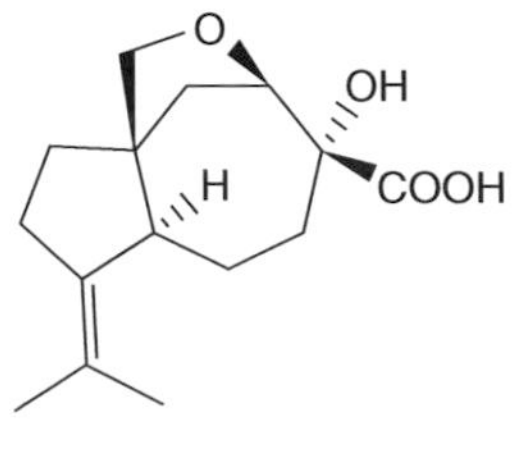

Figure 9.12 Structure of aspterric acid, a carotane-type sesquiterpene, an inhibitor of dihydroxy-acid dehydratase.

9.5.4   *How treated plants die*

It is not clearly understood how treated plants die following ALS inhibition. The diminution of the branched-chain amino acid pool will contribute to a cessation of protein synthesis, but although nanomolar concentrations of herbicide can inhibit ALS within minutes *in vitro*, it may take up to 2 months for the death of intact weeds in the field. Addition of the three branched-chain amino acids to culture media alleviates symptoms of ALS-inhibiting herbicides (Ray, 1984; Shaner and Reider, 1986; Usui *et al.*, 1991; Shimizu *et al.*, 1994; Yamashita *et al.*, 1994). Several physiological and metabolic alterations have been proposed to contribute to weed death. For example, LaRossa *et al.* (cited by Ray, 1989) have found that ALS inhibition results in an accumulation of its substrate, 2-ketobutyrate, which is toxic to *S. typhimurium*. 2-Aminobutyrate has been shown to accumulate in plants treated with ALS inhibitors, but symptoms of ALS inhibition do not appear to be due to 2-aminobutyrate toxicity. Treatment with ALS inhibitors results in a drop in the concentration of valine and isoleucine and a rise in the concentrations of threonine, alanine and norvaline as well as 2-aminobutyrate. The accumulation of abnormal amino acids (e.g. norvaline) owing to increased 2-ketobutyrate concentrations cannot, therefore, be discounted as a possible cause of herbicide-related phytotoxicity.

Accumulation of singlet oxygen has also been observed after treatment with ALS inhibitors, possibly as a result of an oxygen-consuming side-reaction exhibited by ALS when its main enzymic process is inhibited (Durner *et al.*, 1994). However, whole plant symptoms differ markedly from those with other herbicides, the mode of action of which is the production of active oxygen species, suggesting this is not the primary cause of plant death with ALS inhibitors.

One common observation following treatment with ALS inhibitors is a very rapid and potent inhibition of cell division, with the result that an inhibition of elongation of young roots and leaves is evident within 3 h after application. Rost (1984) found that chlorsulfuron blocked the progression of the cell cycle in dividing root cells from peas within 24 h, from G_2 to mitosis (M) and reduced movement from G_1 to DNA synthesis (S). This is a rapid effect on the cell cycle with no direct effect on the mitotic apparatus, which can be overcome by inclusion of isoleucine and valine in the incubation medium. Separate reports that ALS inhibitors may indirectly inhibit DNA synthesis may explain these observations. Furthermore, the involvement of polyamines has been implicated in sulphonylurea action, since the discovery by Giardini and Carosi (1990) that chlorsulfuron causes a reduction in spermidine concentration in *Zea* root tips, which could be responsible for this effect in the cell cycle. These findings have created a new interest in cell-cycle research since they imply a possible regulatory role of branched-chain amino acids in the control of plant cell division.

Growth is therefore retarded or inhibited within hours of foliar treatment, but physical symptoms may take days to develop, first appearing as chlorosis and necrosis in young meristematic regions of both shoots and roots. Young leaves appear wilted, and these effects spread to the rest of the plant. Leaf veins typically develop increased anthocyanin formation (reddening), and leaf abscission is commonly observed, both symptoms being typical responses to stress ethylene production. Under optimum growth conditions plant death may follow within 10 days, although up to 2 months may elapse when weed growth is slow. When ALS inhibitors are applied before the crop is planted or has emerged, susceptible weeds will germinate and grow, presumably utilising stored seed reserves. However, further growth of broadleaf weeds stops at the cotyledon stage, and before the two-leaf stage in grasses.

Note, however, the label recommendations regarding sulphonylurea persistence in subsequent crops, especially in alkaline soils. It is suggested that that a minimum gap of 3 months should be followed between spraying a sulphonylurea herbicide on, for example, spring cereals, to an application on oil-seed rape seedlings in the late summer/early autumn, to avoid herbicide damage to subsequent crops.

9.5.5 Selectivity

A striking feature of the ALS inhibitors is their highly selective action at low dosage. Indeed, some cereals have been reported to tolerate up to 4000 times more chlorsulfuron than some susceptible broadleaf species. Various studies have shown that this extreme species sensitivity is not due to herbicide uptake, movement or sensitivity to ALS, but is correlated to very rapid rates of metabolism in the tolerant crop. In the tolerant crop soybean, the degradation half-life of triazolopyrimidines has been shown to be 49 h compared with 165.3 h in the susceptible species pitted morning glory (Swisher *et al.*, 1991). Several sites on the sulphonylurea molecule are locations for enzyme attack and more than one enzyme system has been demonstrated to be active in their detoxification, as summarised in Figure 9.13 (Beyer *et al.*, 1987).

This extreme species sensitivity may have dramatic consequences to a following crop in the same land. For example, since wheat is more than 1000 times more tolerant to chlorsulfuron than the extremely sensitive sugar beet, which may be inhibited by as little as 0.1 part herbicide per billion parts of soil, then low sulphonylurea residues are crucial to the success of this rotational crop. Chlorsulfuron detoxification in the soil is principally due to microbial action which may degrade the herbicide within 1–2 months under optimum conditions.

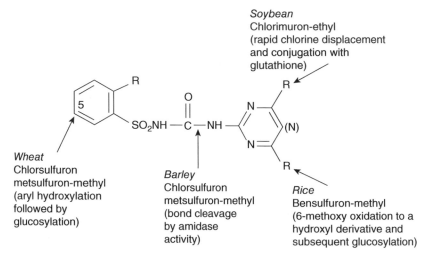

Figure 9.13 Sulphonylurea metabolism in various crops. Source: Beyer, E.M., Duffy, M.J., Hay, J.V. and Schlueter, D.D. (1987) Sulfonylureas. In: Kearney, P.C. and Kaufman, D.D. (eds) *Herbicides: Chemistry, Degradation and Mode of Action*, vol. 3, New York: Marcel Dekker, pp. 117–189.

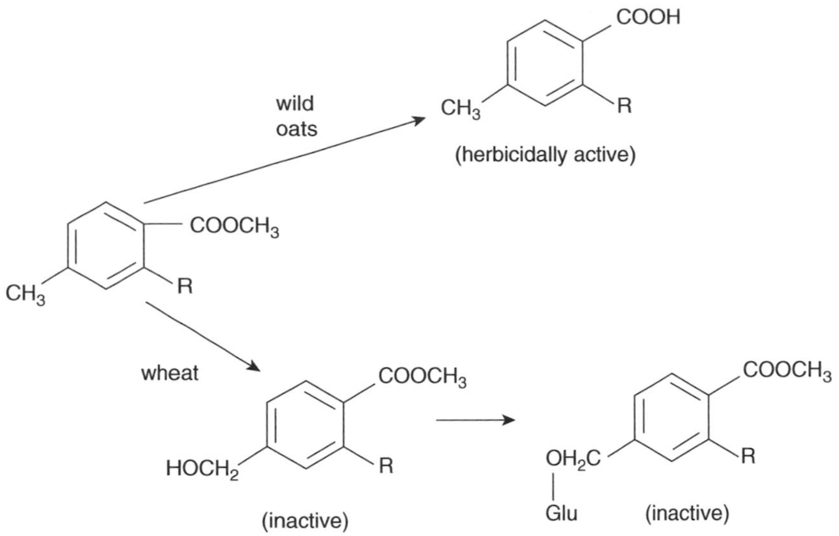

Figure 9.14   Differences in the metabolism of imidazolinones between wheat (tolerant) and wild oats (susceptible).

However, when conditions of low temperature, high rainfall and high soil pH arise, microbial breakdown is greatly reduced and, if prolonged, sufficient residues of chlorsulfuron may remain to cause major damage in succeeding crops for up to a year after the original treatment. The most sensitive crops to chlorsulfuron residues appear to be lentils, sugar beet and onions, but flax, maize, sunflower, mustard, oilseed rape, potatoes and lucerne are also sensitive (Beyer *et al.*, 1987; Blair and Martin 1988). Similar sensitivity to imidazolinone and triazolopyrimidine residues have not yet been reported.

Rapid differential metabolism also provides an explanation for the selectivity of the imidazolinones. Thus, de-esterification of imazamethabenz to phytotoxic acids is observed in susceptible species such as wild oat, and ring methyl hydroxylation followed by glucosylation inactivates the molecule in tolerant maize and wheat (Figure 9.14).

A further aspect of imidazolinone action is noteworthy and remains to be explained. Namely, about 1000 times more imidazolinone is needed to inhibit ALS *in vitro* than an equivalent sulphonylurea, but only five times more herbicide is needed to kill the weed.

Since the sulphonylureas and imidazolinones are such potent inhibitors of ALS and may retain residual activity in soils under certain conditions, it may be predicted that resistant plants will eventually evolve. However, resistance has developed in a surprisingly short time and at present resistance cases to ALS inhibitors are more prevalent than cases of resistance to other herbicides that have been used for many more years (Heap, 2007; Chapter 13). The use of chlorsulfuron and metsulfuron-methyl in minimum tillage winter wheat monoculture for four to five consecutive years has resulted in resistant biotypes of prickly lettuce (*Lactuca serriola*), Kochia (*Kochia scoparia*), Russian thistle (*Salsola iberica*) and chickweed (*Stellaria media*). Although the degree of resistance varies with both biotype and herbicide, it is clear that cross-resistance to the imidazolinones is evident, and a modified, less sensitive ALS is thought to be present in resistant biotypes (Reed *et al.*,

1989). Furthermore, resistant biotypes of *L. serriola* have been crossed with domestic lettuce (*L. sativa*) by Mallory-Smith *et al.* (1990), with the result that inherited resistance is controlled by a single gene with incomplete dominance.

9.6   Inhibition of histidine biosynthesis

The biosynthesis of the amino acid histidine has received extensive study in micro-organisms and is a highly conserved process among species. It is only in recent years that our understanding of histidine biosynthesis in higher plants has begun to increase (Stepansky and Leustek, 2006), largely owing to genetic studies of *A. thaliana* mutants (Muralla *et al.*, 2007).

Histidine biosynthesis in plants is summarised in Figure 9.15. Phosphoribosyl-pyrophosphate reacts with ATP to form phosphoribosyl AMP. This is converted to phospori-bulosyl formimino-5-aminoimidazole-4-carboxamide ribonucleotide, which in turn is converted to imidazole glycerol phosphate (IGP). Histidine is then synthesised from IGP via imidazoleacetol phosphate, histidinol phosphate and histidinol. The enzyme IGP dehydratase (E.C. 4.2.1.19) had long been proposed to be the site of action of the non-selective herbicide 3-amino-1,2,4-triazole (aminotriazole; amitrol; Figure 9.16). However, observations made by Sandmann and Böger (1992) (among others) that amitrol inhibits ξ-carotene desaturase and lycopene cyclase, in addition to 'bleaching' symptoms in treated plants, have now resulted in aminotriazole being classified as a pigment biosynthesis inhibitor. May and Leaver (1993) have also reported that aminotriazole inhibits both catalase and aconitase. Inhibition of the former results in large increases in reduced glutathione pool size. Aminotriazole is a relatively

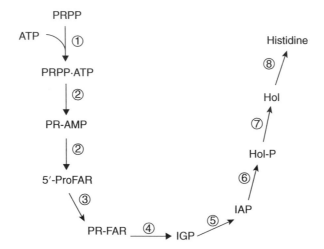

Figure 9.15 Proposed histidine biosynthesis pathway in plants. Source: modified from Muralla, R., Sweeney, C., Stepansky, A., Leustek, T. and Meinke, D. (2007) Genetic dissection of histidine biosynthesis in *Arabidopsis*. *Plant Physiology* **144**, 890–903. For abbreviations see text. Enzymes: 1, phosphorybosyl transferase;2,cyclohydrolase/pyrophosphohydrolase;3,*N*-[(5-phoshoribosyl)-formimino]-5-aminoimadazole-4-carboxamide-ribonucleotide isomerase (BBMI); 4, phosporibulosyl formimino-5-aminoimidazole-4-carboxamide ribonucleotide aminotransferase/cyclase; 5, imidazole glycerol phosphate (IGP) dehydratase; 6, imidazoleacetol phosphate aminotransferase; 7, histidinol phosphatase; 8, histidinol dehydrogenase.

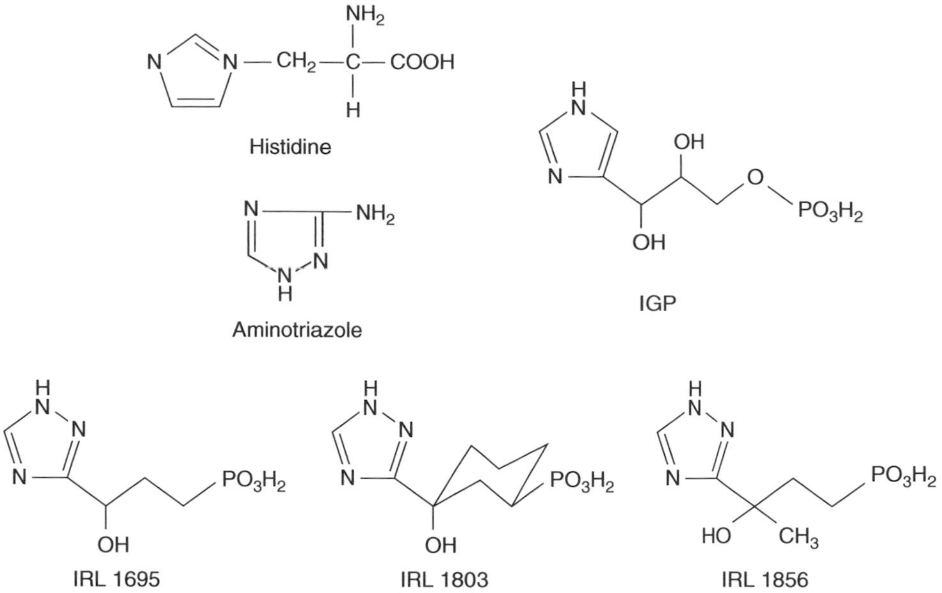

Figure 9.16 The structures of histidine, aminotriazole, IGP and three triazole phosphates reported by Mori *et al*. (1995) to be inhibitors of IGP dehydratase. Source: Mori, I., Fonné-Pfister, R., Matsunaga, S-I., *et al*. (1995) A new class of herbicides: specific inhibitors of imidazoleglycerol phosphate dehydratase. *Plant Physiology* **107**, 719–723.

simple molecule in comparison with many other herbicides and it is perhaps not surprising that it appears to affect a number of metabolic processes in higher plants.

Imidazole glycerol phosphate dehydratase (IGPD) has been targeted by Syngenta for many years, since this enzyme is unique to plants. Mori *et al*. (1995) generated a family of triazole phosphonate inhibitors which are structurally similar to the intermediate in IGPD catalysis (Figure 9.16). The triazole phosphonates are potent inhibitors of IGPD and are phloem-mobile herbicides with activities similar to glyphosate. Structurally, they consist of a triazole head, a hydroxylated link and a phosphate tail. The position of the hydroxyl group affects their binding to the active site by forming chelates with a manganese atom. Glynn *et al*. (2005) have since determined the structure of the enzyme at 3 Å resolution. They suggest that the substrate is bound to a manganese cluster with an imidazolate moiety that subsequently collapses to yield a diazafulvene intermediate. The commercialisation of a triazole phosphonate herbicide as an IGPD inhibitor remains to be demonstrated.

References

Amann, A., Feucht, D. and Wellmann, A. (2000) A new herbicide for grass control in winter wheat, rye and triticale. *Zeitscrift für Pflanzenkrankheiten und Pflanzenschutz* **17**, 545–553.

Babczinski, P. and Zelinski, T. (1991) Mode of action of herbicidal ALS-inhibitors on acetolactate synthase from green plant cell cultures, yeast and *Escherichia coli*. *Pesticide Science* **31**, 305–323.

Beyer, E.M., Duffy, M.J., Hay, J.V. and Schlueter, D.D. (1987) Sulfonylureas. In: Kearney, P.C. and Kaufman, D.D. (eds) *Herbicides: Chemistry, Degradation and Mode of Action*, vol. 3, New York: Marcel Dekker, pp. 117–189.

Blair, A.M. and Martin, T.D. (1988) A review of the activity, fate and mode of action of sulfonylurea herbicides. *Pesticide Science* **22**, 195–219.

Brilisaur, K., Rapp, J., Rath, P., Schollhorn, A., Bleul, L., Weis, E., *et al.* (2019) Cyanobacterial antimetabolite 7-deoxy-sedoheptulose blocks the shikimate pathway to inhibit the growth of phototrophic organisms. *Nature Communications* **10**, 545; doi: org/10.1038/s41467-019-08476-8

Brito, I., Tropaldi, L., Carbonari, C.O. and Velini, E.D. (2018) Hormetic effects of glyphosate on plants. *Pest Management Science* **74**, 1064–1070; doi: org/10.1002/ps.4523

CaJacob, C.A., Feng, P.C.C., Heck, G.R., Alibhai, M.F., Sammons, R.D. and Padgette, S.R. (2004) Engineering resistance to herbicides. In: Christou, P. and Klee, H. (eds) *Handbook of Plant Biotechnology Vol. 1*. Chichester: John Wiley and Sons, pp. 353–372.

Comai, L., Faciotti, D., Hiatt, W.R., Thompson, G., Rose, R.E., and Stalker, D.M. (1985) Expression in a plant of a mutant *aroA* gene from *Salmonella typhimurium* confers tolerance to glyphosate. *Nature* **317**, 741–744.

Denis, M.H. and Delrot, S. (1993) Carrier-mediated uptake of glyphosate in broad bean (*Vicia faba*) via a phosphate transporter. *Physiologia Plantarum* **87**, 569–575.

Duggleby, R.G. and Pang, S.S. (2000) Acetohydroxyacid synthase. *Journal of Biochemistry and Molecular Biology* **33**, 1–36.

Durner, J., Gailus, V. and Böger, P. (1991) New aspects of inhibition of plant acetolactate synthase by chlorsulfuron and imazaquin. *Plant Physiology* **95**, 1144–1149.

Durner, J., Gailus, V. and Boger, P. (1994) The oxygenase reaction of acetolactate synthase detected by chemiluminescence. *FEBS Letters* **354**(1), 71–73.

Franz, J.E., Mao, M.K. and Sikorski, J.A. (1997) *Glyphosate: a Unique Global Herbicide*. ACS Monograph No. 189. Washington, DC: American Chemical Society.

Galili, G., Amir, R. and Fernie, A.R. (2016) The regulation of essential amino acid synthesis and accumulation in plants. *Annual Review of Plant Biology* **67**, 153–178; doi: org/10.1146. annurev-arplant-043015-112213

Giardini, M.C. and Carosi, S. (1990) Effects of chlorsulfuron on polyamine content in maize seedlings. *Pesticide Biochemistry and Physiology* **36**, 229–236.

Glynn, S.E., Baker, P.J., Sedelnikova, S.E., Davies, C.L., Eadsforth, T.C., Levy, C.W., *et al.* (2005) Structure and mechanism of imidazoleglycerol-phosphate dehydrogenase. *Structure* **13**, 1809–1817.

Heap, I. (2007) *The International Survey of Herbicide Resistant Weeds*. Available at www. weedscience.com (accessed 13 November 2007).

Hershey, H.P., Schwartz, L.J., Gale, J.P. and Abel, L.M. (1999) Cloning and functional expression of the small subunit of acetylacetate synthase from *Nicotiana plumbaginifolia*. *Plant Molecular Biology* **40**, 795–806.

Hildebrandt, T.M., Nesi, A.N., Araujo, W.L. and Braun, H.-P. (2015) Amino acid catabolism in plants. *Molecular Plant* **8**, 1563–1579; doi: org/10.1016/j.molp.2015.09.005

Jalal, A., de Oliveira, J.C., Ribeiro, J.S., Fernandes, G.C., Mariano, G.G., Trindade, V.D.R. and dos Reis, A.R. (2021) Hormesis in plants: Physiological and biochemical responses. *Ecotoxicology and Environmental Safety* **207**, 111225; doi: org/10.1016/j.ecoenv.2020.111225

Jaworski, E.G. (1972) The mode of action of *N*-phosphonomethylglycine: inhibition of aromatic amino acid biosynthesis. *Journal of Agriculture and Food Chemistry* **20**, 1195–1198.

LaRossa, R.A. and Schloss, J.V. (1984) The sulphonylurea herbicide sulfometuronmethyl is an extremely potent and selective inhibitor of acetolactate synthase in *Salmonella typhimurium*. *Journal of Biological Chemistry* **259**, 8753–8757.

Leason, M., Cunlife, D., Parkin, D., Lea, P.J. and Miflin, B.J. (1982) Inhibition of pea leaf glutamate synthase by methionine sulfoximine, phosphinothricin and other glutamate analogues. *Phytochemistry* **21**, 855–857.

Los, M. (1986) Synthesis and biology of the imidazolinone herbicides. In: Greenhalgh, P. and Roberts, T.R. (eds) *Pesticide Science and Biotechnology. Proceedings of the Sixth International Congress of Pesticide Chemistry*, Ottawa, Canada. Oxford: Blackwell, pp. 35–42.

Maeda, H. and Dudareva, N. (2012) The shikimate pathway and aromatic amino acid biosynthesis in plants. *Annual Review of Plant Biology* **63**, 73–105; doi: org/10.1146/annurev-arplant-042811-105439

Mallory-Smith, C., Thill, D.C., Dial, M.J. and Zemetra, R.S. (1990) Inheritance of sulphonylurea herbicide resistance in prickly lettuce (*Lactuca serriola*) and domestic lettuce (*Lactuca sativa*) In: Caseley, J.C., Cussans, G. and Atkin, R. (eds) *Herbicide Resistance in Weeds and Crops*. Guildford: Butterworth, pp. 452–453.

Manderscheid, R. and Wild, A. (1986) Studies on the mechanism of inhibition by phosphinothricin of glutamine synthetase isolated from *Triticum aestivum* L. *Journal of Plant Physiology* **123**, 135–142.

May, M.J. and Leaver, C.J. (1993) Oxidative stimulation of glutathione synthesis in *Arabidopsis thaliana* suspension cultures. *Plant Physiology* **103**, 621–627.

McCourt, J.A., Pang, S.S., King-Scot, J., Guddat, L.W. and Duggleby, R.G. (2006) Herbicide-binding sites revealed in the structure of plant acetohydroxyacid synthase. *Proceedings of the National Academy of Sciences, USA* **103**, 569–573.

Mori, I., Fonné-Pfister, R., Matsunaga, S-I., Tada, S., Kimura, Y., Iwaski, G. *et al.* (1995) A new class of herbicides: specific inhibitors of imidazoleglycerol phosphate dehydratase. *Plant Physiology* **107**, 719–723.

Müller, K.H., Koning, K., Kluth, J., Lürssen, K., Santel, H.J. and Schmidt, R.R. (1992) Sulfonylaminocarbonyltriazolinones with oxygen-bound substitutes. EP507171, Bayer AG.

Muralla, R., Sweeney, C., Stepansky, A., Leustek, T. and Meinke, D. (2007) Genetic dissection of histidine biosynthesis in *Arabidopsis*. *Plant Physiology* **144**, 890–903.

Omura, S., Murata, M., Hanaki, H., Hinotozawa, K., Oiwa, R. and Fanaka, H. (1984) Phosalacine, a new herbicidal antibiotic containing phosphinothricin. Fermentation, isolation, biological activity and mechanism of action. *Journal of Antibiotics* **37**, 829–835.

Ray, T.B. (1984) Site of action of chlorsulfuron: inhibition of valine and isoleucine synthesis in plants. *Plant Physiology* **75**, 827–831.

Ray, T.B. (1989) Herbicides as inhibitors of amino acid biosynthesis. In: Böger, P. and Sandmann, G. (eds) *Target Sites of Herbicide Action*. Boca Raton, FL: CRC Press, pp. 105–125.

Reed, W.T., Saladini, J.L., Cotterman, J.C., Primiani, M.M. and Saari, L.L. (1989) Resistance in weeds to sulphonylurea herbicides. *Brighton Crop Protection Conference, Weeds* **1**, 295–300.

Rost, T. (1984) The comparative cell cycle and metabolic effects of chemical treatments on root tip meristems. III. Chlorsulfuron. *Journal of Plant Growth Regulation* **3**, 51–63.

Sandmann, G. and Böger, P. (1992) Chemical structures and activity of herbicidal inhibitors of phytoene desaturase, In: Draber, W. and Fujita, T. (eds) *Rational Approaches to Structure, Activity and Ecotoxicity of Agrochemicals*. Boca Raton, FL: CRC Press, pp. 357–371.

Schloss, J.V., Ciskanik, L.M. and Van Dyk, D.E. (1988) Origin of the herbicide binding site of acetolactate synthase. *Nature* **331**, 360–362.

Schöenbrunn, E., Eschenburg, S., Schuttleworth, W.A., Schloss, J.V., Amrhein, N., Evans, J.N. and Kabsch, W. (2001) Interaction of the herbicide glyphosate with its target enzyme 5-enolpyruvylshikimate-3-phosphate synthase in atomic detail. *Proceedings of the National Academy of Sciences, USA* **98**, 1376–1380.

Schuktz, A., Munder, T., Holländer-Czytko, H. and Amrhein, N. (1990) Glyphosate transport and early effects on shikimate metabolism and its compartmentation in sink leaves of tomato and spinach plants. *Zeitschrift für Naturforschung* **45**, 529–534.

Shaner, D.L. and Reider, M.L. (1986) Physiological responses of corn (*Zea mays*) to AC 243,997 in combination with valine, leucine and isoleucine. *Pesticide Biochemistry and Physiology* **25**, 248–257.

Shaner, D.L., Anderson, P.C. and Stidham, M.A. (1984) Imidazolinones: potent inhibitors of acetohydroxyacid synthase. *Plant Physiology* **76**, 545–546.

Shimada, A., Kusano, M., Takeuchi,S., Fujioka, S., Inokuchi, T. and Kimura, Y. (2002) Aspterric acid and 6-hydroxymellein, inhibitors of pollen development in *Arabidopsis thaliana*, produced by *Aspergillus terreus*. *Zeitschrift für Naturforschung* **57c**, 459–464.

Shimizu, T., Kaku, K., Takahashi, S. and Nagayama, K. (1994) Sensitivities of ALS prepared from SU- and IMI-resistant weeds against PC herbicides. *Journal of Weed Science and Technology* **46**, S32–S33.

Shimizu, T., Nakayama, I., Nagayama, K., Miyazawa, T. and Nezu, Y. (2002) Acetolactate synthase inhibitors, In: Boger, P., Wakabayashi, K. and Hirai, K. (eds) *Herbicide Classes in Development*. Berlin: Springer, pp. 1–41.

Silva, J.R.O., Marquez, J.N.R., Godoy, C.V.C., Batista, L.B. and Ronchi, C.P. (2019) 2,4-D hormesis effect on soybean. *Planta Daninha* **37**, e019216022; doi: org/10.1590/S0100-83582019370100146

Steinrucken, H.C. and Amrhein, P. (1980) The herbicide glyphosate is a potent inhibitor of 5-enolpyruvylshikimic acid-3-phosphate synthase. *Biochemical and Biophysical Research Communications* **94**, 1207–1212.

Stepansky, A. and Leustek, T. (2006) Histidine biosynthesis in plants. *Amino Acids* **30**, 127–142.

Stilling, W.C., Abdel-Meguid, S.S., Lim, L.W., Shien, H.-S., Dayringer, H.E., Leimgruber, N.K., *et al*. (1991) Structure and topological symmetry of the glyphosate target 5-enolpyruvylshikimate-3-phosphate synthase – a distinctive protein fold. *Proceedings of the National Academy of Sciences, USA* **88**, 5046–5050.

Subramanian, M.V. and Gerwick, B.C. (1989) Inhibition of acetolactate synthase by triazolopyrimidines: a review of recent developments. In: Whitaker, J.R. and Sonnet, B.E. (eds) *Biocatalysis in Agricultural Biotechnology*. ACS Symposium Series No. 389. Washington, DC: American Chemical Society, pp. 277–288.

Swisher, B.A., de Boar, G.L., Ouse, D., Geselius, C., Jachetta, J.J. and Miner, V.W. (1991) Metabolism of selected triazolo (1,5-c) pyrimidine sulfonamides in plants. Internal report of DowAgro Sci. (Cited in Dorich and Schultz, 1997. *Down to Earth* **52**, 1–10.)

Szamosi, I.T., Shaner, D.L. and Singh, B.K. (1994) Inhibition of threonine dehydratase is herbicidal. *Plant Physiology* **106**, 1257–1260.

Takano, H.K. and Dayan, F.E. (2020) Glufosinate-ammonium: a review of the current state of knowledge. *Pest Management Science* **76**, 3911–3925; doi: 10.1002/ps.5965

Takano, H.K., Beffa, R., Preston, C., Westra, P. and Dayan, F.E. (2019) Reactive oxygen species trigger the fast action of glufosinate. *Planta* **249**, 1837–1849.

Takano, H.K. and four others (2020) A novel insight into the mode of action of glufosinate:how reactive oxygen species are formed. *Photosynthesis Research* **144**, 361–372; doi: org/10.1007/s11120-020-00749-4

Ullrich, W.R., Ullrich-Eberius, C.I. and Köcher, H. (1990) Uptake of glufosinate and concomitant membrane potential changes in *Lemna gibba* G1. *Pesticide Biochemistry and Physiology* **37**, 1–11.

Usui, K., Suwangwong, S., Watanabe, H. and Ishizuka, K. (1991) Effect of bensulfuron methyl, glyphosate and glufosinate on amino acid and ammonia levels in carrot cells. *Weed Research Japan* **36**, 126–134.

Velini, E.D., Alves, E., Godoy, M.C., Meschede, D.K., Souza, R.Y. and Duke, S.O. (2008) Glyphosate applied at low doses can stimulate plant growth. *Pest Management Science* **64**, 489–496; doi: org/10.1002/ps.1562

Velini, E.D. and three others (2010) Growth regulation and other secondary effects of herbicides. *Weed Science* **58**, 351–354; doi: org/10.1614/WS-D-09-00028.1

Wild, A. and Ziegler, C. (1989) The effect of bialaphos on ammonia assimilation and photosynthesis. I. Effect on the enzymes of ammonium assimilation. *Zeitschrift für Naturforschung* **44c**, 97–102.

Wild, A., Sauer, H. and Ruehle, W. (1987) The effect of phosphinothricin on photosynthesis. I. Inhibition of photosynthesis and accumulation of ammonia. *Zeitschrift für Naturforschung* **42c**, 263–269.

Xu, J.-J., Fang, X., Li, C.-Y., Yang, L. and Chen, X.-Y. (2019) General and specialized tyrosine metabolism pathways in plants. aBIOTECH; doi: org/10.1007/s42994-019-00006-w

Yamashita, K., Nagayama, K., Wada, N. and Abe, H. (1994) A novel ALS inhibitor produced by *Streptomyces hygroscopicus*. *Nippon Nogeikagaka Kaishi* **65**, 658.

Yan, Y., Liu, Q., Zang, X., Yuan, S., Bat-Erdene, U., Nguyen, C. *et al.* (2018) Resistant-gene-directed discovery of a natural-product herbicide with a new mode of action. *Nature* **559**, 415–418.

Zhou, Q., Liu, W., Zhang, Y. and Liu, K.K. (2007) Action mechanisms of acetolactate synthase-inhibiting herbicides. *Pesticide Biochemistry and Physiology* **89**, 89–96.

Chapter 10
The Disruption of Cell Division

It is not so much that the cells make the plant; it is rather that the plant makes the cells.

Heinrich Anton deBary (quoted in the *New Statesman*, 17 April 1920)

10.1   Introduction

According to Vaughn and Lehnen (1991), approximately 25% of all herbicides that have been marketed affect plant cell division, or mitosis, as a primary mechanism of action. For example, the dinitroanilines have been extensively used for over 50 years to selectively inhibit the growth of many annual grasses and some dicot weeds, especially in cotton, soybeans and wheat. They are applied prior to weed emergence, and herbicide absorption occurs through the roots or shoot tissues growing through the treated soil. In this way, they can provide good control of target weeds, allowing the crop seedlings to establish.

Many cell division inhibitors now have a restricted use or have been replaced by more effective, modern, low-dose herbicides. Furthermore, their continuous use has also led to reports of resistant weeds. They will, however, remain an important chemical weapon for weed control in several crops and have in recent years become resurgent in the control strategies for weeds resistant to the photosynthetic inhibitor herbicides, especially in the control of the important grass weeds *Setaria viridis* and *Eleusine indica*, two of the world's worst weeds.

10.2   The cell cycle

A major difference between plants and animals is that plants contain clearly defined regions, termed meristems, which remain embryonic throughout the life of the plant and are responsible for continued growth. Root and shoot meristems generate axial growth: the cambium is an example of a lateral meristem; the pericycle is a potential meristem for the generation of lateral roots; and grasses possess an additional intercalary meristem at the base of nodes. Cells in a meristematic region undergo a precise sequence of events, known as the cell cycle, which is divided into four discrete periods, each lasting for a time specific to a particular

Herbicides and Plant Physiology, Third Edition. Andrew H. Cobb.
© 2022 John Wiley & Sons Ltd. Published 2022 by John Wiley & Sons Ltd.

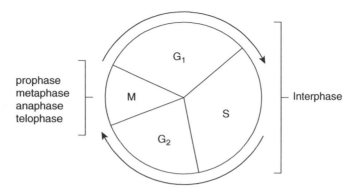

Figure 10.1 The mitotic cell cycle.

species. In G_1 (first gap phase) each nuclear chromosome is a single chromatid containing one DNA molecule and S phase (DNA synthesis) is characterised by a doubling of the DNA content of the nucleus, so that in G_2 (second gap phase) each chromosome now consists of two identical chromatids with identical DNA molecules. The gap phases allow the operation of controls that ensure that the previous phase has been accurately completed. The main regulatory steps operate at the G_1/S and G_2/M boundaries, and both are sensitive to environmental conditions. G_1, S and G_2 are collectively termed interphase, to be followed by cell division (mitosis, or M), which is itself subdivided into prophase, metaphase, anaphase and telophase (Figure 10.1). During prophase the nuclear envelope disintegrates, and pairs of distinctly shorter and fatter chromosomes are observed. In metaphase the chromosome pairs move to the centre of the cell along a structure known as the spindle which consists of proteinaceous microtubules extending between the poles of the cell. At anaphase the chromosome pairs separate and migrate to opposite ends of the cell, so that two identical sets of chromosomes are evident at each pole of the spindle. Mitosis is completed in telophase by the formation of a nuclear envelope around each chromosome set, and the creation of a new cell wall by the fusion of the Golgi apparatus at the equatorial region of the cell. A completed cell cycle may take between 12 and 18 h, and the full metabolic machinery of the cell is required to ensure progression from one stage to the next. It therefore follows that any chemical interference with cell cycle progression will cause growth inhibition and have herbicidal potential. Indeed, herbicides are known that either prevent mitotic entry or disrupt the mitotic sequence.

10.3   Control of the cell cycle

Substantial progress has been made in recent years in our understanding of the molecular mechanisms and control processes of cell division in plants. Ensuring that each daughter cell receives the full complement of DNA requires close regulation of the S and M phases. It seems that the basic control mechanisms are highly conserved in all eukaryotes, the main drivers being a class of serine/threonine kinases known as cyclin-dependent kinases (CDKs). Many regulatory features have evolved in plants that influence CDK activity, including metabolic, hormonal and environmental signals.

Cyclin-dependent kinases are a large family of serine/threonine kinases that ensure that cell division proceeds in an orderly manner. To be active, CDKs need to bind a regulatory protein, known as a cyclin, and cyclin concentration varies depending on the phase of the cell cycle. In *Arabidopsis* there are eight classes of CDKs, named CDKA–G and L. There are also 15 CDK-like proteins (CDKL1–15) whose function is currently unknown. The CDKB family is unique to plants and contains two subfamilies, B1 and B2. CDKB1s are expressed from the outset of S phase until mitosis, while CDKB2s are expressed during the G2 to M transition.

Phosphorylation of a threonine residue (Thr160) is essential for CDK activity, and is achieved by a CDK-activating kinase. Further regulation of CDK activity occurs by inhibitory phosphorylation at the amino terminal residues at Thr14 and Tyr15, catalysed by the WEE 1 kinase.

The CDKs have diverse roles in cell cycle progression and in regulating gene expression. They phosphorylate key proteins, such as histone proteins and cytoskeleton proteins, to drive the cell cycle from one phase to the next. Inhibitors of the CDKs (ICK) and Kip-related proteins (KRPs) bind to the cyclin–CDK complex to negatively control CDK activity. The ICKs are involved at cell cycle 'checkpoints'.

Plants have two families of ICKs, namely the INTER-ACTOR/INHIBITOR OF CDK/ KIP RELATED PROTEINS (ICK/KRPs) and the SIAMESE-RELATED PROTEINS (SMRs). The *Arabidopsis* genome contains seven ICK/KRP and 17 SMR genes, far more than in animals, insects and yeasts. The question has been posed as to why plants have so many of these inhibitors (Kumar and Larkin, 2017)? These authors conclude that both families coordinate cell division, cell expansion and organ growth, with developmental and environmental cues. Evidence suggests that although the role of each ICK is not yet known, SMRs only inhibit the M-phase, at the G2 checkpoint, while KRPs can block entry into both M- and S-phases at the G1–S transition.

Five types of cyclin exist in plants, of which three have clearly defined roles in the cell cycle. The A-type is observed at the beginning of S-phase and is destroyed at the G2/transition. The B-type appears during G2 and is destroyed at anaphase and the D-type controls progression from G1 to S, depending on extracellular signals. As the cell cycle can only be progressed by the synthesis and breakdown of cyclins, they are rapidly degraded via the ubiquitin–proteosome pathway. Ubiquitinylation is followed by DNA duplication and cell division. These two processes are regulated by two complexes: the Skp1/Cullin/F-box (SCF)-related complex and the anaphase promoting complex (APC/C). Thus, the SCF complex allows the cell to progress from G1 to S phase, while the APC/C permits G2 to M.

The circadian clock sets the time of the cell cycle to accurately regulate the number and size of plant cells for synchronising with the environment. The interplay of the clock and the cell cycle therefore controls plant growth. This is achieved by the regulatory component of the clock, TOC1. The overexpression of TOC1 slows down the clock, and so fewer cells are generated in smaller leaves. Conversely, the underexpression of TOC1 leads to a faster clock and so the cell cycle speeds up. TOC1 binds to the promoter of the CDC6 gene to repress its expression. CDC6 is upregulated at the G1-S transition, normally reaching peak activity early in S phase (Fung-Uceda *et al.*, 2018).

10.3.1   *Transition from G_2 to M phase*

Once the cell has duplicated its DNA in S phase, the next step is to generate the mitotic spindle, disassemble the nuclear envelope, condense the chromosomes and align each pair of sister chromatids.

The transition is controlled by CDK–CYCB kinase activity. During G_2, the amount of CYCB (cyclin B) increases and becomes associated with cyclin-dependent protein kinases A (CDKA) and B (CDKB). However, both KRP proteins and inhibitory phosphorylation by WEE kinases render the complex inactive. It is thought that the specific protein phosphatase, PP2A, reverses protein phosphorylation to allow entry into M phase. Thus, cyclin-dependent kinases and phosphatases, acting independently, regulate progress through the cell cycle by the reversible phosphorylation of key proteins. In essence, progression through the cell cycle requires successive phosphorylation and de-phosphorylation events, and the removal of regulatory and structural proteins by proteolytic degradation.

Progression through mitosis involves the association of the CDK complexes with microtubule and chromatin proteins. CDK activity is therefore likely to play an important part in microtubule dynamics and stability. The kinesins are a class of microtubule-associated proteins that have a motor domain that allows movement along the microtubules. They have key roles in spindle formulation and cell plate dynamics (Vanstraelen *et al.*, 2006). At least 23 kinesins have been implicated in mitosis using the energy of ATP to control microtubule organisation, polymerisation, depolymerisation and chromosome movement. The 23 kinesins are upregulated during mitosis and their proposed involvement in spindle dynamics is reviewed by Vanstraelen *et al.* (2006).

The purpose of the cell cycle is to distribute complete and accurate copies of the genome to daughter cells. If for some reason the DNA is damaged, or the chromosomes are only partially replicated, apoptosis (programmed cell death) is initiated. In such cases, cell cycle 'check-points', which are in effect 'surveillance mechanisms', are in place that link cell cycle progression to the correct completion of the previous phase. In this event, CDK-inhibitor proteins stop cell cycle progression until the DNA is repaired, by binding to the cyclin–CDK complex and inactivating it. PP2A controls cell cycle progression by reversing phosphorylation of the cyclin-dependent kinases at the G2–M transition. How the phosphatases themselves are regulated is currently not known.

According to Atkins and Cross (2018), CDKA initiates cell division and activates the transcription of CDKB, which forms a complex with cyclin B. CDKB–cyclin B is not only required for spindle formation, but also represses the activity of the CDKA–cyclin A complex, suggesting a negative feedback loop in the regulation of the cell cycle.

In the prophase of mitosis the cyclin B–CDKB complex enters the nucleus, chromosomes condense and the spindle forms. When mitosis is completed, the APC/C targets a ubiquitin molecule, causing cyclin degradation in the proteosome and inactivating CDK activity, allowing the newly formed daughter cells to enter G1 phase. Figure 10.2 provides an overview of the events in the plant cell cycle.

10.3.2   *After mitosis*

The last step of the cell cycle is when the parent cell separates into two daughter cells with the development of a new cell wall. This complex process is termed cytokinesis and involves the Golgi complex and membrane trafficking. Firstly, a temporary protein 'scaffold' forms in the centre of the parent cell, termed the phragmoplast. This is an array of microtubules and actin filaments which guide the transport of building materials that form the new cell wall, and delivers the Golgi-derived secretory vesicles to the plane of cell division, i.e. at the centre of the dividing cell. The fusion of these vesicles gives rise to the cell

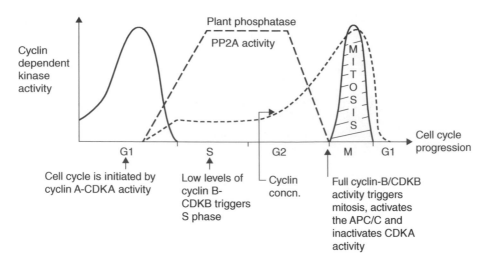

Figure 10.2   An overview of the main events in the plant cell cycle (see text for details).

plate – a membrane-bound compartment which expands outwards from the centre of the parent cell and eventually fuses with the cell membrane. The cell plate then gains celluloses and pectin, and the new cell wall separates the two daughter cells.

Lin *et al.* (2019) have demonstrated how the fusion of the vesicles to the cell plate is controlled, and ensures that the cell wall building materials are transported and assembled in the correct direction. The phosphoinositide kinases Pi4K β1 and β2 integrate these processes in *Arabidopsis*. It is proposed that these kinases act synergistically to control dynamics at the phragmoplast by interacting with a cell-plate-associated MPK4 and regulating the correct positioning of the microtubule-binding protein PLEIADE/MAP65-3.

10.3.3   Hormones and cell division

Plants are adept in their response to changes in their external environment by alterations in physiology, redirecting growth and development. Environmental and internal physiological signals are transmitted via hormones to yield appropriate morphogenetic responses. Auxins and cytokinins are the hormones most directly linked to cell division. Auxin increases both CDKA and cyclin mRNA levels, but not cell division. Cytokinins and brassinosteroids induce expression of cyclin D3, which promotes entry into the mitotic cell cycle. Cytokinins have also been linked to CDK activation at the G_2/M transition, either by direct activation of a phosphatase or by downregulation of the WEE1 kinase. Abscisic acid inhibits cell division by inducing the expression of CDK inhibitor proteins, which result in a decrease in CDKA kinase activity.

A major challenge for future study will be to establish how all the different signal transduction pathways are regulated and integrated in the control of the cell cycle and hence plant growth and development (Stals and Inzé, 2001).

Inhibition of mitotic entry is experimentally observed by an absence or reduction of mitotic chromosomal features in meristematic tissues. Nanomolar concentrations of sulphonylureas and imidazolinones block cell cycle progression at G_1 and G_2 within minutes of application, even though the tissues are capable of S and M phases. This inhibition can be overcome by the addition of branched-chain amino acids and suggests that these compounds are worthy of further study as possible regulators of cell cycle progression.

10.4   Microtubule structure and function

Microtubules appear in electron micrographs as elongated, hollow cylinders occasionally up to 200 µm long and 25 µm in diameter. Chemically, they consist of the dimeric protein tubulin which is composed of similar but distinct subunits of 55 kDa molecular weight. In cross-section they are seen to consist of 13 units composed of the tubulin subunits arranged in a helical fashion. They are found as groups in different parts of the cell and at different times of the cell cycle. They perform a range of cellular functions as four distinct microtubule arrays. Cortical microtubules control the orientation of cellulose microfibrils in the young, developing cell and so determine the final shape of the cell. Pre-prophase microtubules control tissue morphogenesis, since they determine the planes of new cell divisions. Spindle microtubules, which form the spindle that is visible at metaphase during mitosis, enable the movement of chromosomes. After cell division, phragmoplast microtubules organise the new cell plate between the daughter cells. How these microtubules are formed into distinct arrays, however, remains uncertain. In animal cells, centrioles exist at the poles of cells to control spindle organisation and these are termed microtubule organising centres. Higher plant cells do not possess a centriole. The microtubules commonly originate at the endoplasmic reticulum and the nuclear envelope, but how is unknown.

Microtubule assembly is a very dynamic process. A pool of free tubulin subunits exists in the cytoplasm, which can be reversibly polymerised to form microtubules depending on the stage of the cell cycle. The proportion of tubulin assembled into microtubules can vary from 0 to 90%. Mitchison and Kirschner (1984) envisaged the addition of tubulin subunits to the growing end of a microtubule, their loss at the opposite end and the possibility of a loss of the whole structure or parts of it. The biochemistry of tubulin polymerisation and de-polymerisation appears complex (Figure 10.3).

The process appears to be very sensitive *in vitro* to the concentration of calcium and magnesium ions, and GTP binding appears essential for polymerisation to take place (Gunning and Hardham, 1982; Dawson and Lloyd, 1987). Microtubule-associated proteins (MAPs) have also been identified in plants (Cyr and Palevitz, 1989), which are also believed to influence and interact with microtubule assembly and crosslinking.

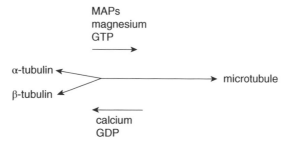

Figure 10.3 Factors affecting tubulin polymerisation and depolymerisation. MAPs, Microtubule-associated proteins.

10.5   Herbicidal interference with microtubules

It has been known for some time that members of the dinitroaniline and *N*-phenylcarbamate classes of herbicides interfere with the structure and functions of microtubules. Representative structures of these herbicides are presented in Table 10.1.

10.5.1   Dinitroanilines

Trifluralin, oryzalin and pendimethalin are widely used pre-emergence herbicides in dicot crops such as cotton and soybean for the control of grasses, and useful selectivity in wheat is evident. At concentrations as low as 1 µM, weeds show characteristic reductions in root length and swollen root tips, an effect identical to that obtained when treating seedlings with the well-known mitotic disruptor colchicine. Ultrastructural analysis reveals cells arrested at prometaphase (Vaughn and Lehnen, 1991) and so no metaphase or later stages are observed. Instead, a nuclear membrane reforms around the chromosomes and the nucleus appears highly lobed. No spindle microtubules are evident in treated cells and cortical microtubules are also affected, with the result that cells appear square-shaped rather than rectangular, explaining the swollen appearance of the root tips.

Both dinitroanilines and colchicine inhibit microtubule assembly by forming a tubule–herbicide complex that disrupts polymerisation and hence microtubule assembly. In so doing, the de-polymerisation process continues, shortening the microtubules until they are eventually undetectable. Unlike colchicine, the dinitroaniline herbicides do not inhibit tubulin polymerisation in animals, nor disrupt cell division in animal cells. At higher concentrations the dinitroaniline can also inhibit photosynthetic electron flow and oxidative phosphorylation.

These herbicides act by binding to α,β-tubulin dimers. The incorporation of dimer–herbicide complexes into a polymerising tubulin filament is thought to block further polymerisation and to cause microtubule disruption. It follows that any mutation causing changes in the amino acid content of tubulin may alter herbicide binding and could lead to herbicide resistance (Anthony *et al.*, 1998).

Resistance to trifluralin and other dinitroaniline herbicides has been documented in several weed species in important crops, and has reached a level of concern in *Lolium rigidum* in Australia. Resistance is due to a mutation in the α-tubulin gene, where Arg 243 is substituted by methionine or lysine. These mutations reduce herbicide binding and confer resistance in the field, implying that this site is also important for microtubule function.

10.5.2   N-Phenylcarbamates

The phenylcarbamates also disrupt mitosis and inhibit photosynthetic electron flow and oxidative phosphorylation at high concentrations. In this case, microscopy reveals intact microtubules, but chromosomal abnormalities. Chromosome movement during anaphase normally generates two sites at opposite poles of the cell, but in tissues treated with herbicide, three or more chromosome sites are observed after anaphase. Nuclear membranes then form around each group of chromosomes, and abnormal phragmoplasts organise irregularly shaped, abnormal cell walls. The proposed mechanism is the interference of the spindle microtubule-organising centres, fragmenting them throughout the cell, giving rise to the symptoms of multipolar cell division. How this is achieved and the sites of action are unknown. Barban, carbetamide, propham and chlorpropham all interfere with mitosis in this manner. These herbicides have been known since the 1950s and show useful activity

Table 10.1 Structures of herbicides that interfere with microtubule assembly or function.

(A) Dinitroanilines

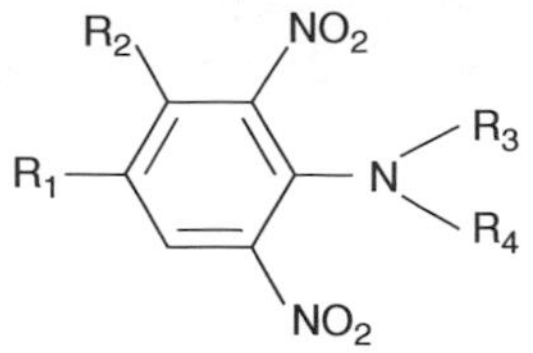

R_1	R_2	R_3	R_4	Common name
CF_3	H	C_3H_7	C_3H_7	Trifluralin
SO_2NH_2	H	C_3H_7	C_3H_7	Oryzalin
CH_3	CH_3	H	$CH(C_2H_5)_2$	Pendimethalin

(B) N-Phenylcarbamates

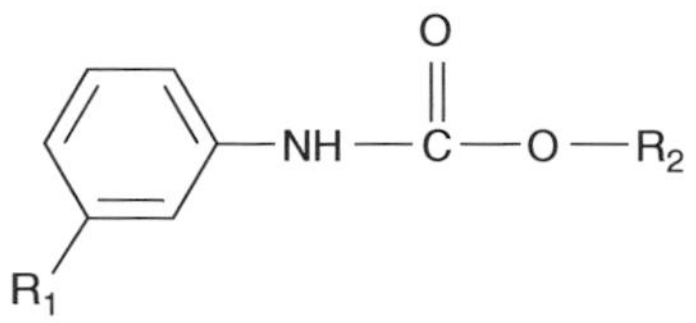

R_1	R_2	Common name
Cl	$CH_2C{=}CCH_2Cl$	Barban
H	$CHCH_3CONHC_2H_5$	Carbetamide
Cl	$CH(CH_3)_2$	Chlorpropham
H	$CH(CH_3)_2$	Propham

(C) Others

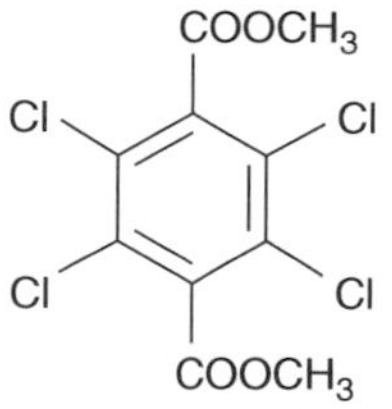

chlorthal-dimethyl(DCPA)

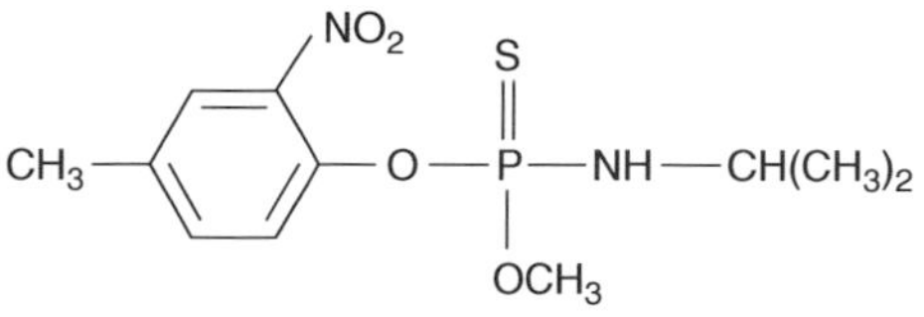

amiprophos-methyl

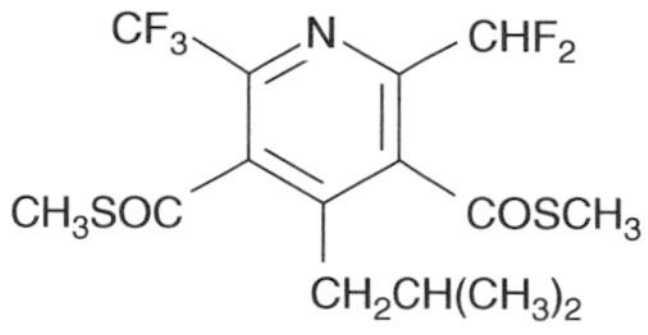

dithiopyr

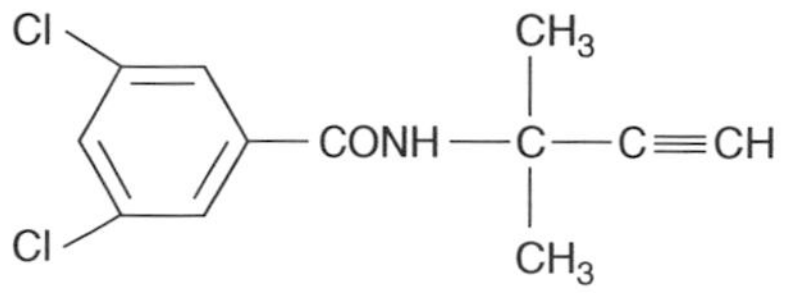

propyzamide

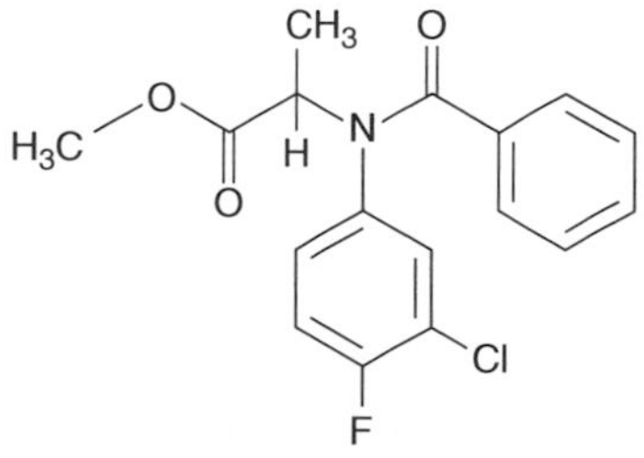

flamprop-methyl

against grasses. Nowadays, chlorpropham is principally used for the suppression of sprout growth in stored potato tubers, owing to the inhibition of sprout cell division.

10.5.3   Others

Chlorthal-dimethyl (DCPA) is widely used in turf grasses. It appears to block cell plate formation in susceptible species, via the disruption of phragmoplast microtubule organisation and production, by an unknown mechanism. There is some indication that CDK phosphatase activity is inhibited by endothal, thus preventing G_2/M transition (Ayaydin *et al.*, 2000). Amiprophos-methyl results in a loss of microtubules and gives similar symptoms to the dinitroanilines. Propyzamide acts in a different fashion to the compounds described so far, such that small microtubules rather than none at all result from treatment with this herbicide. It binds directly to the tubulin, preventing microtubule assembly (Akashi *et al.*, 1988). Dithiopyr acts similarly to propyzamide, but binds to a MAP of molecular weight 65 kDa, rather than to tubulin. Vaughn and Lehnen (1991) are of the view that dithiopyr interacts with a MAP involved in microtubule stability, resulting in shortened microtubules.

Flamprop-methyl was developed in the early 1970s by Shell for the post-emergence control of wild oats in barley. While it was known to inhibit cell elongation, its mode of action was not established until 2008, when Tresch *et al.* demonstrated by microscopy that seedling roots treated with 50 μM flamprop-methyl had a severely disrupted orientation of the spindle and phragmoblast microtubules, such that the chromosomes remained in a condensed state at metaphase. The active metabolite, flamprop-acid did not inhibit soybean tubulin polymerisation, and so the authors suggested that this was due to a novel mode of action, perhaps via microtubule disassembly.

Tresch *et al.* (2005), have demonstrated that the cyanoacrylates (CA1 and CA2: Figure 10.4) are a new chemical class of herbicides with the same mechanism of action as the dinitroanilines, interacting with the α-tubulin binding site to prevent tubulin polymerisation.

Clearly, there remain many unanswered questions regarding microtubule assembly and function in plant cell division. The aforementioned herbicides may provide useful probes to study microtubule biochemistry and further work may yield additional novel herbicide targets.

Intriguingly, it has also been observed that members of the imidazolinone and sulphony-lurea herbicide families inhibit entry into mitotic cell division. Rost (1984) found that chlorsulfuron inhibited the cell cycle progression at G_2 and G_1 transitions and also demonstrated a similar effect of imidazolinones (Rost *et al.* 1990). Both herbicide families inhibit

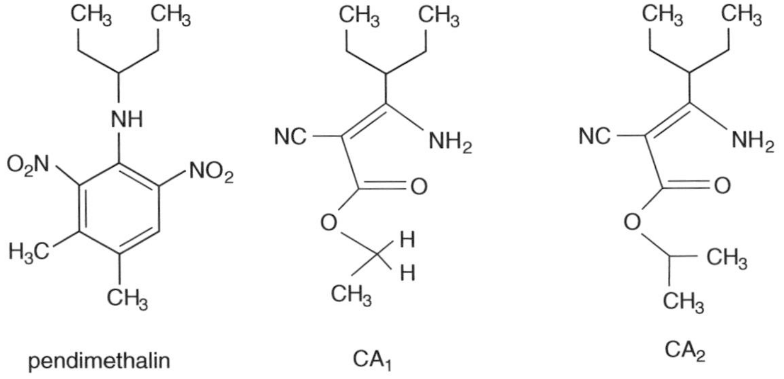

Figure 10.4   The structures of pendimethalin, CA$_1$ and CA$_2$.

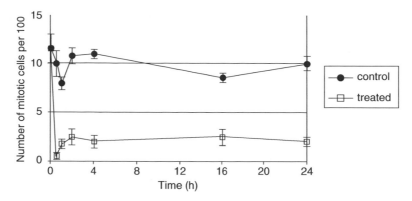

Figure 10.5 The effect of treatment with 0.1 g active ingredient (a.i.) L⁻¹ imazethapyr on mitotic divisions of potato (*Solanum tuberosum*) cv. Cara tuber root tips using aerated liquid media. Data are means of six replicates ± standard error. Treated and control samples were significantly different at every point ($p < 0.01$). Source: Spackman, V.M.T. and Cobb, A.H. (1999) Cell cycle inhibition of potato root tips treated with imazethapyr. *Annals of Applied Biology* **135**, 585–587. Reproduced with permission of John Wiley & Sons.

acetolactate synthase (Chapter 9) in the synthesis of the branched-chain amino acids (valine, leucine, isoleucine) and cell cycle inhibition can be reversed by the addition of these amino acids. This may imply a link between branched-chain amino acid biosynthesis and cell cycle entry, although the precise nature of this remains to be established.

Spackman and Cobb (1999), observed a rapid inhibition of cell division in potato root tips treated with imazethapyr. Significantly lower numbers of mitotically dividing cells were recorded after only 30 min incubation with 0.35 μM imazethapyr (Figure 10.5). The commercialisation of this response as a novel, low-dose potato-sprout suppressant remains to be demonstrated.

10.6   Selectivity

These herbicides inhibit cell division and so all meristematic regions are affected. Symptoms include short, swollen coleoptiles in grass weeds and swollen hypocotyls in broad-leaf plants. Short and stubby secondary roots are also commonly observed leading to a poorly developed root system and a weak plant.

The dinitroanilines are more phytotoxic to grasses than dicots and it has been proposed that selectivity is based on the lipid content of the seeds (Hilton and Christiansen, 1972). The dinitroanilines are highly lipophilic, while the lipid content of grass seeds is lower than that of dicot seeds. Thus, it is believed that these lipophilic molecules are partitioned into the dicot seed lipid reserves, and so sequestered away from their site of action. These authors also demonstrated that coating the seeds of normally susceptible species with α-tocopherol acetate prior to planting gave some protection to the presence of trifluralin in the growth medium.

Depth protection also offers a more selective action for crops. The dinitroanilines are usually incorporated into the top 10 cm of soil and do not readily leach downwards through the soil column. Hence, large-seeded, deep-sown crops with large lipid reserves can grow through the treated zone, whereas small-seeded, shallow-germinating weed seedlings cannot fare as well.

References

Akashi, T., Izumi, K., Nagano, E., Enomoto, E., Mizuno, K. and Shibaoka, H. (1988) Effects of pro-pyzamide on tobacco cell microtubules *in vivo* and *in vitro*. *Plant Cell Physiology* **29**, 1053–1062.

Anthony, R.G., Walden, T.R., Ray, J.A., Bright, S.W.J. and Hussey, P.J. (1998) Herbicide resistance caused by spontaneous mutation of the cytoskeletal protein tubulin. *Nature* **393**, 260–263.

Atkins, K.C. and Cross, F. (2018) Inter-regulation of CDKA/CDK1 and the plant-specific cyclin-dependent kinase CDKB in control of the Chlamydomonas cell cycle. *The Plant Cell* **30**, 429–446.

Ayaydin, F., Vissi, E., Meszaros, T., Miskokzi, P., Kovacs, I., Feher, A., *et al.* (2000) Inhibition of serine/threonine-specific protein phosphatases causes premature activation of cdcMsF kinase at G_2/M transition and early mitotic microtubule organisation in alfalfa. *Plant Journal* **23**, 85–96.

Chu, Z., Chen, J., Nporko, A., Han, H., Yu, Q. and Powles, S. (2018) Novel alpha-tubulin mutations con-ferring resistance to dinitroaniline herbicides in *Lolium rigidum*. *Frontiers in Plant Science* **9**, 1–12.

Cyr, R.J. and Palevitz, B.A (1989) Microtubule-binding proteins from carrot. I. Initial characterisa-tion and microtubule bundling. *Planta* **177**, 245–260.

Dawson, P.J. and Lloyd, C.W. (1987) Comparative biochemistry of plant and animal tubulins. In: Davies, D.D. (ed.), *The Biochemistry of Plants. A Comprehensive Treatise*, vol. 12. New York: Academic Press, pp. 3–47.

Fung-Uceda, J., Lee, K., Seo, P.J., Polyn, S., de Velder, L. and Mas, P. (2018) The circadian clock sets the time of DNA replication licensing to regulate growth in *Arabidopsis*. *Developmental Cell* **45**, 101–113; doi: org/10.1016/j.devcel.2018.o2.022

Gunning, B.E.S. and Hardham, A.R. (1982) Microtubules. *Annual Review of Plant Physiology* **33**, 651–698.

Hilton, J.L. and Christiansen, M.N. (1972) Lipid contribution to selective action of trifluralin. *Weed Science* **20**, 290–294.

Kumar, N. and Larkin, J.C. (2017) Why do plants need so many cyclin-dependent kinase inhibitors? *Plant Signalling and Behaviour* **12**, e1282021. dx.doi.org/10.1080/15592324.2017.1282021

Lin, F. and five others (2019) A dual role for cell plate-associated Pi4K beta in endocytosis and phragmoplast dynamics during plant somatic kinesis. *The EMBO Journal* **38**, e100303.

Mitchison, T.J. and Kirschner, M. (1984) Dynamic instability of microtubule growth. *Nature* **312**, 237–242.

Rost, T.L. (1984) The comparative cell cycle and metabolic effects of chemical treatment on root tip meristems. III. Chlorsulfuron. *Journal of Plant Growth Regulation* **3**, 51–63.

Rost, T.L., Gladdish, D., Steffen, J. and Robbins, J. (1990) Is there a relationship between branched chain amino acid pool size and cell cycle inhibition in roots treated with imidazolinone herbicides? *Journal of Plant Growth Regulation* **9**, 227–232.

Spackman, V.M.T. and Cobb, A.H. (1999) Cell cycle inhibition of potato root tips treated with imazethapyr. *Annals of Applied Biology* **135**, 585–587.

Stals, A. and Inzé, D. (2001) When plant cells decide to divide. *Trends in Plant Science* **8**, 359–364.

Tresch, S., Plath, P. and Grossmann, K. (2005) Herbicidal cyanoacrylates with antimicrotubule mechanism of action. *Pest Management Science* **61**, 1052–1059.

Tresch, S., Niggeweg, R. and Grossman, K. (2008) The herbicide flamprop-M-methyl has a new anti-microtubule action. *Pest Management Science* **64**, 195–203.

Vanstraelen, M., Inzé, D. and Geelen, D. (2006) Mitosis-specific kinesins in *Arabidopsis*. *Trends in Plant Science* **11**, 167–174.

Vaughn, K.C. and Lehnen, L.P. Jr. (1991) Mitotic disrupter herbicides. *Weed Science* **39**, 450–457.

Chapter 11
The Inhibition of Cellulose Biosynthesis

The cosmopolitan character of many weeds is perhaps a tribute both to man's modification of environmental conditions and his efficiency as an agent of dispersal.

E.J. Salisbury (1886–1978)

11.1 Introduction

The plant cell wall is a very complex yet highly organised network of polysaccharides, proteins and phenylpropanoid polymers (e.g. lignin), set in a slightly acidic solution containing several enzymes and many organic and inorganic substances. It is not a static structure but a dynamic, metabolic compartment of the cell, sharing a molecular continuity with the plasmalemma and the cytoskeleton. It also changes throughout the life of the cell, from cell division to expansion and elongation, expanding its original area by orders of magnitude. Readers are referred elsewhere for a more detailed understanding and review of the cell wall (e.g. Carpita, 1997).

Cellulose is the most abundant bio-polymer on earth, and constitutes about 95% of paper and 90% of cotton. It provides the framework for the higher plant cell wall. Its biosynthesis is therefore an important target for herbicide development, especially as no animal counterpart exists that could, perhaps, lead to issues of toxicity. The cellulose biosynthesis inhibitors are used as pre-emergent herbicides in lawns, golf courses, vineyards and railway tracks. As far as we currently know, there is no known weed resistance to these herbicides, and so they can be an attractive addition to weed control strategies. The fact that resistance in the field is yet to be observed is probably due to the complexity of cellulose biosynthesis in plants.

11.2 Cellulose biosynthesis

While we know that cellulose biosynthesis takes place at the plasma membrane, cell-free synthesis has not yet been demonstrated. It is thought to be catalysed by multimeric terminal complexes that have been visualised by freeze-fracture images of the plasmalemma. These are, however, irreversibly disrupted and lost on membrane isolation. UDP-glucose is considered to be the primary substrate and progress is awaited on identifying all of the

Herbicides and Plant Physiology, Third Edition. Andrew H. Cobb.
© 2022 John Wiley & Sons Ltd. Published 2022 by John Wiley & Sons Ltd.

genes and enzymes involved. *Arabidopsis* mutants have been generated and their further characterisation should shed light on this important process.

Cellulose, the fundamental unit characterising the plant cell wall, consists of microfibrils that are themselves a bundle of parallel $\beta \rightarrow 4$ glucan chains, hydrogen-bound to form a microcrystalline array. Cellulose synthase genes (CesA) that code for the catalytic subunit of cellulose synthase (E.C. 2.4.1.12) were first isolated from cellulose-producing bacteria in the early 1990s and have since been characterised in other prokaryotes and higher plants. All cellulose synthase-like proteins share transmembrane helices and a cytoplasmic loop consisting of four conserved regions involved in substrate binding and catalysis. In *Arabidopsis* 10 CesA genes have been identified. Eckardt (2003) has suggested that three CesA proteins (A_1, A_3 and A_6) interact as subunits within the cellulose synthase complex in primary cell wall biosynthesis, whereas $CesA_4$, A_7 and A_8 form the complex in secondary cell walls.

The transmembrane nature of the cellulose synthase-like proteins is supported by freeze-facture electron microscopy studies, which show hexagonal arrays of nascent cellulose microfibrils and terminal complexes, termed rosettes, on the cell wall face of the plasma membrane. More conclusive evidence that these rosettes are involved in cellulose synthesis has come from Kimura *et al.* (1999), who labelled freeze-facture replicas of mung bean rosettes with antibodies raised against a CesA gene product. It is now generally accepted that the rosettes are functional cellulose synthase complexes.

The Golgi apparatus is the known cell site for the synthesis and processing of glycoproteins and targetting them to the plasma membrane. Glycans and pectins are also polymerised at this location, along with the synthesis of hexose- and pentose-containing polysaccharides. It appears that all non-cellulosic polysaccharides are synthesised at the Golgi apparatus.

Cellulose synthase subunits are synthesised at the endoplasmic reticulum and move to the plasma-membrane via the trans-Golgi network. Once delivered to the plasma membrane, the synthesis of cellulose microfibrils begins, and the direction of micro-fibril synthesis is guided by cortical microtubules. Zhang *et al.* (2016) have shown that STELLIO 1 and STELLIO 2 are Golgi-localised proteins that can interact with cellulose synthases and control cellulose quantity (Figure 11.1). The authors propose that these proteins are glucosyltransferases, facing the Golgi lumen. (*Stello* is from the Greek, meaning 'to set in place and deliver'.)

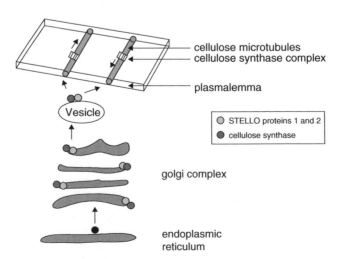

Figure 11.1   A simplified scheme for cellulose synthesis in plants.

More recently, Zhu *et al.* (2018), have proposed a docking site for cellulose synthase-containing vesicles for access to the plasma membrane. The plant-specific protein PATROL1 (a Cellulose Synthase Interacting Protein) assists in exocytosis. Both PATROL1 and the exocyst complex determine the rate of delivery of the cellulose synthase complex to the plasma membrane.

The organising system that controls a cellulose cell wall is the plant cortical microtubule array. They regulate the insertion of cellulose synthase complexes into the plasma membrane and guide their movement during the deposition of cellulose microfibrils. Microtubules interact with each other through collisions and can rapidly change in response to internal and external cues. The microtubule-severing enzyme, katanin, is required for normal alignment and orientation of the microtubule array (Deinum *et al.*, 2017). If two microtubules cross each other without colliding (i.e. they are not growing in the same direction), katanin severs the microtubule lying on top (i.e. the one that arrives last). In this way, all of the microtubules grow in the same direction.

Several additional proteins are required for cellulose biosynthesis, namely:

- KORRIGAN – an endoglucanase, thought to edit the new cellulose strands;
- COBRA – a glycosyl-phosphatidyl inositol-anchored protein, thought to be crucial to maintaining cellulose structure; and
- CELLULAR SYNTHASE INTERACTING PROTEINS CSI-1/POM2 and CSI-3 – thought to act as molecular cross-linking agents.

Furthermore, the phosphorylation and de-phosphorylation of proteins is probably involved in post-translational modification, suggesting that protein kinases and phosphatases have a role in this process (Tateno *et al.*, 2016). A complex process, indeed!

The cell wall also contains four major structural proteins: extensins, the repetitive proline-rich protein, the glycine-rich protein and the threonine-hydroxyproline-rich protein. Extensins contain repeating Ser-$(Hyp)_4$ and Tyr–Lys–Tyr sequences that are important for wall structure. They are rod-shaped and may cross-link to other extensins. The 33 kDa proline-rich proteins are similar to extensins and are more highly expressed in later stages of cell development. The glycine-rich protein may have over 70% glycine residues and is thought to form a plate-like structure at the interface of the plasma membrane and the cell wall, which is thought to serve as an initiation site for lignification. The threonine–hydroxyproline-rich proteins predominate in grasses and are thought to be related to extensins. They accumulate early in the cell cycle and become insoluble during cell elongation and differentiation.

11.3   Cellulose biosynthesis inhibitors

The cellulose biosynthesis inhibitors (CBIs) are a small group of chemically unrelated compounds including the herbicides dichlobenil, indaziflam, isoxaben and flupoxam. They are all effective CBIs and, because this site of action is not shared with mammals, they may be ideal choices for the purposes of registration and selective herbicidal activity. A lack of observed resistance in the field may enable them to become useful agents in the control of resistant weeds. Furthermore, they have become valuable probes in our understanding of cellulose biosynthesis and cell wall formation in plants.

Cellulose is an integral part of the plant cell wall and hence cellulose synthesis is vital in order for the cell wall to be synthesised. The cell wall determines to some extent both the morphology and the function of a cell, but even more importantly it controls the degree to which a cell can expand. It is suggested that if cellulose biosynthesis is inhibited then weakened cell walls will result, causing expansion of the cell and disruption of cellular processes. This in turn leads to abnormal or restricted growth and subsequent plant death. As most CBIs are used pre-emergence then seedling growth is inhibited. Post-emergence use of CBIs leads to stunted growth and swollen roots in treated plants.

Tateno *et al.* (2016) have tentatively characterised CBIs into three groups, based on livecell confocal microscopy of cellulose synthase A proteins tagged to a fluorescent marker following treatment with inhibitors. See Figure 11.2 for chemical structures.

Group 1: Cause clearance of cellulose synthase A particles from the plasma membrane. Examples include isoxaben, quinoxyphen, AEF 150944, CGA 325,615, thaxtomin-A, flupoxam, triazofenamide, CESTRIN and acetobixan.

Group 2: Cause increase in cellulose synthase A particles in the plasma membrane, but movement is slowed or arrested. Examples include dichlorbenil, indaziflam and triaziflam.

Group 3: Modify the trajectory of cellulose synthase A particles at the plasma membrane. Examples include morlin and cobtorin.

These different processes of disruption suggest that members of each group may target a different aspect of cellulose biosynthesis.

Research has identified at least two separate points at which CBIs inhibit. Dichlobenil analogues have been shown to bind to an 18 kDa polypeptide associated with the multienzyme complex. It is postulated that this may be a regulatory subunit associated with cellulose synthase, as it appears to be too small to be the enzyme itself. Reported effects of dichlobenil treatment include synthesis of callose in place of cellulose. Callose is a β-1,3-glucan, often formed in plants as a wounding response. It is not usually present in the cell walls of undamaged cells. It is postulated that dichlobenil inhibits the incorporation of UDP-glucose into cellulose and this is therefore shunted into callose (and xyloglucan) synthesis.

Isoxaben has been demonstrated to inhibit the incorporation of ^{14}C-labelled glucose into the cell wall. Other studies have revealed that the synthesis of both cellulose and callose is reduced by this herbicide. This suggests an inhibition of cellulose synthesis at an earlier stage than dichlobenil, preventing incorporation of glucose into both cellulose and callose. It has been postulated that the step where isoxaben acts may be at the point where UDP-glucose is formed from sucrose. Flupoxam appears to inhibit at a similar point, although less information on this herbicide's mode of action is available. Quinclorac, an auxin-type herbicide for broad-leaf weed control, also appears to inhibit cellulose biosynthesis in susceptible grasses, although the mechanism by which it accomplishes this has yet to be deduced. This implies that quinclorac has a second site of action in monocots, unrelated to its primary mechanism in dicots (Koo *et al.*, 1997).

A new herbicide, 5-*tert*-butyl-carbamoyloxy-3-(3-trifluoromethyl)phenyl-4-thiazolidinone (Figure 11.3), also appears to inhibit cellulose biosynthesis at the same place as isoxaben (Sharples *et al.*, 1999). This is a representative of a novel class of *N*-phenyl-lactam-carbamate herbicides patented by Syngenta in 1994. It shows potential for the pre-emergence control of a range of grass weeds (including *Setaria viridis*, *Echinochloa crus-galli* and *Sorghum halepense*) and small-seeded broad-leaf weeds (including *Amaranthus retroflexus* and *Chenopodium album*) in soybean, at an application rate of approximately 125–250 g ha^{-1}.

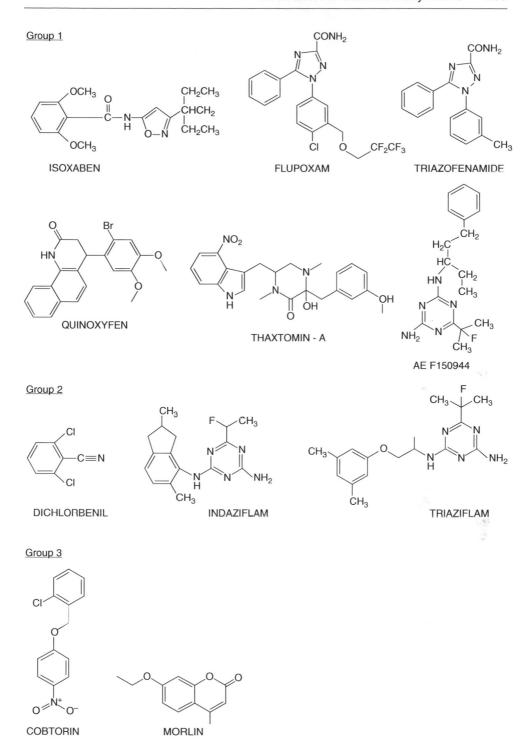

Figure 11.2 Categorisation of cellulose biosynthesis inhibitors according to Tateno, Brabham and DeBolt, 2016.

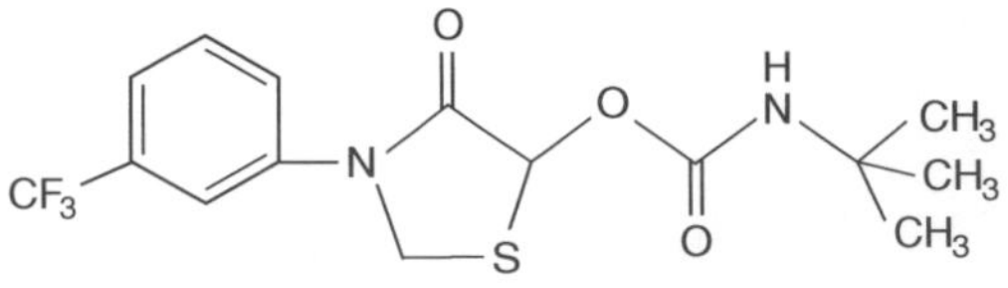

Figure 11.3   5-*tert*-Butyl-carbamoyloxy-3-(3-trifluoromethyl)phenyl-4-thiazolidonone (compound 1).

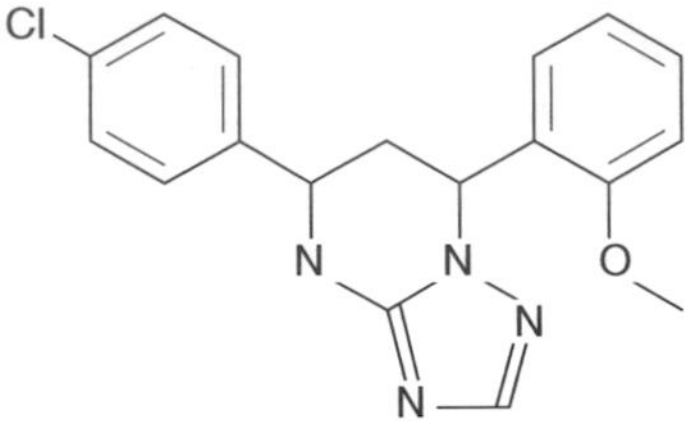

Figure 11.4   Chemical structure of the novel cellulose biosynthesis inhibitor, termed C17 (Hu *et al.*, 2019).

Root growth was completely inhibited at 2 μM and visual symptoms typically included stunted and swollen roots, with no lateral branches. It was shown by Sharples *et al.* (1999) to be a potent inhibitor (IC_{50} of 50 μM) of the incorporation of ^{3}H-labelled glucose into the acid-insoluble polysaccharide cell wall fraction of *Zea mays* roots – which is assumed to be cellulose. Similar inhibition was observed by isoxaben and dichlobenil, and cross-resistance was demonstrated to an isoxaben-resistant *Arabidopsis* mutant (Ixr1-1), which implies a shared molecular binding site with isoxaben. These authors conclude that the final identification of this site will require a combination of molecular approaches and traditional biochemistry.

Hu *et al.* (2019) reported the finding of a novel CBI, termed C17, from a compound-screening approach (Figure 11.4), which results from the depletion of CESA from the plasmalemma. They suggest that it acts differently from other CBIs by a direct interaction with CESA subunits, resulting in a weaker cell wall. C17 was shown to inhibit the growth of both monocot and dicot weeds, although the inhibition of monocot weed growth required a much higher dosage, suggesting potential selectivity in rice and maize crops. Furthermore, an additive growth inhibition was observed when mixed with isoxaben or indaziflam, which may be of value in slowing the evolution of herbicide-resistant weeds.

The formation of cell plates in the presence and absence of CBIs has received detailed scrutiny in both light and electron microscope studies. Cell plates are the cell walls formed to separate daughter nuclei after cell division. In the presence of dichlobenil, cell plates develop more undulated and thicker regions than in untreated tissues, resulting in cell walls that are incomplete or attached to only one existing cell wall. Treated cell walls also are enriched in callose compared with controls. This may imply that the inhibition of cellulose biosynthesis has increased the amount of UDP-glucose available for callose or xyloglucan synthesis. Conversely, treatment with either isoxaben or flupoxam results in the production of a thin cell plate, with no callose nor xyloglucan enrichment.

Another useful experimental system for the study of CBIs (Vaughn, 2002) is the developing cotton fibre, whose secondary wall is composed almost exclusively of cellulose. Fibre development is severely inhibited by the CBIs. The cell wall of control fibres consists of two layers, an outer wall area enriched in pectin and an inner layer enriched in cellulose–xyloglucan. After treatment with diclobenil, the inner layer became enriched in callose and cellulose was absent. On the other hand, treatment with isoxaben and flupoxam generated relatively more outer wall, highly enriched in de-esterified pectin.

In summary, dichlobenil appears to divert cellulose biosynthesis into callose, while isoxaben and flupoxam inhibit the production of both cellulose and callose in both cell plates and cotton fibres. The further identification of the enzymes involved in the presence of CBIs is awaited with interest.

The most recently discovered cellulose biosynthesis inhibitors, flupoxam and triaziflam, are active at much lower rates (100–250 g ha^{-1}) than their predecessors, such as dichlobenil (3000 g ha^{-1}). This may open up the possibility of more imitative chemistry in future to lower the dosage used in this herbicide class. According to Grossman *et al.* (2001), triaziflam may also affect photosynthetic electron transport and microtubule formation. This multiple activity may be a positive feature and may delay the onset of herbicide-resistant weeds.

The relationship between cellulose microfibrils and cortical microtubules has long been debated. Paradez *et al.* (2006) have fluorescently labelled cellulose synthases that assemble into functional cellulose synthase complexes in *Arabidopsis* hypocotyl cells. They have demonstrated that the cellulose synthase complexes move in the plasma membrane along tracks that appear to be defined by co-labelled microtubules. It appears that the force for the movement of the cellulose synthase complex is derived from cellulose microfibril production itself. According to Emons *et al.* (2007), this discovery provides the experimental tools that may allow us to answer the following questions:

- How do the cellulose synthase complexes enter the plasma membrane, via exocytosis?
- Where are they inserted into the plasma membrane?
- How are they activated, for how long and how are they regulated?
- Which gene products influence microfibril production?
- How does the network of molecules involved in cellulose production function?

Note that in these studies oryzalin was added to depolymerise the cortical microtubules, showing the value of this herbicide as a tool in unravelling this complex process.

11.4   How treated plants die

Inhibitors of cellulose biosynthesis are pre-emergence herbicides that prevent cell division, especially in root tips, and control some germinating broad-leaf weeds and grasses. In grasses, symptoms are typically short and swollen coleoptiles, while broad-leaf weeds show swollen hypocotyls. Consequently, treated plants are stunted or seedlings fail to germinate or emerge.

11.5   Selectivity

Cellulose synthesis inhibitors generally inhibit root elongation in dicots at lower rates than in monocots. As yet, there is no indication that metabolism can explain this selectivity. Perhaps selectivity is due to structural differences in cell walls. Garcia-Angelo *et al.* (2012) suggest that some cellulose biosynthesis inhibitors may divert radiolabelled glucose added to cell cultures into a pectin-rich fraction. This could serve as a compensation mechanism, so that instead of cellulose, hemicelluloses are produced in grasses and pectins in dicots. More recently, Brabham *et al.* (2017) examined this idea further and conclude that indeed that structure of the cell wall in grasses may be a factor in tolerance to these herbicides. Clearly, more work is needed to explain selectivity to these herbicides.

Various studies on the metabolic fate of isoxaben in both cell cultures and *Arabidopsis* mutants have implied that selectivity is not due to altered uptake or metabolism, but instead results from an alteration at the target site. Furthermore, the observation (Sharples *et al.* 1999), that compound 1 is metabolised very slowly *in vivo* also implies that the selectivity of this molecule in soybean is due to a species-dependent variability at the target site. This is presumed to be the multimeric complex, cellulose synthase. Further details of this enzyme complex are eagerly awaited.

References

Brabham, C., Stork, J., Barrett, M. and DeBolt, S. (2017) Grass cell walls have a role in the inherent tolerance of grasses to the cellulose biosynthesis inhibitor Isoxaben. *Pest Management Science* **74**, 878–884.

Carpita, N. (1997) Structure and biosynthesis of plant cell walls. In: Dennis, D.T., Turpin, D.H., Lefebvre, D.B. and Layzell, D.D. (eds) *Plant Metabolism*, 2nd edn. Harlow: Longman, Ch. 9, pp. 124–147.

Deinum, E.E., Tindermans, S.H., Lindeboom, J.L and Mulder, B.M. (2017) How selective severing by katanin promotes order in the plant cortical microtubule array. *Proceedings of the National Academy of Science* **114**, 6942–6947.

Eckardt, N.A. (2003) Cellulose synthesis takes the CesA train. *Plant Cell* **15**, 1685–1687.

Emons, A.M.C., Höfte, H. and Mulder, B.M. (2007) Microtubules and cellulose microfibrils: how intimate is their relationship? *Trends in Plant Science* **12**, 279–281.

Garcia-Angulo, P., Alonso-Simon, A., Encina, A., Alavarez, J.M. and Acebes, J. (2012) Cellulose biosynthesis inhibitors: comparative effect on bean cell cultures. *International Journal of Molecular Sciences* **13**, 3685–3702.

Grossman, K., Tresch, S. and Plath P. (2001) Triaziflam and diamino-triazine derivatives affect enantioselectivity multiple herbicide target sites. *Zeitschrift für Naturforschung* **56c**, 559–569.

Hu, Z., Zhang, T., Rombaut, D., Decaestecker, W., Xing, A., de Haeyer, S. *et al.* (2019) Genome editing-based engineering of CESA3 dual cellulose-inhibitor-resistant plants. *Plant Physiology* **180**, 827–836.

Kimura, A., Laosinchai, W., Itoh, T., Cui, X., Linder, C.R. and Brown, R.M.(1999) Immunogold labelling of rosette terminal cellulose-synthesising complexes in the vascular plant *Vigna angularis*. *Plant Cell* **11**, 2075–2086.

Koo, S.J., Neil, J.C. and Ditomaso, J.M. (1997) Mechanism of action of quinclorac in grass roots. *Pesticide Biochemistry and Physiology* **57**, 44–53.

Paradez, A.R., Somerville, C.R. and Ehardt, D.W. (2006) Visualisation of cellulose synthase demonstrates functional association with microtubules. *Science* **312**, 1491–1495.

Sharples, K.R., Hawkes, T.R., Mitchell G., Edwards, L.S., Langford, M.P., Langton, D.W., *et al.* (1999) A novel thiazolidinone herbicide is a potent inhibitor of glucose incorporation into cell wall material. *Pesticide Science* **54**, 368–376.

Tateno, M., Brabham, C. and De Bolt, S. (2016) Cellulose biosynthesis inhibitors – a multifunctional toolbox. *Journal of Experimental Botany* **67**, 553–542.

Vaughn, K.C. (2002) Cellulose biosynthesis inhibitor herbicides. In: Böger, P., Wakabayashi, K. and Harai, K. (eds), *Herbicide Classes in Development*. Berlin: Springer, pp. 139–150.

Zhang, Y., Nikolovski, N., Sorieul, M., Vellosillo, T., McFarlane, H.E., Dupree, R., *et al.* (2016) Golgi-localised STELLO proteins regulate the assembly and trafficking of cellulose synthase complexes in Arabidopsis. *Nature Communications* 7, article number 11656.

Zhu, X., Li, S., Pan, S., Xin, X. and Gu, Y. (2018) CST1, PATROL and exocyst complex cooperate in the delivery of cellulose synthase complexes to the plasma membrane. *Proceedings of the National Academy of Science* **115**, 3578–3587.

Chapter 12
Plant Kinases, Phosphatases and Stress Signalling

The plant is blind, but it knows enough to keep pushing upwards towards the light, and will continue to do this in the face of endless discouragements.

George Orwell (1903–1950)

12.1   Introduction

Kinases are enzymes that transfer the gamma phosphate group of ATP to serine, threonine or tyrosine amino acids of proteins in a process termed phosphorylation. On the other hand, phosphatases are members of a broad group of phospho-hydrolase enzymes that remove the phosphate group (Figure 12.1). Thus, the phosphorylation state of a protein is governed by these mutually antagonistic enzymes. Of course, if they were both operational at the same time, they would cancel each other out. Therefore, it is thought that they work alternately, while in communication with each other at a regulatory level.

Approximately 5% of the *Arabidopsis* genome codes for these enzymes, although only a small number have been studied. The serine/threonine kinases and phosphatases play key roles in the regulation of plant growth, but their regulation appears complex or currently unknown.

Since kinase activity can often become abnormal in certain human diseases and especially cancer, kinase inhibitors have become one of the most sought-after targets in medicinal chemistry research. Kinase inhibitors can manipulate the human immune system to target and kill cancer cells. As an example, palbociclib is a promising drug for breast cancer therapy. It inhibits cyclin-dependent kinase (CDK) 4 and 6 activity, which normally enables entry into the cell cycle of cancer cells. This kinase inhibitor therefore helps to slow cancer growth and spread (e.g. Guiley *et al.*, 2019). Other kinases are showing activity in the treatment of conditions such as diabetic nephropathy, a kidney disease that is a common complication of diabetes. There are now approaching 50 small molecules approved as drugs in the pharmaceutical sector for the inhibition of kinases. Many are showing promise as treatments for oncology and inflammatory disorders. Furthermore, about 200 crystal structures of kinases are now available, with details of active sites enabling the rational design of inhibitors. For example, see www.icr.ac.uk, and Klaeger *et al.* (2017), for further details.

Herbicides and Plant Physiology, Third Edition. Andrew H. Cobb.
© 2022 John Wiley & Sons Ltd. Published 2022 by John Wiley & Sons Ltd.

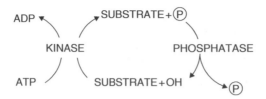

Figure 12.1 A simplified scheme of phosphorylation and de-phosphorylation. P denotes a phosphate group, PO_4.

Currently, however, our understanding of plant kinases lags behind our knowledge of animal kinases. As a result of publications in the last decade, their roles in plants are becoming increasingly understood, and they are certain to become important targets for herbicide discovery and development.

12.2   Plant kinases

Plant protein kinases phosphorylate serine, threonine or tyrosine residues in proteins. They all have a 250–300 amino acid domain that is responsible for phosphate transfer. Those in flowering plants, termed the Kinome, are more abundant than in other eukaryotes, ranging from 600 to 2500 members. When the *Arabidopsis* genome was sequenced in 2000, over 1000 protein kinases were identified, compared with 518 identified in the human genome in 2002. Almost 600 are membrane-located receptor kinases and almost 400 are soluble kinases with distinct functions in signal transduction. Furthermore, there are an estimated 1500 kinases in rice and 2000 in soybean. Thus, 1–2% of functional genes encode protein kinases, indicating their importance in all aspects of cell regulation in both animals and plants (Lehti-Shiu and Shiu, 2012).

In recent years, plant protein kinases have been found in signalling networks, such as the perception of biotic agents (i.e. bacteria, fungi, viruses, animals and weeds) and abiotic stressors (such as extremes of temperature, UV light, drought and salinity) that affect plant growth and development. Indeed, they also function in diurnal and circadian rhythms, developmental processes, modulation of vesicle transport, chemical activities and regulation of cell metabolism. They are highly conserved in all organisms.

Lehti-Shiu and Shiu (2012) have published a comprehensive and fully referenced account of the evolution and functional diversity of the plant kinome. The following is a brief account of their classification and functional diversity that will no doubt be added to in time, with apologies for an excess of acronyms.

(A)   *Central roles in metabolic signalling and stress responses*
  AMPK/SNF1 and LKB1: The AMP-activated protein kinase (AMPK)/SNF1 (sucrose non-fermenting 1) kinases play central roles in sensing energy status and maintaining energy homeostasis under stress conditions, such as during starvation and long periods of darkness.
  PDK1: Phosphoinositide-dependent kinase 1 is a master regulator of AGC kinase activity. AGC kinases are among the most well studied kinases, and are named after cyclic <u>A</u>MP-dependent kinases, cyclic <u>G</u>MP-dependent kinases and

the diacyglycerol-activated, phospholipid-dependent kinase. Nine of the 39 AGC kinases in *Arabidopsis* have been shown to be involved in polar auxin transport and regulating photo-autotrophic growth (Rademacher and Offringa, 2012). Ribosomal S6 kinases are also phosphorylated by PDK1 in response to nutrient availability.

IRE1: Inositol-requiring kinase 1 senses stress at the endoplasmic reticulum and activates the unfolded protein response, regulating the expression of stress responsive genes.

GCN2: General control, non-repressed 2 regulates translation in response to nutrient stress, such as amino acid starvation. In yeast, it is also activated by UV light and is required as a checkpoint, delaying progression from G1 to S phase in the cell cycle.

(B) *Central roles in the regulation of mitosis and cytokinesis*
Protein kinases in the Aurora and cyclin-dependent families play central roles in the regulation of mitosis and the control of the cell cycle.

Aurora: Plants have two Aurora kinases which are found together with mitotic structures, including the centromeres and the cell plate, and phosphorylate histone protein 3.

CDKA and CDKB: These cyclin-dependent kinases regulate the G2/M transition in the cell-cycle.

TLK: First identified as a regulator of flower initiation and development in *Arabidopsis*, they can also phosphorylate histones and a histone chaperone protein that functions in chromatin assembly.

IRK: The inflorescence and root apices receptor kinase is located at the outer plasmalemma surface of root endodermal cells and functions to repress specific cell divisions (Campos *et al.*, 2019).

(C) *Protein kinase families with diverse functions, but with conserved network signalling components*
MAPK: Mitogen-activated protein kinases link extra- and intracellular signals to downstream processes. In *Arabidopsis*, 110 genes have been identified which code for MAPKs involved as cascade kinases. Comparable numbers are found in other plant species. They play key roles in the transduction of environmental and developmental signals via the phosphorylation of kinases, enzymes, cytoskeletal proteins and transcription factors. Jagodzik *et al.* (2018) review the involvement of MAPK in hormone signalling in plants. As an example, ethylene signal transduction involves a MAPK cascade (Merchante *et al.*, 2013). In this instance, the receptors CTR1 and EIN2 are located at the endoplasmic reticulum. In the absence of ethylene, CTR1 inhibits EIN2 by direct phosphorylation. In the presence of ethylene, the *C*-terminus is cleaved off and migrates to the nucleus where it activates a transcriptional cascade.

STE 7 and STE 11: Like the MAPKs, the STE 7 and 11 (sterility) comprise a large kinase subfamily in *Arabidopsis* that function as MAP2K and MAP3K, respectively. Their activity leads to the downstream transcription of defence-response genes in response to stresses, such as pathogen attack. The signal transduction of auxin, ethylene, abscisic acid and cytokinins all utilise a MAPK pathway (Mishra *et al.*, 2006).

(D) *Divergent functions, but conserved regulatory mechanisms*
OSTL: Open stomata-like, SnRK2 kinases are involved in osmotic stress, abscisic acid responses and signalling of sugar metabolism.

CHK1: These kinases play a role in response to DNA damage, the regulation of ion homeostasis and confer salt tolerance.

CK1: Casein kinase 1 is involved in microtubule organisation. The Rice Early Flowering 1 gene encodes a CK1 that negatively regulates gibberellic acid signalling by phosphorylating a gibberellic acid receptor.

(E)   *Involvement in plant-specific processes*

GSK3: Glycogen synthase kinase 3 regulates cell migration, metabolism and proliferation. GSK3s also function in brassinosteroid signalling, stress pathways and flower development.

IPK: Isopentenyl phosphate kinases (IPKs) re-activate isopentenyl phosphate by ATP-dependent phosphorylation to form the key primary metabolite isopentenyl diphosphate used in isoprenoid and terpenoid biosynthesis (Henry *et al.*, 2018).

NEK: NimA-related kinases regulate microtubules and have roles in cell-cycle regulation. NEC6 in *Arabidopsis* interacts with microtubules to regulate epidermal cell morphogenesis and is also involved in stress responses.

DRK1 and DRK2: Downstream receptor kinases regulate salicylic acid-inducible defence signalling, ethylene responses and drought responses.

TKL-Pl4: These kinases are expressed in the guard cells and regulate stomatal movement in response to carbon dioxide concentrations. They regulate transit peptides on chloroplast-targeted pre-proteins and are required for chloroplast differentiation. The Pl5 subfamily is also involved in leaf venation and response to auxins and brassinosteroids.

WNK/NRBP: With-no-lysine and nuclear-receptor binding protein kinases, WNK genes are transcriptionally regulated by the circadian clock and abiotic stress. Soybean WNK1 kinase regulates root architecture in response to abscisic acid and osmotic signals. These kinases also appear to protect plant cells against water loss. WNK kinases are serine/threonine kinases characterised by the placement of a catalytic lysine residue, critical for ATP binding, located at a different domain. They are present in all animals, although the plant genome encodes more WNK kinases than animal genomes do. They have been well studied in *Arabidopsis* and are associated with a range of physiological processes ranging from hormonal regulation and ionic stress to an involvement in circadian rhythms and flowering time, as reviewed by Cao-Pham *et al.* (2018). The detailed mechanisms involved, however, remain to be elucidated and there is much extrapolation from animal to plant physiology. Nonetheless, it is clear that plant WNKs are central to how plants modulate their responses to environmental stress.

(F)   *Kinases that allow plants to perceive and respond to various environmental signals*

CDPKs: The calcium-dependent protein kinases (CDPKs) are a large family. Different members bind to calcium with different affinities. They play roles in regulating transcription in response to hormone levels. Several are known to function in abiotic stress responses and abscisic acid signalling via phosphorylation of target sites, such as ion channels and transporters. Activity in guard cells controls stomatal movement in response to abscisic acid and methyl jasmonate.

RLKs: Receptor-like kinases are signalling proteins that have an extracellular domain connected via a trans-membrane domain to a cytoplasmic kinase. In this way, RLKs perceive external signals and transduce them in the cell. They are a major gene family, with more than 600 encoded in the *Arabidopsis* genome and 1100 in the rice genome, which are involved in responses to pathogen attack (Morillo and Tax, 2006). More than half are located at plasma membranes, most likely functioning in response to extracellular signals. Uemura *et al.* (2020), have investigated soybean RLKs as mediators of herbivore danger signals in response to oral polysaccharide secretions extracted from larvae of herbivores, which result in ethylene signalling as a plant defence response.

Intriguingly, few protein kinases have been located in chloroplasts and mitochondria, perhaps owing to their endosymbiotic ancestry. Lundquist *et al.* (2012), have suggested that the ABC1K family of protein kinases are located there, functioning in the regulation of quinone synthesis and perhaps prenyl-lipid metabolism.

In conclusion, the plant protein kinase super-family is larger than in other eukaryotes and plays many central roles in the regulation of plant metabolism and the cell cycle. However, why are there so many of them? Lehti-Shiu and Shiu (2012) suggest that selection pressure to respond to an ever-changing environment may be the main reason. Furthermore, plants are literally 'rooted to the spot' and so have had to evolve means of perception and signalling when exposed to environmental change, which can lead to altered gene expression and so alter plant growth and development.

At the time of writing (summer 2020), the author is not aware of any candidates in the literature in development for the herbicidal inhibition of plant protein kinases. Perhaps it is only a matter of time before such discoveries are made?

12.3   Plant phosphatases

In eukaryotes, two major complexes, PP1 and PP2, account for more than 90% of plant protein phosphatase activity. They function to selectively de-phosphorylate plant kinases, and it is the balance of protein kinase and phosphatase activities that orchestrates gene expression and cell cycle progression.

While great attention has been devoted to the protein kinases, protein phosphatases have received relatively limited study. The plant phosphatases are categorised into the serine/threonine phosphatases and the tyrosine phosphatases, depending on the amino acid residue that they de-phosphorylate. The former category is subdivided into two families: PP1 and PP2. PP1, PP2A and PP2B are the main ones, while PP2C require a metal ion (magnesium or manganese) for catalysis. PP2C represents the main group in plants, with 65% of all protein phosphatases. There are 126 in the *Arabidopsis* genome, in comparison with over 1000 protein kinases (Fuchs *et al.*, 2013). The PP2Cs counteract MAPK-activated protein kinase pathways, while others function as co-receptors for abscisic acid.

The plant tyrosine phosphatases are also divided into two groups: the tyrosine-specific plant tyrosine phosphatases and those termed 'dual specific', i.e. they can de-phosphorylate phospho-tyrosine in addition to phospho-serine and phospho-threonine (Singh *et al.*, 2015). The PP2Cs have taken centre-stage in recent years, with their involvement in the action of abscisic acid (ABA), which is involved in seed development, seed dormancy, the control of stomatal apertures and plant responses to abiotic stress.

A novel, ABA-soluble receptor-protein (ABAR) has been discovered in the cytoplasm and the nucleus of plant cells. This protein family has 14 members in *Arabidopsis*, most of which are capable of forming an ABA–receptor complex able to activate the transcription of ABA-responsive genes. At low or no ABA concentrations, PP2C phosphatases interact with SnRK2 kinases to inhibit their auto-phosphorylation and activation. In the presence of ABA, however, ABA binds to the receptor protein and a conformational change results so that the PP2C phosphatases are inactivated. The consequence is that the SnRK2 kinases are now phosphorylated and activated, which in turn phosphorylates and activates a transcription factor, resulting in the transcription of ABA responsive genes (Figure 12.2; Kline et al., 2010; Singh *et al.*, 2015).

The herbicidal properties of endothall were first reported in 1951, and it was commercialised for the selective, post-emergence control of broad-leaf and grass weeds in sugar beet. It also found use as a pre-harvest desiccant in potatoes. Early studies on mode of action showed effects in mitosis and membrane dysfunction, but its mechanism of action was not reported until 2011 by Tresch *et al.* These workers concluded that endothall blocked checkpoints in the cell cycle, implying that PP1 and PP2A protein phosphatases were involved in the regulation of the cell cycle. The following year Bajsa *et al.* (2012) demonstrated that endothall was a close analogue of the natural secondary metabolite cantharidin and that both were potent inhibitors of serine/threonine protein phosphatases, acting as slow, irreversible inactivators of the enzyme (Figure 12.3).

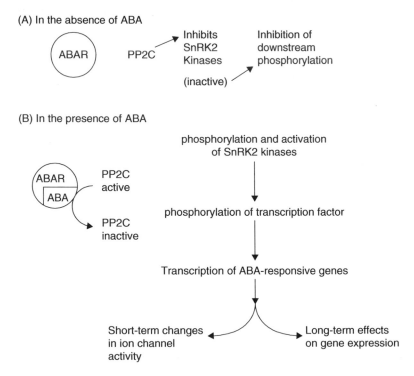

Figure 12.2   A model of ABAR signalling. Source: based on Kline, K.G., Sussman, M.R. and Jones, A.M. (2010) Abscisic acid receptors. *Plant Physiology* **154**, 479–482.

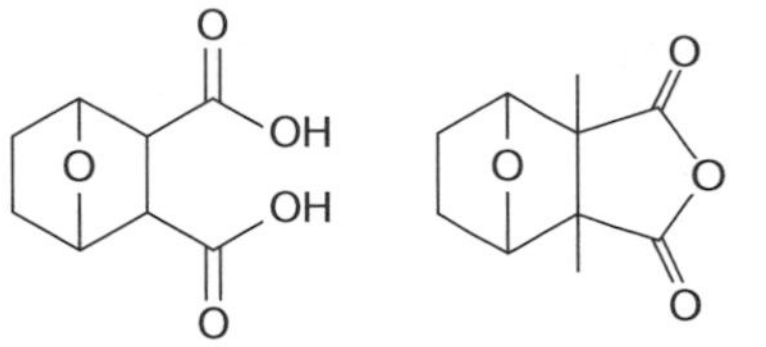

Figure 12.3 Structures of endothall and cantharidin.

12.4 Cyclin-dependent kinases and plant stress

Since plants are immobile and 'rooted to the spot', they have evolved mechanisms to enable them to perceive biotic and abiotic stressors and to alter gene expression, thereby enabling adaptation to these external factors in terms of growth and development. Typical biotic stressors are attack or competition by bacteria, fungi, viruses, animals and weeds, while abiotic stressors include temperature, light quality and quantity, drought and salinity.

Cyclin-dependent protein kinases are increasingly regarded as playing important roles in stress perception and transduction. Perception involves the activation of signalling cascades that result in a prolonged S-phase in the cell-cycle and so a delayed entry into mitosis is commonly observed.

Response to abiotic (environmental) stress involves the production of stress-responsive genes, including those that encode the late-embryogenesis-abundant (LEA) proteins and the protein kinases. The former act to protect cell membranes and proteins, assist in the re-naturation of damaged proteins and may act as hydration buffers. On the other hand, the protein kinases act as regulatory molecules in signal transduction cascades.

CDPKs and SnRKs (sucrose non-fermentation 1-related kinases) regulate stress-responsive gene expression, including the LEA genes. Overexpression of these genes stimulate root development and water and nutrient uptake, thus favouring tolerance to the stress. Once perceived, the signal is transduced via MAPK cascades, e.g. Figure 12.4.

As a case study, the molecular basis of plant adaptation to cold has become better understood in recent years. The expression of cold-regulated genes (COR) is crucial to cold tolerance. These include the dehydrins (proteins that confer membrane stabilisation and prevent protein aggregation by re-folding denatured proteins), pathogen-related proteins, chitinases and lipid-transfer proteins. Some also have anti-freeze activity (e.g. β-1,3-glucanase), which prevents ice-crystal formation. An increase in unsaturated fatty-acids in cell membranes also enhances cold tolerance.

Approximately 40% of COR genes and 46% of cold-regulated transcription factor genes (CBFs, C-repeat binding factors) are regulated by ICE1, a transcription factor that binds to the promotor CBF3/DREB1A (dehydration-responsive element binding protein 1A). ICE1 is therefore regarded as a master 'switch' in the regulation of transcription, that is itself regulated by ubiquitinylation and sumoylation. An overview is presented in Figure 12.5.

Temperature is a major environmental factor that affects plant growth and development. For a plant growing at, say, 25°C, cold stress is classified as chilling stress (0–15°C) and freezing stress (less than 0°C). In temperate climatic regions, chilling stress can induce cold acclimation, thereby generating tolerance to freezing.

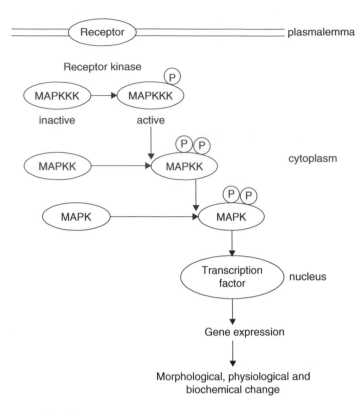

Figure 12.4   A typical MAPK cascade.

Plants perceive cold temperature at a protein receptor at the plasma membrane. Fluctuations in temperature cause changes in plasma membrane fluidity and rearrangement of the cytoskeleton to form bundles. De-polymerisation of the cytoskeleton is thought to be a pre-requisite for low temperature-induction of gene expression.

In *Arabidopsis*, a plasma membrane-localised, cold-responsive protein kinase (CRPK1) phosphorylates 13-3-3 proteins which destabilise the CBF (binding factor), preventing a cold-stress response during freezing.

Exposure to low temperatures, however, results in an influx of calcium ions from the apoplast and the vacuoles that leads to the activation of transcription factors, such as dehydration-responsive element binding (DREB). Calmodulin (CaM), CaM-like proteins (CML), calcium-dependent protein kinases and calcineurin B-like proteins (CBLs) are all calcium-binding proteins which can act as calcium sensors, to rapidly transduce the cold signal leading to the expression of COR.

MAPKs (mitogen-activated protein kinases) mediate the transduction of the cold signal. Typically, a MAPK cascade contains three protein kinases:

- MAP kinase kinase kinase (MAPKKK, MAP3K or MEKK);
- MAP kinase kinase (MAPKK, MKK or MEK); and
- MAP kinase (MAPK).

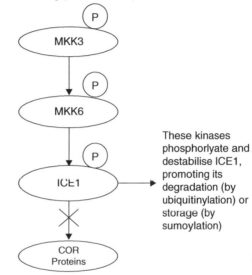

Figure 12.5   An overview of the cold signalling pathway (modified from Li *et al.*, 2017).

Induced by cold, MAPKKKs activate MKKs at serine/threonine residues; the activated MAKKs activate MAPKs by phosphorylating their threonine and tyrosine residues, leading to the activation of various effector proteins. In this way MKK2 interacts with MPK4 and MPK6 to regulate their activity, leading to the expression of COR genes to enhance tolerance to chill stress (Figure 12.6).

The reader is referred to the comprehensive review by Guo *et al.* (2018) for further details and recent findings, including how CBF gene expression is additionally regulated by other positive and negative factors, such as the perception of temperature by phytochrome.

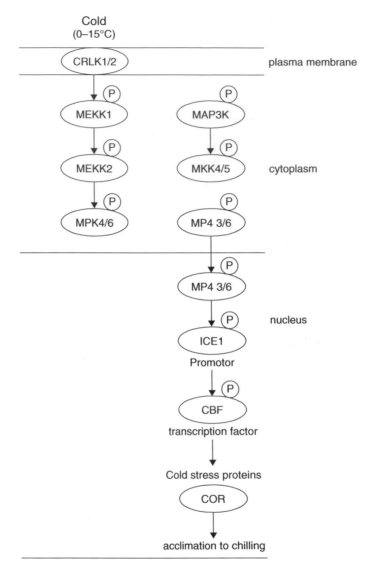

Figure 12.6 Cold acclimation in *Arabidopsis* (modified from Zhao *et al.*, 2017).

12.5   Post-translational modification of proteins

Phosphorylation is a common post-translational modification (PTM) for regulating enzyme activity, protein stability and protein–protein interactions. In doing so, plant kinases have now been documented as playing essential roles in regulating plant growth and development, metabolism, the cell cycle and responses to both biotic and abiotic stress. Precisely how this is achieved will no doubt be unravelled in the decades to come.

Table 12.1   The diversity and complexity of post-translational modifications.

Modification	Examples
Addition of chemical groups (reversible)	Phosphorylation, acetylation, methylation, redox-based modifications
Addition of polypeptides (enzymatically reversible)	Ubiquitination, sumoylation
Addition of complex molecules (reversible)	Acylation, prenylation, glycosylation, ADP-ribosylation, AMPylation
Direct modification of amino acids or of polypeptides (irreversible)	De-amidation, eliminylation
Protein cleavage (irreversible)	Proteolysis

Source: modified from Spoel, S.H. (2018) Orchestrating the proteome with post-translational modifications. *Journal of Experimental Botany* **69**, 4499–4503.

In recent years, additional post-translational modifications have been identified. They have advanced our understanding of how a relatively small genome can generate massive complexity in the output of a single gene. Indeed, the sheer number of possible post-translational modifications, combined with the number of possible amino acid residues, leads to an almost astronomical number of possible protein forms that can each support a different function (Spoel, 2018; Table 12.1).

Spoel (2018) envisages the employment of these modifications to enable a rapid and dynamic signalling by enzymes that then function as 'writers' and 'erasers' of post-translational modifications that are vast in number. Thus, in *Arabidopsis*, phosphorylation is regulated by more than 1100 genes encoding protein kinases and phosphatases, and over 1600 genes encode ubiquitin ligases and de-ubiquitinases. It is then argued that this variation and complexity allow plants to adapt at the molecular level to their ever-changing external environment.

In addition to the dynamic phospho-proteome, we can now add a range of regulatory possibilities, which are only now beginning to be unravelled. Post-translational modifications can now be assumed to act as 'directors' of proteome function and so act as major determinants of plant phenotype. The challenge is then to make physiological sense of these countless interactions, and how they might be manipulated in the search for new plant protection products.

The small ubiquitin-like modifier (SUMO) polypeptides appear to have important roles in plant development and responses to the environment, such as tolerance to drought. The conjugation and de-conjugation of SUMO polypeptides are tightly regulated and numerous SUMO proteases have been identified, as part of the cysteine protease family.

Acetylation of the protein *N*-terminus is also an abundant PTM, catalysed by *N*-acetyltransferases, with acetyl-coenzyme A acting as the C2 donor.

A recent paper by Borg *et al.* (2020) indicates that, via PTM, plants also have memories! Histones, which are important proteins for packaging and indexing DNA, can be modified to mark genes that are turned off. One such histone modification, termed H3K27me3, accumulates in the cold at genes that control flowering. This nomenclature indicates that the modification is a tri-methylation at the 27th lysine residue of the histone H3 protein. This 'epigenetic memory' is transmitted from cell to cell so that, when winter is over, plants will flower at the right time. Yet how do the plants forget this 'memory' of the cold when they set seeds? These researchers found that H3K27me3 disappeared in pollen, so the histone modification is erased from hundreds of genes, not only those that prevent flowering, but also ones which control many functions in seed function, which are produced when the pollen fuses with the plant egg cell. This phenomenon is termed 'epigenetic resetting', and is akin to erasing and reformatting data on a hard-drive.

Lysine methylation of non-histone proteins is also regarded as a common PTM that regulates molecular and cellular functions in plants by lysine methyltransferases, although the precise roles remain to be elucidated.

DNA methylation and de-methylation are important epigenetic markers in plants. DNA methylation is the addition of a methyl group to the fifth carbon of the cytosine ring by the enzyme activities of DOMAINS REARRANGED METHYLASE 2 (DRM2), METHYLTRANSFERASE 1 (MET1), CHROMOMETHYLASE 3 (CMT3) and CHROMOMETHYLASE 2 (CMT2). The methyl group cannot be directly removed from methyl cytosine. Instead, the base is removed by DNA glycosylases/lyases that belong to the RELEASE OF SILENCING 1 (ROS1) enzyme family. DNA de-methylation can be passively lost or actively removed, and regulatory factors for active de-methylation have been identified (Lin and Lang, 2020). DNA de-methylation is involved in fruit development and ripening, as demonstrated in many species, stress responses, nitrogen fixation and the development of epidermal and aleurone layer cells.

PTM of intracellular proteins by O-linked-N-acetyl glucosamine modulates protein activities and thereby controls plant cell function according to nutrient and energy status. In animals, serine and threonine residues are modified by O-linked-N-acetyl glucosamine acylation catalysed by a specific transferase. Acylation plays key roles in cellular homeostasis and responses to nutritional and stress factors. There are 971 O-linked-N-acylated peptides in 262 *Arabidopsis* proteins, one of which is AtACINUS, an analogue of the mammalian apoptopic chromatin inducer in the nucleus (ACINUS). Bi *et al.* (2021), have shown that the functions of AtACINUS are regulated by O-linked glycosylation and are involved in regulating seed germination, ABA sensitivity and flowering via the ABA-signalling pathway and the transcriptional regulation of the floral repressor, FLC. AtACINUS is modified by both O-linked-N-acetylglucosamine and also by O-fucose.

It is probable that more examples of PTM in plants will be discovered in the coming years and more functions discovered.

References

Bajsa, J., Pan., Z., Dayan, F.E., Owens, D.K. and Duke, S.O. (2012) Validation of serine/threonine protein phosphatase as the herbicide target site of Endothall. *Pesticide Biochemistry and Physiology* **102**, 38–44.

Bi, Y., Deng, Z., Ni, W., Shrestha, R., Davage, D., Hartwig, T., *et al.* (2021) *Arabidopsis* ACINUS is O-glycosylated and regulates transcription and alternative splicing of regulators of reproductive transitions. *Nature Communications* **12**, 945; doi: org/10.1038/s41467-021-20929-7

Borg, M., Jacob, Y., Susaki, D., LeBlanc, C., Buendia, D., Axelsson, E., *et al*. (2020) Targetted reprogramming of H3K27me3 resets epigenetic memory in plant paternal chromatin. *Nature Cell Biology*, May 2020; doi: org/10.1038/s41556-020-0515-y

Campos, R., Goff, J., Rodriguez-Furlan, C. and van Norman, J.M. (2020) The Arabidopsis receptor kinase IRK is polarised and represses specific cell divisions in roots. *Developmental Cell* **52**, 183–195; doi: org/10.1016/j.devcel.2019.12.001

Cao-Pham, A.H., Urano, D., Ross-Elliot, T.J. and Jones, A.M. (2018) Nudge, nudge, WNK–WNK (kinases), say no more? *The New Phytologist* **220**, 35–48.

Fuchs, S., Grill, E., Meskiene, I. and Schweighofer, A. (2013) Type 2C protein phosphatases in plants. *FEBS Journal* **280**, 681–693.

Guiley, K.Z., Stephenson, J.W., Lou, K., Barkovitch, K.J., Kumarasamy, V., Wijeratne, T.U., *et al*. (2019) P27 allosterically activates cyclin-dependent kinase 4 and antagonises Palbociclib inhibition. *Science* **366**, 1330.

Guo, Z., Liu, D. and Chong, K. (2018) Cold signalling in plants: insights into mechanisms and regulation. *Journal of Integrative Plant Biology* **60**, 745–756.

Henry, L.K., Thomas, S.T., Widhalm, J.R., Lynch, J.H., Davis, T.C., Kessler, S.A., *et al*. (2018) Contribution of isopentenyl phosphate to plant terpenoid metabolism *Nature Plants* **4**, 721–729.

Jadodzik, P., Tajdel-Zielinska, M., Ciesla, A., Marczak, M. and Ludwikow, A. (2018) Mitogen-activated protein kinase cascades in plant hormone signalling. *Frontiers in Plant Science*, October 2018; doi: org/10.3389/fpls.2018.01387

Klaeger, S., Heinzlmeir, S., Wilhelm, M., Polzer, B., Vick, B., Koenig, P.-A., *et al*. (2017) The target landscape of clinical kinase drugs. *Science* **358**, 1148.

Kline, K.G., Sussman, M.R. and Jones, A.M. (2010) Abscisic acid receptors. *Plant Physiology* **154**, 479–482.

Lehti-Shiu, M.D. and Shiu, S.-H. (2012) Diversity, classification and function of the plant kinase super-family. *Philosophical Transactions of the Royal Society B*, **367**, 2619–2639.

Li, H., Ding, Y., Shi, Y., Zhang, X., Zhang, S., Gong, Z. and Yang, S. (2017) MPK3- and MPK6-mediated ICE1 phosphorylation negatively regulates ICE1 stability and freezing tolerance in *Arabidopsis*. *Developmental Cell* **43**, 630–642.

Lin, R. and Lang, Z. (2020) The mechanism and function of active DNA de-methylation in plants. *Journal of Integrative Plant Biology* **62**, 148–159.

Lundquist, P.K., Davis, J.I. and van Wijk, K.J. (2012) ABC1K atypical kinases in plants: filling the organellar kinase void. *Trends in Plant Science* **17**, 546–555.

Merchante, C., Alonso, J.M. and Stepanova, A.N. (2013) Ethylene signalling: simple ligand, complex regulation. *Current Opinions in Plant Biology* **16**, 554–560.

Mishra, N.S., Tuteja, R. and Tuteja, N. (2006) Signalling through MAP Kinase networks in plants. *Archives of Biochemistry and Biophysics* **452**, 55–68.

Miura, K. and Furumoto, T. (2013) Cold signalling and cold tolerance in plants. *International Journal of Molecular Sciences* **14**, 5312–5337.

Morillo, S.A. and Tax, F.E. (2006) Functional analysis of receptor-like kinases in monocots and dicots. *Current Opinion in Plant Biology* **9**, 460–469.

Rademacher, E.H. and Offringa, R. (2012) Evolutionary adaptations of plant AGC kinases: from light signalling to cell polarity regulation. *Frontiers of Plant Science* **3**, 250.

Singh, A., Pandey, A., Srivastava, A.K., Phan Tran, L.-S. and Pandey, G.K. (2015) Plant protein phosphatases 2C: from genomic diversity to functional multiplicity and importance in stress management. *Critical Reviews in Biotechnology* **36**, 1023–1035.

Spoel, S.H. (2018) Orchestrating the proteome with post-translational modifications. *Journal of Experimental Botany* **69**, 4499–4503.

Tresch, S., Schmotz, J. and Grossmann, K. (2011) Probing mode of action in the plant cell cycle by the herbicide Endothall, a protein phosphatase inhibitor. *Pesticide Biochemistry and Physiology* **99**, 86–95.

Uemura, T., Hachisu, M., Desaki, Y., Ito, A., Hoshino, R. and Sano, Y. (2020) Soy and Arabidopsis receptor-like kinases respond to polysaccharide signals from Spodoptera species and mediate herbivore resistance. *Nature, Communications Biology* **3**, 224; doi: org/10.1038/s42003-020-0959-4

Zhao, C., Wang, P., Si, T., Hsu, C.-C., Wang, L.W., Zayed, O., *et al.* (2017) MAP kinase cascades regulate the cold response by modulating ICE1. *Developmental Cell* **43**, 618–629.

Chapter 13
Herbicide Resistance

Farmers only worry during the growing season, but townspeople worry all the time.

E.W. Howe (1853–1937)

13.1   Introduction

At 8 April 2020, there were 512 unique cases (i.e. species × target site) and 262 examples known of herbicide-resistant weed species in the world, made up of 152 dicots and 110 monocots, reported in 92 crops in 70 countries. Moreover, weeds have now evolved resistance to 23 of the 26 commercialised herbicide target sites and to 167 different herbicides (Heap, 2020). For comparison, in the summer of 2010, Heap reported 194 herbicide-resistant weed species in the world, towards 17 of the 26 commercialised targets. We must conclude that there has been a dramatic increase in herbicide-resistance in the last decade. Indeed, as an example, *Alopecurus myosuroides* (black-grass), is now regarded as the worst herbicide-resistant weed in Europe. Some level of black-grass resistance is now found in most of the 20,000 or so arable farms in the UK where herbicides are routinely used for weed control (Section 13.6). It follows that as long as herbicides play a part in global weed management, then resistance will remain a problem. The integration of herbicide usage and non-chemical weed management practices can reduce this risk.

In this chapter, herbicide resistance will be defined; reasons will be presented why it is so widespread; underlying mechanisms will be considered; examples of resistance to commercial herbicide group will be given; and strategies for the control of herbicide-resistant weeds will be highlighted, for now and the future.

13.2   Definition of herbicide resistance

Herbicide resistance is the inheritable ability of a plant to survive a herbicide dose that would be lethal to a member of a normal population of that plant species. Central to this definition are three important points:

1   Herbicide resistance is inheritable, so it is a characteristic coded for in the plant genome. At least some of the progeny of that plant will either be resistant or will carry the resistance

---

Herbicides and Plant Physiology, Third Edition. Andrew H. Cobb.
© 2022 John Wiley & Sons Ltd. Published 2022 by John Wiley & Sons Ltd.

trait. This distinguishes herbicide resistance from other causes of poor herbicide efficacy, perhaps caused by environmental factors including their effect on spray effectiveness, plant physiology and biochemistry.

2   Herbicide-resistant plants survive herbicide treatment and can successfully complete their life-cycle by flowering and producing seed. This does not mean that resistant individuals will not show symptoms of herbicide damage, but that they are not killed by herbicides. In many cases some herbicide damage may be observed, but it does not lead to plant death.

3   A normal population of the species is one in which, when treated with an optimum dose of a herbicide under ideal conditions, all individuals are killed. This population will be one that has never been exposed to a herbicide. Such 'wild-type' populations are not always available to the researcher, so populations that have demonstrated 100% susceptibility are often used in research, regardless of their field history. In order to determine baseline sensitivity for a particular herbicide acting on a distinct species, several normal populations are ideally used.

Resistance should not be confused with tolerance, which is the term used to describe an individual species that is not controlled by a particular herbicide. For example, cleavers (*Galium aparine*) is described as tolerant to isoproturon because the species as a whole is naturally tolerant to this active ingredient.

13.3   How resistance occurs

Herbicide resistance occurs through imposing selection pressure on a weed population. The selection pressure is the application of herbicides, which select for survival of members of the population that contain a mechanism imparting resistance to them. These individuals will survive and reproduce, so year upon year, as long as the same herbicides are used, the percentage of such individuals in a population will increase (Figure 13.1). In practice, population size plays a very important part in the occurrence of herbicide resistance. If selection pressure is exerted, but a population is kept small by other weed management methods, then the number of resistant individuals in the population may increase, but the population size will remain small, resulting in less of a problem. Conversely, where the population size remains large, the number of surviving resistant individuals will be large, and so greater problems are encountered. Therefore, resistance problems in practice are a result of repeated imposition of selection pressure coupled with agricultural practices that result in large weed populations.

Factors affecting the speed at which resistance develops include:

- the initial frequency of resistant individuals in the population;
- size of the seed population in the seed bank and the length of time the seeds remain viable;
- the susceptibility of individual weed plants;
- growth rate, fecundity and method of seed dispersal;
- repeated use of a single herbicide active ingredient or single mode of action, year upon year;
- the use of herbicides with long periods of residual activity in the soil;
- the use of herbicides with highly specific single modes of action.

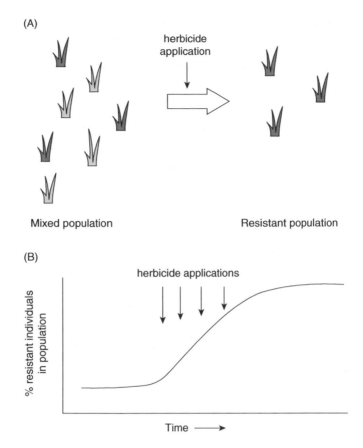

Figure 13.1 How resistance occurs. (A) Herbicide application imposing a selection pressure on a mixed population. Light grey, herbicide sensitive; dark grey, herbicide resistant. (B) Repeated use of the same herbicide will repeat the selection process.

13.4   A chronology of herbicide resistance

Herbicide resistance was first predicted by J.L. Harper in 1956. In 1957, the first cases of resistance to 2,4-D were reported in climbing dayflower (*Commelina diffusa* Burm.f.) in Hawaii and wild carrot (*Daucus carota* L.) in Canada (cited in Busi *et al.*, 2017). Ryan (1970) reported resistance to simazine and atrazine in groundsel (*Senecio vulgaris*) in 1968. Since then, the number of reports of herbicide-resistant weeds has increased every decade, and is most prevalent against acetolactate synthase (ALS) inhibitors, triazines and acetyl-CoA carboxylase (ACCase) inhibitors. Although they were only introduced to the market in 1982, resistance to ALS inhibitors is the most widespread mode of action for which weed resistance has been reported. This is probably a combined result of the extensive use of these herbicides, the number of ALS inhibitors on the market (Table 9.1) and the specificity of the site of action.

Table 13.1   Number of herbicide-resistant weeds reported by country.

Rank	Country	Number
1	USA	165
2	Australia	95
3	Canada	68
4	France	55
5	Brazil	50
6	China	45
7	Spain	39
8	Israel	37
9	Japan	36
10	Germany	30
11	Italy	30
12	UK	29

Source: modified from Heap, I. (2020) *The International Herbicide-Resistance Database*. www.weedscience.org.

As noted by Heap (2020), five weed families account for about 70% of all cases of herbicide resistance. The grasses (Poaceae), crucifers (Brassicaceae) and the pigweeds (Amaranthaceae) appear to be very prone to the development of resistance, compared with other plant families. Both farmers and the agrochemical industry must be concerned that weeds have now evolved resistance to 23 of the 26 commercialised herbicide target sites and to 167 different herbicides, up from the 17 target sites noted by Heap in 2010.

Annual ryegrass (*Lolium rigidum*) is probably the world's worst herbicide-resistant weed, having evolved resistance to 11 herbicide target sites in 12 countries in millions of hectares of arable land, especially in Australia, where it was originally introduced as a grass feed for livestock. It has a high degree of genetic variability and rapidly evolves resistance to almost any herbicide that it is exposed to. It is especially troublesome because it has often evolved cross-resistance (both target site and non-target site) and rapidly evolves multiple resistance to a wide variety of herbicides by outcrossing. At least one herbicide-resistant weed has now been reported in 70 countries, according to a list currently led by the USA (Table 13.1).

Herbicide resistance will probably continue to increase, especially where the number of herbicide groups with different modes of action is reduced, where fewer novel target sites are being introduced, where increasing production costs reduce herbicide choice and where genetically modified, herbicide-tolerant crops (Chapter 14) encourage the repeated use of a single herbicide.

An example of the latter is the rapid spread of glyphosate-resistant *Digitaria insularis* (sourgrass) in Brazil and Paraguay (Lopez-Ovejero *et al.*, 2017). Sourgrass can reduce yields of maize and soybean by up to 32 and 44%, respectively. Glyphosate-resistant soybean was 93% of the Brazilian crop in 2014, equivalent to 30 million hectares. Lopez-Ovejero *et al.* (2017) performed a comprehensive study of the frequency and dispersion of glyphosate-resistant sourgrass throughout Brazil and found it in all soybean production areas, suggesting the need to implement integrated weed management strategies for the effective control of this species (Figure 13.2).

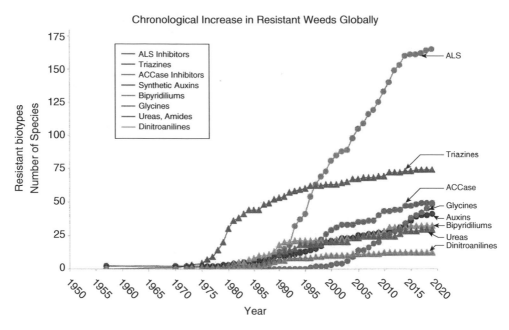

Figure 13.2 Development of herbicide resistance in weeds, from Ian Heap, accessed 18th April 2020. *See color inset section for the color representation of this figure.*

13.5   Mechanisms of herbicide resistance

13.5.1   *Target site resistance*

Herbicides have distinct target sites where they act to disrupt biochemical processes leading to cell, tissue and plant death. The majority of target sites are enzymes and the interaction between herbicide and target site can be disrupted if there is a change in the primary structure of the enzyme protein molecule. Where this occurs, the herbicide may no longer be effective in blocking the action of the target site and the plant will not die, but exhibit herbicide resistance (Figure 13.3).

Target site resistance is often the result of a single point mutation in the gene coding for the target site, so it is the result of a change in one nucleic acid in the target site gene, resulting in one amino acid change in the final target site protein.

13.5.1.1   *Target site-mediated resistance to ACCase-inhibiting herbicides*

A number of mutations in the ACCase gene have been reported in grasses, which result in target site resistance to ACCase-inhibiting herbicides. Resistance was first observed in 1982 in *L. rigidum*, only four years after the introduction of diclofop-methyl. It is fascinating to note that target site mutations have been found in herbarium samples of grasses collected between 1788 and 1975, i.e. before their commercial release! This implies a long-standing genetic variation in grass weed populations rather than *de novo* mutations in the field.

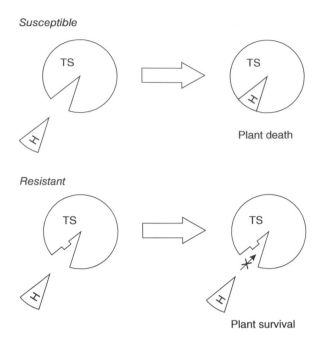

Susceptible

TS

TS

H

Plant death

Resistant

TS

TS

H

H

Plant survival

Figure 13.3   Diagrammatic representation of target site resistance. TS, Target site; H, herbicide.

Target-site mutations in ACCase are well characterised, with 16 listed by Takano *et al.* (2020a, b) at carboxytransferase sites 1781, 1999, 2007, 2041, 2078, 2088 and 2096. Different cross-resistance patterns are noted for each mutation (Devine, 1997; Devine and Shukla, 2000; Délye, 2005). An isoleucine-to-leucine substitution within the carboxytransferase region of ACCase confers resistance in *Avena fatua* (Christoffers *et al.*, 2002), *L. rigidum* Gaud. (Zagnitko *et al.*, 2001; Délye *et al.*, 2002a; Tal and Rubin, 2004), *Setaria viridis* (L.) Beauv. (Délye *et al.*, 2002b) and *A. myosuroides* Huds (Délye *et al.*, 2002a). In *A. myosuroides* this mutation is designated I1781L (Délye *et al.* 2002a), indicating the position where the amino acid substitution has taken place. In all cases this mutation confers cross-resistance to all fops and most dims. Another mutation (Ile 2041 → Asn), also in the carboxytransferase region of ACCase, has been reported in *L. rigidum* and *A. myosuroides*, which confers resistance to fops, but not to dims (Délye *et al.*, 2003). These observations may increase our understanding of the differences in herbicide binding to ACCase between fops and dims.

In Argentina, *Lolium* species occur in 40% of winter-cereal crops and ACCase inhibitors have been applied for the control of grass weeds. Resistance to pinoxaden was due to single point mutation in ACCase at Asp-2078-Gly, and resistant plants also required much greater concentrations of clethodim and quizalofop for control (Yanniccari and Gigon, 2020).

In South America *D. insularis* (sourgrass) is a problematic weed owing to glyphosate resistance and has also evolved resistance to ACCase inhibitors. Takano *et al.* (2020a, b) have recently demonstrated that resistance is due to a target site mutation, Trp-2027-Cys, conferring a high resistance to haloxyfop, a low resistance to pinoxaden and no resistance to clethodim. The authors conclude that the mutation obstructs access to the binding of fops, while permitting the binding of dims and dens.

Since climate projections indicate an increase in mean global temperature and carbon dioxide concentrations in the atmosphere by the end of the twenty-first century, it is important to understand and predict how climate change might influence herbicide resistance in years to come. Refatti *et al.* (2019) examined the effects of increasing temperature and carbon dioxide concentration on the absorption, translocation and efficacy of cyhalofop-butyl on multiple resistant and susceptible *Echinochloa colonum* under simulated climate change conditions. The multiresistant biotype had extreme resistance to quinclorac and propanil, which involves non-target site resistance mechanisms. Herbicide efficacy was reduced by about 50% in multiple resistant plants compared with plants grown at ambient conditions. This observation implies that a changing climate might increase selection pressure in favour of resistance, which is a warning call to weed management practitioners.

13.5.1.2   Target site-mediated resistance to ALS-inhibiting herbicides

Many of the cases of herbicide resistance to ALS inhibitors reported to date are caused by an altered target site. Five naturally occurring mutations are found in native populations that give rise to resistance. Many others are found in bacteria and yeast systems. Mutations are found in two regions: A, amino acid sequences 124–205; and B, amino acid sequences 574–653 (Gressel, 2002; Table 13.2). This is not the catalytic site of the enzyme, but a

Table 13.2   A selection of mutations that confer target site resistance to ALS, after Gressel, 2002. In some cases different cross-resistance patterns have been reported for a single mutation by different researchers.

Mutation	Gene name	Selector and cross-resistances	Little or no cross-resistance
$Met_{124} \rightarrow Glu$		SU, IM	
$Met_{124} \rightarrow Ile$		IM	
$Ala_{155} \rightarrow Thr$		SU, IM, TP, PC	SU, TP
$Pro_{197} \rightarrow His$		SU, some IM	
$Pro_{197} \rightarrow Gln$	C3	SU, some IM	
$Pro_{197} \rightarrow Ala$	S4	SU, some IM	
$Pro_{197} \rightarrow Ser$	Crs-1	SU, TP	IM, PC
$Pro_{197} \rightarrow Arg$		SU, some IM	
$Pro_{197} \rightarrow Leu$		SU, some IM	
$Pro_{197} \rightarrow Thr$		SU, some IM	
$Arg_{199} \rightarrow Ala$			
$Arg_{199} \rightarrow Glu$			
$Ala_{205} \rightarrow Asp$			
$Gln_{269} \rightarrow His$		IM, TP, PC, SU	
$Asn_{522} \rightarrow Ser$			
$Tro_{574} \rightarrow Leu$	ALS3	IM, TP, PC, SU	
$Trp_{574} \rightarrow Ser$		IM > SU	
$Trp_{574} \rightarrow Phe$		IM, TP, SU	
$Ser_{653} \rightarrow Asn$	Csr l-2, imr	IM, PC	SU, TP
$Ser_{653} \rightarrow Thr$		IM	SU
$Ser_{653} \rightarrow Phe$		IM > SU	

SU, Sulphonylurea; TP, triazolopyrimidine sulfonamide; IM, imidazolinone; PC, pyrimidyloxybenzoate.
Source: modified from Gressel, J. (2002) *Molecular Biology of Weed Control*. London: Taylor & Francis.

separate herbicide-binding site. Different patterns of cross-resistance are noted for each of these amino acid substitutions, and the flexibility of substitution while still maintaining enzymic activity has probably contributed to the rapid rise in resistance to this class of herbicide. Resistance to imidazolinones, conferred by Ser670 → Asp or Ala122 → Thr, does not confer very high levels of resistance to the other classes of ALS inhibitor (Sathasivan *et al.*, 1990, 1991; Bernasconi *et al.*, 1995). Studies with ALS-resistant *Arabidopsis thaliana* mutants *csr*1-1 (Pro197 → Ser) and *csr*1-2 (Ser653 → Asn) identified complex cross-resistance patterns with respect to 22 ALS-inhibiting herbicides (Roux *et al.*, 2005). This suggests complex and non-ubiquitous binding characteristics for this herbicide family.

More recent investigations reveal further altered ALS amino acid sequences in resistant plants. Examples include an Asp-376 substitution in *Sorghum halepense* (Panozzo *et al.*, 2017), Pro-197-Ser in *Amaranthus palmeri* (Nakka *et al.*, 2017), Pro-197-Ala in *Myosoton aquaticum* (Bai *et al.*, 2019) and Trp-574-Arg in *Echinochloa crus-galli* (Fang *et al.*, 2019). The latter authors note 28 types of amino acid substitutions at eight conserved positions in ALS, namely Ala122, Pro197, Ala205, Asp376, Arg377, Trp574, Ser653 and Gly654. Mutations at Ala122, Trp574 and Ser653 also confer cross-resistance to ALS inhibitors in *Echinochloa* species.

It is also worth noting that resistance to ALS inhibitors continues to evolve, and resistance patterns have changed in the last decade (e.g. Li *et al.*, 2019).

13.5.1.3   *Target site-mediated resistance to Photosystem II-inhibiting herbicides*

In 1970 Ryan reported the failure to control groundsel (*S. vulgaris*) with simazine and atrazine in a nursery in Washington State, USA. These residual herbicides had been used once or twice a year in the nursery from 1958 to 1968. Susceptible plants grown from seeds collected from a location where triazines had not been in continuous use were adequately controlled by 0.56–1.12 kg ha^{-1}, but seedlings from the nursery were barely affected, even by doses as high as 17.92 kg ha^{-1}. Triazine resistance is now known to have evolved independently throughout the world owing to persistent and prolonged use in monocultures in orchards, vineyards, nurseries and maize crops. Cross-resistance is commonly observed (Table 13.3).

Table 13.3   Photoreduction of DCPIP by thylakoids isolated from atrazine-resistant and -susceptible biotypes of oilseed rape (*Brassica* napus L. cv. Candelle; A.H. Cobb, unpublished observations).

Herbicide	Resistant (R)	Susceptible (S)	R/S
Atrazine	7.5×10^{-4}	1.0×10^{-8}	75,000
Phenmedipham	1.0×10^{-5}	2.0×10^{-7}	50.00
Diuron	1.75×10^{-6}	6.5×10^{-8}	26.90
Metamitron	9.0×10^{-6}	1.8×10^{-6}	5.00
Bentazone	8.0×10^{-5}	6.0×10^{-5}	1.33
Ioxynil	2.6×10^{-7}	3.5×10^{-7}	0.74
Dinoseb	1.1×10^{-7}	0.8×10^{-6}	0.14

Values are the molar concentration of herbicide needed to inhibit the photoreduction of DCPIP by 50% (IC_{50}). The resistance ratio IC_{50R}/IC_{50S} reflects the degree of resistance measured in the assay.

A single mutation in the *psbA* gene coding for a 32 kD protein that forms part of the Photosystem II (PS II) complex in the thylakoids can lead to resistance to the herbicides that inhibit photosynthesis at PS II (Oettmeier, 1999). These single mutations cause resistance to a number of PS II inhibiting herbicides to varying degrees (see Table 5.2). In some instances 1000 times more herbicide is needed to displace Q_B from this site. As this binding site is common to many PS II-inhibiting herbicides, the cross-resistance observed is not surprising. However, cross-resistance does not always extend to all PS II herbicides as is clearly demonstrated in Table 5.2. The mutation Ser264 → Gly results in decreased binding and efficacy of triazines, but not other classes of PS II inhibitors. The mutation Ser264 → Thr results in a broader spectrum of resistance to phenylureas and the triazine herbicides, for example linuron and atrazine, respectively. Resistance to diuron and metribuzin is conferred by the mutation Val219 → Ile. In addition to these mutations of the D1 protein, Ala251 → Arg and Val280 → Leu are also found in weed populations in the wild. In addition, laboratory studies have also identified a further mutation, Ser268 → Pro, that also confers resistance. It is interesting to note that there is a fitness price to pay for these mutations to the D1 protein, with reduced CO_2 fixation, quantum yield and biomass accumulation, all reported for biotypes possessing a mutation to the D1 protein that results in herbicide resistance. The mechanism of this fitness price has not been satisfactorily elucidated to date, but it has been shown to affect electron flow and chlorophyll a:b ratios and to increase damage in high-intensity light conditions (Holt and Thill, 1994).

Lu *et al.* (2018) collected two atrazine-resistant populations of wild radish (*Raphanus raphanistrum*) in Western Australia and investigated the mechanisms of resistance. They sequenced the psbA gene and found the expected Ser-264-Gly mutation in population R1, which conferred a high level of resistance to atrazine, but sensitivity to bromoxynil. On the other hand, population R2 had a novel Phe-274-Val mutation, giving moderate resistance to atrazine, metribuzin and diuron, but also sensitivity to bromoxynil.

Further novel mutations at PS II have been discovered in the control of pigweeds (*Amaranthus* spp.) in carrot production in Ontario, Canada (Davis *et al.*, 2020). Resistance to linuron was due to several mutations in psbA, including Val-251-Ile, which was most common, but also populations with Ala-251-Ile and Phe-274-Val. All substitutions gave varying levels of resistance to linuron. The authors conclude that such widespread failure of linuron to control these weeds in carrot crops is due to the selection of the weeds with the mutated psbA sequences, suggesting the need for a change in weed management strategies.

13.5.1.4   Target site-mediated resistance to Photosystem I-diverting herbicides

There have been only limited reports of resistance to this class of herbicide and little is known regarding the resistance mechanisms responsible. No evidence of target site resistance involvement has yet been found and it is thought that most cases of resistance are a result of enhanced herbicide metabolism or sequestration away from the site of action. In *Lolium perenne* and *Conyza bonariensis* increased levels and activities of enzymes that detoxify active oxygen species have been measured in resistant biotypes. These are superoxide dismutase, ascorbate reductase and glutathione reductase (Shaaltiel and Gressel, 1986). Other studies with *C. bonariensis* have suggested the immobilisation of paraquat in resistant biotypes, possibly by binding to cell wall components, so that less of the herbicide can reach the thylakoid. Alternatively, paraquat uptake and movement may be reduced in resistant biotypes of *Hordeum glaucum*.

No evidence of target site resistance has been found to this class of herbicide. Most cases of reported resistance are due to reduced translocation and/or vacuolar sequestration. Interestingly, Li *et al.* (2013) reported an *Arabidopsis* paraquat-resistant mutant (PAR1) that showed strong resistance to paraquat. PAR1 encodes an amino acid transporter localised to the Golgi body. This protein is not responsible for the intercellular uptake of paraquat, but caused a reduction of paraquat accumulation in the chloroplasts of the mutant. Over-expression of the gene in rice (OsPAR1) resulted in an increased sensitivity to the herbicide.

13.5.1.5   *Target site-mediated resistance to cell division inhibitors*

Dinitroanilines have been used annually for approaching four decades in the cotton fields of the USA and on oilseeds and small-grain cereals in the Canadian prairies. It was therefore not unexpected when resistance was reported in 1984 for *Eleusine indica* (L.) Gaertn. (goosegrass), in 1989 for *S. viridis* (green foxtail) and in 1992 for *S. halepense* (Johnsongrass), as detailed by Smeda and Vaughn (1994). In *E. indica* the resistant biotype has shown cross-resistance to all dinitroanilines, dithiopyr and amiprophos-methyl, but sensitivity to propyzamide and chlorthal-dimethyl, and enhanced sensitivity to the carbamates (Vaughn *et al.*, 1987). Furthermore, there seem to be no differences in fitness between the resistant and susceptible biotypes.

In resistant biotypes of *S. viridis* and *E. indica* a specific base change in the sequence of the *TUA1* gene causes an amino acid substitution in α-tubulin from threonine to isoleucine at position 239. This results in a conformational change to the surface of the α-tubulin molecule, which prevents herbicide binding.

In both species, resistance is controlled by nuclear, recessive genes that are mutant alleles of an α-tubulin gene that causes a substitution of either Thr239 to Ile, or Met268 to Thr in *E. indica* or Leu136 to Phe in *S. viridis*. Thr239 → Ile is reported to give rise to resistance to the dinitroanaline herbicides, whereas Met268 → Thr gives rise to low levels of trifluralin resistance (Vaughn *et al.*, 1990; Yamamoto *et al.*, 1998). The recessive nature of the resistant gene probably accounts for the relative rarity of target site resistance to these herbicides. Indeed, resistance to the microtubule inhibitors has only been reported in 12 weed species.

13.5.1.6   *Target site-mediated resistance to glyphosate*

The mode of action of glyphosate is to inhibit the enzyme EPSP synthase (Chapter 9). A single mutation in the gene encoding this enzyme is reported to have resulted in glyphosate resistance in *E. indica* in Malaysia (Lee and Ngim, 2000). The same mutation (Pro106 → Thr) has also been reported in the EPSP synthase gene from herbicide-resistant *L. rigidum* in Australia (Wakelin and Preston, 2006). A mutation has also been documented at Ala 103. A further case of glyphosate resistance appears to be due to an increase in the transcription of EPSP synthase, resulting in twice the amount of enzyme (Gruys *et al.*, 1999). Wiersma *et al.* (2015) investigated the mechanism of resistance to glyphosate of *Kochia scoparia* and found that the EPSPS copy number and transcript abundance were elevated 3- to 10-fold in resistant plants.

For such an extensively used herbicide, there were few reports of glyphosate-resistant weed populations between 1970 and 1995. The withdrawal of paraquat, a reduction in the cost of glyphosate following the expiry of its patent and the introduction of glyphosate-tolerant crops in 1996 has had a profound effect on weed control (Chapter 14). The expansion of glyphosate-tolerant maize, cotton and soybean in the USA and South America in particular has selected for a range of glyphosate-resistant weeds, with 43 reported so far. In particular, glyphosate-resistant *Amaranthus hybridus* and *palmeri* have been observed in South America and a triple mutation in the EPSPS gene of Thr-102-Ile, Ala-103-Val and Pro-106-Ser was first observed in Argentina in *A. hybridus* (Perotti *et al.*, 2018). Resistant plants completed a normal life-cycle even after a 32-fold increased dose of glyphosate. This is the first report of a naturally occurring triple mutation at the EPSPS target site.

According to Duke (2019), there are many mechanisms now known whereby resistance to glyphosate has evolved, largely owing to massive selection pressure, which is greater than any other herbicide or herbicide class. In addition to target site mutations, there are examples of EPSPS gene duplication and amplification.

13.5.1.7 Target site-mediated resistance to auxin-type herbicides

Although auxin-type herbicides have been extensively used for approaching 80 years there are few reported examples of weeds that are resistant to them. This is not too surprising, as these herbicides are generally foliar-applied and non-persistent, and it is difficult to envisage the generation of the major selection pressures needed to favour the evolution of resistant weeds. However, in 1985 Lutman and Lovegrove demonstrated an approximately 10-fold difference in sensitivity to mecoprop in two populations of chickweed (*Stellaria media*), the more resistant population being found in grassland where mecoprop was seldom used. In subsequent studies Lutman *et al.* suggested that resistance was not due to differences in the uptake or movement of mecoprop, but because more of the herbicide was apparently immobile in resistant plants, presumably bound to structural polymers (Coupland *et al.*, 1990). An alternative hypothesis has been proposed by Barnwell and Cobb (1989) and Cobb *et al.* (1990). Their studies confirmed the original observations of resistance by Lutman and Lovegrove (1985) and reported reduced vigour in resistant plants (Figure 13.4). Furthermore, an investigation of H^+-efflux induced by mecoprop in etiolated *S. media* stems revealed that 170,000 times more mecoprop was needed in resistant plants to induce an amount of efflux equivalent to that obtained from susceptible plants. Such differences in mecoprop binding could imply an alteration in an auxin receptor in resistant plants that may also contribute in mecoprop resistance in this important weed species.

Further examples of increased tolerance of weeds to auxin-type herbicides have been observed in New Zealand and reported by Popay *et al.* (1990). These include populations of nodding thistle (*Caruns nutans*) requiring 5–30 times more MCPA and 2,4-D than expected, and populations of giant buttercup (*Ranunculus aris*) being tolerant of 4.8 times the dose of MCPA compared with susceptible populations. Since uptake, translocation and enhanced metabolism have been eliminated as resistance mechanisms in some cases of resistance to auxin analogues, then it seems a distinct possibility that mutations in the auxin-binding protein might be involved in the resistance (Peniuk *et al.*, 1992). However, mutation in the auxin-binding protein has only been confirmed in one case, *Sinapis arvensis*. In this case the auxin receptor did not bind auxin-type herbicides. However, it also did not bind endogenous auxins either and this represents a severe fitness price for possessing herbicide resistance. Plants containing the insensitive auxin receptor lacked apical dominance and, in the absence of auxin-type herbicides, were very uncompetitive owing to their heavily branched growth habit and their lack of vertical growth (Deshpande and Hall, 2000; Webb and Hall, 1999).

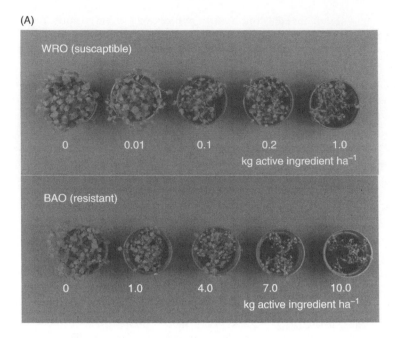

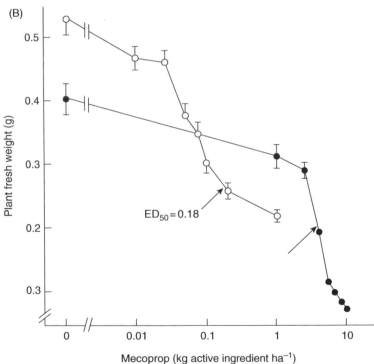

Figure 13.4 Mecoprop resistance in *Stellaria media* (chickweed). (A) Plants photographed from above. (B) ED$_{50}$ graph. ●, WRO (susceptible); ●, BAO (resistant). Source: Barnwell, P. and Cobb, A.H. 1989, Physiological studies of mecoprop-resistance in chickweed (*Stellaria media* L.). *Weed Research* **29**, 135–140. Reproduced with permission of John Wiley & Sons.

Understanding of resistance-mechanisms to auxin-type herbicides remains obscure. Prior to 2018, target site resistance to auxin-type herbicides had not been reported. In that year, LeClere *et al.* reported target site resistance in *Kochia scoparia* resulting from a single amino acid substitution at Gly-73-Asn in an IAA co-receptor that resulted in a loss of interaction between IAA16 and TIRI, a finding which was also confirmed in *A. thaliana*.

Poppy (*Papaver rhoeas*) is a common broad-leaf weed in cereals in Europe. Torra *et al.* (2017) investigated increased resistance to 2,4-D in Spain and concluded that enhanced metabolism of the herbicide to 2,3-D and 2,5-D was evident in resistant populations, which could be reversed by pre-treatment with a cytochrome P450 inhibitor. Reduced translocation may also contribute to herbicide resistance in this species.

Wild radish (*R. raphanistrum*) is an important dicot weed in South Australia, resulting in yield losses and increased weed control costs. Goggin *et al.* (2016) applied radiolabelled 2,4-D to leaves and found that phloem loading and transport were impaired in resistant plants. The auxin transporting ATP-binding cassette (ABC transporters) of the B subfamily facilitates polar auxin transport and is often associated with the transport regulating proteins, TWD1s, and the auxin efflux proteins, PINs, to ensure effective auxin transport at the plasma membrane. It is thought that mutations in these proteins impair the transport of auxin-type herbicides. Sequence analysis from ABCB genes and gene products may resolve the basis of resistance in this and other resistant species (Schulz and Segobye, 2016).

13.5.1.8 *Target site-mediated resistance to cellulose biosynthesis inhibitors*

There have been no reports of resistance in the field to the cellulose biosynthesis inhibitors. Equally, the generation of resistant mutants of *Arabidopsis* in the laboratory has been difficult (Vaughn, 2002). Selection from extensive mutant screens eventually generated isoxaben- and dichlobenil-resistant plants, which were not cross-resistant to either flupoxam or isoxaben, indicating that the three herbicides did not share a common site. Furthermore, none of the mutants exhibited enhanced herbicide metabolism. Vaughan (2002) reports that at least one of the isoxaben-resistant mutants has an alteration in a cellulose synthase gene. Further confirmation of this finding at the gene sequence is awaited.

13.5.1.9 *Target site resistance to phytoene-desaturase inhibitors*

Evolved target site resistance to PDS inhibitors has only been reported in the last decade. Substitutions of Arg 304 to Ser, Cys or His have been noted in six species. A Leu-498-Val mutation, equivalent to the Leu 538 position in rice, was observed in oriental mustard (*Sisymbrium orientale*) in Australia, corresponding to resistance to pyridazinones in several species. Substitution at both Glu-425-Asp and Leu-498-Val revealed resistance to both diflufenican and picolinofen (Murphy and Tranel, 2019). On the other hand, diflufenican resistance in a *R. raphanistrum* population in Western Australia was most likely due to non-target site-enhanced metabolism involving cytochrome P450s (Lu *et al.*, 2019).

13.5.1.10 *Target site resistance to PROTOX inhibitors*

As PROTOX inhibitors have target sites in both the chloroplasts and mitochondria, resistance can be at both locations. Reported mutations include the deletion of a Gly at position 210, and substitutions of Arg 128 to Leu, Gly or Met. An Arg-128-substitution confers resistance to the diphenyl-ethers and triazolinones. A Gly-399-Ala substitution was reported in *A. palmeri* (Rangani *et al.*, 2019). Crystal structure analysis revealed that the substitution decreased the size of the binding-pocket to diphenyl-ethers.

13.5.1.11 Target site resistance to glutamine synthetase inhibitors

Two isoforms of glutamine synthetase exist in plants. One is nuclear-encoded and targeted to the cytoplasm and the other is nuclear-encoded and targeted to the chloroplast. Both forms are inhibited by glufosinate. An Asn-171-Asp substitution reported by Avila-Garcia *et al.* (2012) was the first report of an altered target site for this enzyme.

13.5.1.12 Target site resistance to HPPD inhibitors

Some rice varieties are sensitive to benzobicyclon, a triketone inhibitor of HPPD. Maeda *et al.* (2019) have identified a rice gene HIS1 (HPPD Inhibitor sensitive 1) that confers resistance to benzobicyclon. HIS1 encodes an oxyglutarate-dependent oxygenase that detoxifies triketone herbicides by catalysing their hydroxylation. When expressed in *Arabidopsis* it gave resistance to four additional triketone herbicides. The authors conclude that this gene may prove useful in breeding herbicide-resistant crops.

13.5.2 Non-target site resistance

13.5.2.1 Enhanced metabolism resistance

Plants possess a host of enzymes for the metabolism of xenobiotics and unwanted substances. It is these detoxifying enzymes that modify or break down herbicides once they enter a plant cell. The rate at which these enzymes carry out this task will determine whether a plant lives or dies, and is the main contributor to herbicide selectivity between crops and weed species (see Chapter 4 for more information about the enzyme systems involved).

If an individual weed biotype within a population has the ability to metabolise a herbicide at an increased rate, then it may survive a herbicide treatment. Such biotypes are described as possessing enhanced metabolism resistance (Figure 13.5). This type of resistance has been reviewed by Yuan *et al.* (2006).

The main enzymes involved in metabolism of xenobiotics in plants are the cytochrome P450s and glutathione *S*-transferases (GSTs), esterases and UDP-glucosyl transferases. These enzyme families have both been implicated in herbicide metabolism in tolerant crop and weed species and also in biotypes possessing enhanced metabolism resistance (Table 13.4). Enhanced GST activity has been correlated with resistance to isoproturon, clodinafop-propargyl and fenoxaprop-*p*-ethyl in *A. myosuroides* populations and offers a possible screening test for resistance in this and other species (Reade and Cobb, 2002). Increased GST and P450 activity may be due to a more active enzyme, increased transcription of gene(s) coding for the enzyme, or the presence of more copies of the gene(s).

Cytochrome P450 monooxygenases are a large and ubiquitous family of enzymes that carry out a number of reactions (see Chapter 4). Enhanced P450 activity has been reported in some weed biotypes that demonstrate herbicide resistance, implying that increased herbicide metabolism by P450s is the mechanism of herbicide resistance (Katagi and Mikami, 2000; Mougin *et al.*, 2001). Inhibitors of P450 activity (e.g. 1-aminobenzotriazole) have been shown to reduce herbicide metabolism and also levels of herbicide resistance in some resistant populations (e.g. Singh *et al.*, 1998). It appears that in most cases the P450 is

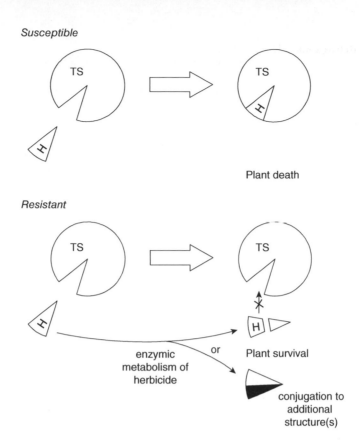

Susceptible

TS

TS

Plant death

Resistant

TS

TS

enzymic
metabolism of
herbicide

or

Plant survival

conjugation to
additional
structure(s)

Figure 13.5 Diagrammatic representation of enhanced metabolism resistance. TS, Target site; H, herbicide.

Table 13.4 Weed species for which glutathione S-transferases (GSTs) and/or P450s have been implicated in herbicide resistance (From Devine and Preston, 2000).

Weed species	Herbicide(s)[a]	Proposed enzymatic system
Alopecurus myosuroides	Chlorotoluron Pendimethalin Diclofop-methyl Fenoxaprop-p-ethyl Propaquizafop Chlorsulfuron	Cytochrome P450 monooxygenases/ glutathione S-transferase
Abutilon theophrasti	Atrazine	Glutathione S-transferase
Avena sterilis	Diclofop-methyl	Cytochrome P450 monooxygenases
Avena fatua	Triallate	Cytochrome P450?
Digitaria sanguinalis	Fluazifop-p-butyl	Unknown
Echinochloa colonum	Propanil	Aryl acylamidase
Echinochloa crus-galli	Propanil	Aryl acylamidase
Hordeum leporinum	Fluazifop-p-butyl	Unknown
Lolium rigidum	Simazine Diclofop-methyl Fluazifop-p-butyl Tralkoxydim Chlorsulfuron Metribuzin Chlorotoluron	Cytochrome P450 monooxygenases/ unknown
Phalaris minor	Isoproturon	Cytochrome P450 monooxygenases

[a]'p' in this column refers to herbicidally active isomer.

Source: Devine, M.D. and Preston, C. (2000) The molecular basis of herbicide resistance. In: Cobb, A.H. and Kirkwood, R.C. (eds) *Herbicides and their Mechanisms of Action*. Sheffield and Boca Raton, FL: Sheffield Academic Press/CRC Press, pp. 72–84.

not induced in resistant biotypes by herbicide treatment but is constitutively expressed. This raises the interesting possibility that the protection mechanisms usually induced by herbicide application, often too late in the case of susceptible biotypes, are already being expressed in the resistant biotypes. Where P450s are responsible for resistance, then wide-ranging and unpredictable levels of cross-resistance can be encountered. This will be determined by the means by which a herbicide is metabolised rather than by its mode of action.

According to Gaines *et al.* (2020), cytochrome p450-mediated herbicide resistance has now been identified in broad-leaf weeds, including to auxin-type herbicides, and inhibitors of microtubule assembly, phytoene desaturase, deoxy-xylulose 5-phosphate synthase (DXS), 4-hydroxyphenypyruvate dioxygenase and PPO inhibitors.

Glutathione *S*-transferases (E.C. 2.5.1.18) are a superfamily of enzymes which catalyse the conjugation of a wide range of substrates, including some herbicides and xenobiotics, to the tripeptide glutathione (GSH, γ-Glu-Cys-Gly; see Chapter 4). Researchers have suggested a role for GSTs in the metabolism of atrazine (Anderson and Gronwald, 1991; Gray *et al.*, 1996), alachlor, metolachlor and fluorodifen (Hatton *et al.*, 1996). GST activity in a herbicide-resistant black-grass biotype has been reported to be approximately double that of herbicide-susceptible biotypes (Reade *et al.*, 1997). GSTs may also play a role in protecting plants from damage from active oxygen species that result from the action of certain herbicides. In black-grass a GST possessing glutathione peroxidase activity has been identified which may undertake this role in the metabolism of organic hydroperoxides (Cummins *et al.*, 1999). This raises the possibility of GSTs not only aiding in the metabolism of certain herbicides, but also in protecting the plant from the damage resulting from herbicide treatment. The diverse roles of GSTs in toxin metabolism, stress responses and secondary metabolism have been reviewed by Marrs (1996).

Resistance to propanil in a biotype of *Echinocloa colonum* is due to raised activities of aryl acylamidase. Inhibition of this enzyme in resistant biotypes increases their susceptibility and, as some of these inhibitors are also herbicides, has proved useful in the control of resistant *E. colonum* populations.

Amaranthus tuberculatus (waterhemp) is a problematic dicot weed in maize, soybean and cotton crops in the USA that has evolved resistance to HPPD-inhibiting herbicides owing to rapid oxidative metabolism. While maize rapidly metabolised topramezone to desmethyl and benzoic acid derivatives, hydroxy-topramezone metabolites were found in some multiresistant biotypes, indicating the ability of other mechanisms to evolve in this weed (Lygin *et al.*, 2018).

Conyza bonariensis (L.) Cronq. (hairy fleabane) is in the family Asteraceae and is native to the Americas. It has become one of the world's worst weeds because it has evolved herbicide resistance in both crops and fallow land. Target site resistance has not yet been demonstrated in *C. bonariensis*, so one or more non-target site resistance mechanisms must be responsible for the resistance observed. The molecular mechanisms for non-target site glyphosate resistance, however, remain unknown. Piasecki *et al.* (2019) have performed a transcriptome analysis of resistant and susceptible biotypes of this species. A total of 9622 genes were differentially expressed as a result of glyphosate treatment in the two biotypes. These authors identified 41 new candidate target genes related to herbicide transport and metabolism, namely 19 ABC transporters, 10 cytochrome P450s, one glutathione *S*-transferase, five glucosyltransferases and six genes related to the antioxidant enzymes catalase, peroxidase and superoxide dismutase. As the authors point out, these target genes may inform further functional genome studies to elucidate glyphosate-resistance mechanisms in this and other species. Interestingly, Piasecki *et al.* (2020) have tested the hypothesis that adding the PROTOX inhibitor saflufenacil to glyphosate may control glyphosate-resistant *C. bonariensis*. This combination resulted in more lipid peroxidation and oxidative stress compared with the effects of either herbicide alone.

Glyphosate is readily metabolised in microorganisms by glyphosate oxidoreductase, GOX (Figure 14.5), to amino methyl phosphonic acid (AMPA) and glyoxylate. A high level of expression of aldo-keto reductase has been reported to confer glyphosate resistance in *E. colonum*, yielding the same products (Pan *et al.*, 2019). Duke (2019) lists reduced movement of glyphosate into the plant, reduced translocation of glyphosate, vacuolar sequestration of glyphosate, rapid necrosis followed by regeneration and enhanced glyphosate degradation as further causes of resistance to glyphosate.

The genus *Lolium* spp. can thrive in agricultural and non-agricultural areas, and populations have evolved multi- and cross-resistance to at least 14 herbicide modes of action in more than 21 countries (Suzukawa *et al.*, 2021). Non-target site resistance has been reported in *Lolium* spp. to ACCase, ALS, microtubule assembly, PS II, EPSPS, GS, Very long chain fatty acid inhibitors (VLCFA) and PS I electron diverters. Resistance mechanisms include enhanced herbicide metabolism, reduced herbicide uptake and translocation, vacuolar sequestration and protection-based resistance, such as protection against oxidative damage by active oxygen species. Note also the findings of Duker *et al.* (2019) that glutathione transferase is implicated in resistance to flufenacet in *Lolium* sp.

Very long chain fatty acid inhibitors, such as flufenacet, metolachlor and pyroxasulfone, are metabolised more rapidly in resistant *Lolium* spp. plants by glutathione conjugation via GST (Ducker *et al.*, 2019), and by GST overexpression in *L. rigidum* (Busi *et al.*, 2018).

For additional accounts of enhanced-metabolism resistance, the reader is referred to the detailed reviews by Gaines *et al.* (2020), Jugulam and Shyam (2019) and Suzukawa *et al.* (2021).

13.5.2.2   *Enhanced compartmentalisation*

Little evidence exists to indicate enhanced compartmentalisation (Figure 13.6) as a resistance mechanism and what is known suggests that it may act in conjunction with enhanced metabolism mechanisms. An example of this is the identification of an ATP-binding cassette (ABC) transporter which moves xenobiotic substances from the cytoplasm into the cell vacuole once they have been conjugated to glutathione. One such ATP-dependent glutathione *S*-conjugate pump has been identified in barley (*Hordeum sativum*) (Martinola *et al.*, 1993). Conjugation of the herbicide metolachlor to glutathione in this system results it its movement across the tonoplast into the vacuole. Similar ABC transporters have been reported to be responsible for the movement of chloroacetanilide and triazine herbicides into vacuoles (Ishikawa *et al.*, 1997). Where conjugation to glutathione forms part of the metabolism of a herbicide, then enhanced activity or the presence of such an ABC transporter would remove herbicides from their site of action quicker, perhaps resulting in plant survival. Alternatively, an increase in the number of ABC transporters might also result in resistance. Similar transport systems for substances that have been conjugated to sugars have also been postulated as a means by which herbicides are removed from the cytosol. Yuan *et al.* (2006) have reviewed the role of enhanced compartmentalisation as a mechanism of herbicide resistance.

Lasat *et al.* (2006) studied paraquat resistance in the grass weed *H. glaucum* Steud. using roots as a model system. While paraquat influx was similar in both resistant and susceptible biotypes, they observed more herbicide accumulation in the root vacuoles of resistant seedlings. Furthermore, efflux of paraquat from the vacuoles of resistant seedlings was five times slower than in susceptible plants. After 48 h, almost seven times more herbicide was

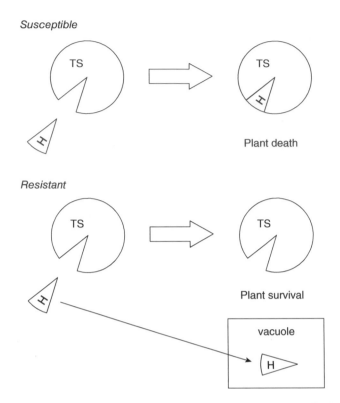

Figure 13.6 Diagrammatic representation of Enhanced Compartmentalisation Resistance. TS, target site; H, herbicide.

retained in the roots of resistant compared with susceptible plants. They concluded that paraquat resistance in this grass may be due to paraquat sequestration in the vacuole.

Ge *et al.* (2010) used ^{31}P-NMR to study the resistance mechanism in glyphosate-resistant horseweed, *Conyza canadensis* L. and found that there was significantly more accumulation of glyphosate in the vacuole of the resistant biotype. The same team (Ge *et al.*, 2014) characterised this vacuolar sequestration of glyphosate as similar to the ATP-binding cassette transporters found in mammalian cells, and more recently used *in vivo* NMR spectrometry to report that the sequestration was rapid (d'Avignon and Ge, 2018).

13.5.3 Cross-resistance

Cross-resistance refers to a plant's resistance to a number of herbicides owing to the presence of only one resistance mechanism. Where a resistant plant has an altered target site, it is likely that resistance to many, if not all, herbicides that act at that target site will be found. In the case of target site resistance to PS II-inhibiting herbicides, three distinct herbicide groups are identified for which cross-resistance occurs. Cross-resistance also occurs where resistance is due to enhanced metabolism. It is less easy to predict which herbicides will be included in such cross-resistance situations, as a detailed knowledge of herbicide metabolism is often lacking.

13.5.4   *Multiple resistance*

Multiple resistance is a plant's resistance to two or more herbicides owing to the presence of more than one resistance mechanism. An individual biotype with target site resistance to ACCase inhibitors and ALS inhibitors would be deemed as possessing multiresistance, as would a biotype possessing both target site and enhanced metabolism resistance. In theory, a biotype could contain many distinct resistance mechanisms, and quite clearly in such cases, predicting which herbicides would still work in controlling such a biotype becomes very difficult. An example of multiresistance, in *L. rigidum*, is presented by Tan *et al.* (2007).

Busi *et al.* (2020) consider that pre-emergence herbicides are key tools for the control of multiple-resistant *L. rigidum* and other major weeds in the Australian no-till cropping system. They observed that cinmethylin controls both sensitive and multiresistant *L. rigidum* when the herbicide is applied to the soil surface and when the wheat seeds are sown at least 1 cm below the soil surface. They concluded that cinmethylin tolerance in wheat is metabolism based and probably mediated by cytochrome P450 mono-oxygenase activity.

Reports of multiple weed resistance to glyphosate are now commonplace in the literature, especially in North America and Australia. Interestingly, species belonging to the genus *Bidens* L. Asteraceae, commonly known as beggarsticks, are widespread in South America in both tropical and subtropical environments, where *B. subalternans* and *B. pilosa* are considered important weeds. These species have a long history of herbicide resistance, with evidence of resistance to ALS inhibitors in soybean crops in the 1990s, multiresistance to ALS inhibitors and atrazine in soybean and maize crops in the mid-2010s, and more recently, multiple-resistance to both glyphosate and imazethapyr in *B. subalternans* in Paraguay (Mendes *et al.*, 2019).

Evolution of multiresistance in Palmer amaranth in the USA has become a serious management challenge. This weed is a summer, annual, C_4 dicot with a fast growth rate. It is highly competitive in maize, soybean, cotton and sorghum crops. Shyam *et al.* (2021) report the evolution of six-way resistance in a Palmer amaranth population in Kansas, mediated by enhanced P450 and GST activity and the amplification of the EPSPS gene by up to 88 copies. This population was able to survive recommended field rates of post-emergence applications of 2,4-D, ALS, PS II, EPSPS, PPO and HPPD inhibitors. Management strategies now need to evolve to control this important weed, such as cultural and mechanical methods, and the use of herbicide mixtures.

The genus *Chloris* in the Poaceae family is found in tropical and warm temperate regions. *Chloris radiata* (radiate fingergrass) is a C4 annual species native to Colombia and is a weed in rice. *Chloris polydactyla* competes with soybean, *C. truncata* decreases wheat yields and *C. virgata* is an important weed in sorghum. Hoyos *et al.* (2021), report a *C. radiata* population resistant to glyphosate and the ALS inhibitor, imazamox. Poor control was also observed with bispyribac-sodium and metsulfuron-methyl, suggesting cross-resistance to other ALS inhibitors, and to quinclorac, indicating multiresistance to auxin-type herbicides.

Although only introduced into the market in 1982, resistance to the ALS inhibiting herbicides is the most widespread mode of action for which resistance is reported. This is probably a mode of action for which resistance is reported. This is probably a combined result of the extensive use of these compounds, the number of different active ingredients of this type on the market and the specificity of the site of action. Additionally, the natural occurrence of a number of mutations of the target enzyme (ALS) that are less sensitive to herbicide, but are still enzymatically active, has undoubtedly added to the problem.

Resistance is likely to continue to increase, especially where the number of available herbicides with different modes of action is reduced, where production costs reduce herbicide choice and where genetically modified, herbicide-tolerant crops encourage the repeated use of a single herbicide in cropping systems.

13.6 Case study – black-grass (*A. myosuroides* Huds)

In the UK at least two-thirds of cereal crops are now sown as winter crops, as early as September. The winter wheat crop successfully overwinters and, with agrochemical inputs, yields at harvest can approach 9 tonnes/hectare. However, the earlier sowing of winter cereals has favoured autumn-emerging annual grass weeds such as black-grass and *Lolium* spp. From an economic perspective, herbicide resistance in black-grass alone may cost £1 billion each year in the UK.

The average winter wheat crop is treated 2.5 times with herbicides, and in 2012 the use of graminicides in arable crops accounted for 9.9 million sprayed hectares, dominated by ALS inhibitors, ACCase inhibitors, oxyacetamides and dinitroanilines (Food and Environment Research Agency, 2013). The inevitable consequence of an overdependence on ALS and ACCase inhibitors has been the evolution of herbicide-resistant weed populations, with target site mutations at ACCase (mainly at site 1781) and ALS (mutations at sites 197 and 574), in addition to enhanced metabolism resistance (Food and Environment Research Agency, 2013). Resistance has also been noted to all pre-emergence herbicides used in the UK, although this resistance has tended to be slower in evolution.

Black-grass is an autumn-germinating annual grass that is a major weed of winter cereal crops in many countries (Figure 13.7). Its presence can result in yield reductions of up to 44% (Moss, 1987) and in some cases results in total crop loss. Black-grass is capable of producing a very high number of seeds that remain viable in the soil for up to 7 years. In

Figure 13.7 Black-grass in a field of wheat. On the left side the field is free of black-grass. On the right the black-grass population is approximately 500 plants per m². Yield will be seriously reduced in the part of this trial where black-grass has been allowed to remain. Source: photograph courtesy of J.P.H. Reade. *See color inset section for the color representation of this figure.*

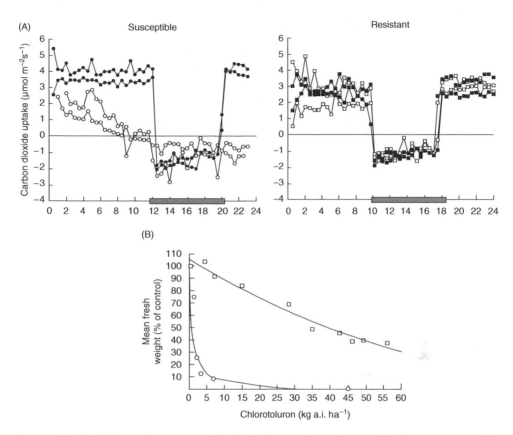

Figure 13.8 (A) CO_2 uptake of susceptible (circles) and resistant (squares) black-grass over a 24 h time period. Plants were initially treated with distilled water (solid circles) or chlorotoluron (open circles) added at the amount recommended in the field (3.5 kg a.i. ha^{-1}; solid symbols). A dark period is indicated by the shaded area on the x-axis. (B) The effect of increasing amounts of chlorotoluron (kg a.i. ha^{-1}) on the fresh weight of susceptible (circles) and resistant (squares) black-grass, 2 weeks after herbicide treatment. The doses calculated to cause a 50% reduction in fresh weight were 0.93 and 39.3 kg ha^{-1}, respectively, giving a resistance factor of 42. Source: Sharples, C.R, Hull, M.R. and Cobb, A.H. (1997) Growth and photosynthetic characteristics of two biotypes of the weed black-grass (*Alopecurus myosuroides* Huds) resistant and susceptible to the herbicide chlorotoluron. *Annals of Botany* **79**, 455–461. Reproduced with permission of Oxford University Press.

the early 1980s poor herbicidal control of black-grass was reported at a farm in Peldon, Essex, UK where the PS II-inhibiting herbicide chlorotoluron (CTU) was routinely used for weed control. Subsequent studies demonstrated that the Peldon biotype was resistant to CTU (Moss and Cussans, 1985). Little or no reduction in photosynthetic rate was noted when field-rate CTU (3500 g active ingredient (a.i.) ha^{-1}) was applied to the Peldon biotype *in vivo* (Sharples, *et al.*, 1997; Figure 13.8).

Since the early 1980s many further cases of resistance to different herbicides have been reported in black-grass. By 1999, as many as 750 herbicide-resistant populations had been identified in the UK alone (Moss *et al.*, 1999). Herbicide resistance in black-grass has also been reported in Belgium, France, Germany, Israel, The Netherlands, Spain, Switzerland

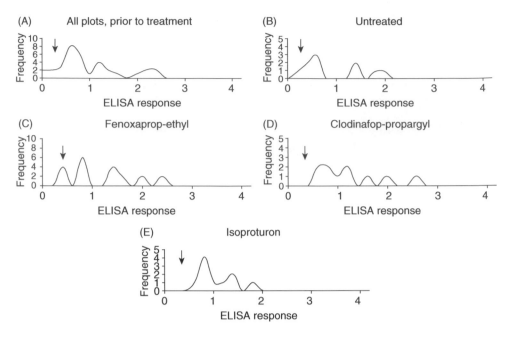

Figure 13.9   Glutathione *S*-transferase (GST) abundance in the field, measured by ELISA using monoclonal anti-*Alopecurus myosuroides* GST antiserum. (A) Prior to herbicide treatment; (b–e) 25 days after treatment. Arrow indicates plants giving a low ELISA response (B) that are absent from treated plots (C–E). Source: Reade, J.P.H. and Cobb, A.H. (2002) New, quick tests for herbicide resistance in black-grass (*Alopecurus myosuroides* Huds) based on increased glutathione S-transferase activity and abundance. *Pest Management Science* **58**, 26–32. Reproduced with permission of John Wiley & Sons.

and Turkey (Heap, 2009). Many of these cases involve cross- or multiresistance, and in the UK herbicide resistance owing to enhanced metabolism is considered endemic.

A number of studies have attempted to identify the mechanisms responsible for herbicide resistance in black-grass. The GST activity in the herbicide-resistant biotype Peldon is approximately double that in herbicide-susceptible biotypes (Reade *et al.*, 1997). Study of further biotypes demonstrating resistance to fenoxaprop-ethyl and clodinafop-propargyl showed correlation between the level of herbicide resistance and level of GST activity (Reade *et al.*, 1997). GST activities were found to be raised in a number of black-grass biotypes that had previously been characterised as possessing enhanced metabolism resistance, but not in biotypes that had been characterised as possessing target site resistance or that were herbicide susceptible (Cocker *et al.*, 1999; Reade and Cobb, 2002). Field studies have further demonstrated that subpopulations of black-grass that have survived herbicide treatment have higher GST activity and abundance than the untreated 'parent' populations (Reade *et al.*, 1999; Figure 13.9), which implies that the individuals with lower GST activity and abundance have been removed by the herbicide treatment. Studies suggest that the increased GST activity in resistant biotypes is not due solely to increased abundance of a GST present in both susceptible and resistant plants, but is also due, in part, to the presence of a GST not detected in susceptible biotypes. Cummins *et al.* (1997) identified GSTs with molecular weights of 27 and 28 kDa in addition to a 25 kDa GST in resistant biotypes: the susceptible biotype only had the 25 kDa GST. Reade and Cobb (1999) purified a 30 kDa GST from resistant black-grass that was absent from susceptible black-grass. This GST, the identity of which was confirmed by polypeptide sequencing (Reade and Cobb, 2002), showed different kinetic properties from those found in both susceptible and resistant biotypes.

Chlorotoluron did not induce GST activity in either biotype (Sharples *et al.*, 1995) but treatment with the herbicide safeners benoxcor, flurazole and oxabetrinil did increase GST activity in susceptible but not in resistant biotypes (Sharples, 1996). This is interesting, as studies in wheat (*Triticum aestivum* L.) have indicated that some herbicide safeners do increase glutathione conjugation of herbicides (Tal *et al.*, 1993, 1995; Edwards and Cole, 1996). GST activity in black-grass increases from winter to spring, which corresponds to a drop in the efficacy of the herbicides fenoxaprop-ethyl and clodinafop-propargyl with respect to black-grass control (Milner *et al.*, 1999, 2001).

Cytochrome P450s have also been implicated in herbicide resistance in black-grass. Kemp and Caseley (1987) and Kemp *et al.* (1988) used the P450 inhibitor 1-aminobenzotriazole to demonstrate that P450s played a role in metabolism of both chlortoluron and isoproturon in a resistant black-grass biotype. Hall *et al.* (1995) also showed that inhibitors of P450 activity reduced tolerance to chlortoluron and reduced herbicide metabolism in a resistant black-grass biotype, although this was not the case for fenoxaprop (Hall *et al.*, 1997). Cell culture studies by Menendez and De Prado (1997) also suggest a role for P450s in chlortoluron metabolism. It is tempting, therefore, to interpret these findings as demonstrating a role for P450s in chlortoluron metabolism, but not in fenoxaprop metabolism, in black-grass. GST activity in resistant black-grass biotypes has been identified against fenoxaprop, fluorodifen and metolachlor (Cummins *et al.*, 1997).

Interestingly, Brazier *et al.* (2002) have also identified an *O*-glucosyltransferase in a resistant black-grass biotype, raising the possibility of a third mechanism of enhanced metabolism resistance in this species.

In a recent publication, Mellado-Sánchez *et al.* (2020) report the use of virus-induced gene silencing (VIGS) and virus-mediated gene overexpression (VOX) to investigate the involvement of specific genes in blackgrass-sensitive (Rothamsted) and a resistant biotype (Peldon, which is multiple-herbicide resistant). They first inserted a gene of interest into a virus and then infected the resistant or sensitive blackgrass with it. Using these techniques, they confirmed that the GST gene AmGSTF1 is required for multiple-herbicide resistance in Peldon. This study suggests that VIGS and VOX approaches provide new opportunities to directly investigate the roles of specific genes in herbicide resistance in this and other important grass weed species.

13.7   Strategies for the control of herbicide-resistant weeds

Farming strategies that will reduce the development and spread of resistance weeds include:

1   using a mixture of herbicides each year, with at least two different modes of action, preferably with a rotation of application sequence.
2   maintaining good weed control over time, using a mixture of weed control methods that include chemical, mechanical and cultivation approaches;
3   considering minimum tillage and sowing a weed-suppressing cover crop;
4   rotating crops grown in the same field each year;
5   purchasing clean crop seed.

6   knowing the weed spectrum in each field, each year, and noting when weed control has been incomplete;
7   establishing the resistance status of problematic weeds each year by testing;
8   trying to kill resistant weeds at the seedling stage, i.e. before they set seed;
9   reducing the soil seed bank by making greater use of residual, pre-plant and pre-emergence herbicides, such as the thiocarbamate triallate, the PS II inhibitor metribuzin, the microtubule inhibitor trifluralin, the PROTOX inhibitors flumioxazin, saflufenacil and sulfentrazole, and the VLCFA inhibitor pyroxasulfone;
10   preventing escapes from the field before seed set and at harvest – cleaning farm machinery and operator clothes.

The reader is also referred to the work of the Herbicide Resistance Action Committee (HRAC, www.HRACGLOBAL.com), Busi *et al.* (2019), Délye *et al.* (2013) and Refatti *et al.* (2019).

13.8   The future development of herbicide resistance

Herbicide resistance is the ultimate evidence of the extraordinary capacity of weeds to evolve under stressful conditions (Vila-Aiub, 2019). Agroecosystems are defined by intense disturbance and stress imposed by humans, and weed traits are selected to maximise reproductive capacity and plant success.

The appearance of herbicide resistance over the last 60 years has undoubtedly encouraged weed control to be viewed as an integrated process in which chemical means of control are used alongside non-chemical means. Using such approaches should reduce the selection pressure for the survival of herbicide-resistant biotypes by using, among other methods, rotations, stale seedbeds and delayed drilling to control weeds. Unfortunately, as crop production costs rise and as the number of active ingredients available continues to fall (or, at least, the number of herbicide target sites remain limited), then resistance is likely to increase for the foreseeable future. The use of genetically modified, herbicide-tolerant crops (Chapter 14) will also increase selection pressure for the survival of resistant biotypes, as well as posing a risk of genes conferring resistance being transferred from crop to weed. It is alarming to consider that even if herbicides with new sites of action are developed and released to market, then the weeds that will be resistant to them are already present in existing weed populations. They simply require the appropriate selection pressure to be applied, which will ensure their survival advantage. Thus, weed populations are in a constant state of flux. Those that succeed will be best adapted to the biotic and abiotic environments presented to them.

On a positive note, recent research (eg, Rigon *et al.*, 2020) indicates that our understanding of herbicide resistance is improving with the identification of specific genes involved in each phase of herbicide metabolism (Table 13.5).

Looking to the future, a risk assessment approach has been developed which attempts to quantify the risk of further herbicide resistance developing in Europe (Moss *et al.*, 2019). This is based on (a) the characteristics of the active ingredient, including currently known cases and mechanisms of resistance, (b) the characteristics of the target weed, including its predisposition to evolve resistance and (c) procedures to minimise the resistant risk, including weed management practices. These factors have been used to classify weed species as having a high, medium, low or very low inherent risk of evolving herbicide resistance (Table 13.6). All other weed genera and species are considered to have a low or very low risk of evolving

Table 13.5 Genes for herbicide metabolism in herbicide-resistant weeds, according to the phases of herbicide metabolism.

Phase 1	Gene family	Herbicide group
	CYP734 A1	ALS
	CYP76 C1	ALS
	CYP86 B1	ALS
	CYP76 B1	ALS
	CYP73 B	ALS
	CYP72 A	ACCase
	CYP81 A6	ALS
	CYP81 A12, A21	ACCase, ALS, PS II, DXS, HPPD
	CYP81 A14, A18	ALS
	CYP81 A15	ALS, PS II, DXS
	CYP81 A24	ALS, PS II, PDS, DXS, PPO, HPPD
Phase II	GST 1 and 2	VLCFA
	GST U1 and U6	ALS
	GST F1	ACCase, ALS, PS II, VLCFA
	GST UF2	PS II
	GT	ACCase, ALS
Phase III	ABC C10	ALS
Activation	AAA	PS II
	EST	ALS
	NMO	ACCase
	AKR	EPSPS

AAA, Aryl acylamidase; AKR, aldo-keto reductase; ABC, ATP-binding cassette; CYP, cytochrome P450 mono-oxygenase; EST, esterase; GST, glutathione *S*-transferase; GT, glucosyltransferase; NMO, nitronate mono-oxygenase. Source: based on Rigon, C.A.G., Galnes, T.A., Kupper, A. and Dayan, F.E. (2020) Metabolism-based herbicide resistance, the major threat among the non-target site resistance mechanisms. *Outlooks on Pest Management* **31**, 162–168; doi: org/10.1564/v31_aug_04.

Table 13.6 Weed species or genera in Europe with inherent risk of evolving herbicide resistance.

High inherent risk	Medium inherent risk
Alopecurus myosuroides	*Avena* spp.
Amaranthus spp.	*Conyza* spp.
Apera spica-venti	*Echinochloa* spp.
Chenopodium spp.	*Matricaria* spp.
Lolium spp.	*Phalaris* spp.
Papaver rhoeas	*Senecio vulgaris*
	Stellaria media

Source: Moss, S.R., Ulber, L. and den Hoad, I. (2019) A herbicide resistance risk matrix. *Crop Protection* **115**, 13–19; doi: org/10.1016/j.cropro.2018.09.005

herbicide resistance in Europe at the present time. This quantitative approach is considered to be robust and realistic by Moss *et al.* (2019), and its use may be recommended to assess the risk of herbicide resistance developing in weeds in other parts of the world.

Finally, it is undoubtedly the case that the rate of herbicide resistance is currently faster than the discovery of new herbicide active ingredients. It follows that to lessen the former, we must optimise and challenge how we use our remaining active ingredients. Otherwise the value and effectiveness of these precious molecules will soon be lost.

References

Anderson, M. and Gronwald, J. (1991) Atrazine resistance in a velvetleaf (*Abutilion theophrasti*) biotype due to enhanced glutathione *S*-transferase activity. *Plant Physiology* **96**, 104–109.

d'Avignon, A.A. and Ge, X. (2018) *In vivo* NMR investigations of glyphosate influences on plant metabolism. *Journal of Magnetic Resonance* **292**, 59–72.

Avila-Garcia, W.V., Sanchez-Olguin, E., Hulting, A.G. and Mallory-Smith, C.A. (2012) Target-site mutation associated with glufosinate resistance in Italian ryegrass (*Lolium perenne* L. ssp. Multiflorum). *Pest Management Science* **68**, 1248–1254.

Bai, S., Zhang, F., Zuren, Li., Wang, H.,Wang, Q., Wang, J., Liu, W. and Bai, L. (2019) Target-site and non-target-site-based resistance to tribenuron-methyl in multiply-resistant *Myosoton aquaticum* L. *Pesticide Biochemistry and Physiology* **155**, 8–14.

Barnwell, P. and Cobb, A.H. (1989) Physiological studies of mecoprop-resistance in chickweed (*Stellaria media* L.). *Weed Research* **29**, 135–140.

Bernasconi, P., Woodworth, A.R., Rosen, B.A., Subramanian, M.V. and Siehl, D.L. (1995) A naturally occurring point mutation confers broad range tolerance to herbicides that target acetolactate synthase. *Journal of Biological Chemistry* **270**, 17381–17385.

Brazier, M., Cole, D.J. and Edwards, R. (2002) *O*-Glucosyltransferase activities towards phenolic natural products and xenobiotics in wheat and herbicide-resistant and herbicide-susceptible black-grass (*Alopecurus myosuroides*). *Phytochemistry* **59**, 149–156.

Busi, R., Goggin, D.E., Heap, I.M., Horak, M.J., Jugulam, M., Masters, R.A., *et al.* (2017) Weed resistance to synthetic auxin herbicides. *Pest Management Science* **74**, 2265–2276.

Busi, R., Porri, A., Gaines, T.A. and Powles, S.B. (2018) Pyroxasulfone resistance in *Lolium rigidum* is metabolism-based. *Pesticide Biochemistry and Physiology*, **148**, 74–80; doi: org/10.1016/j.pestbp.2018.03.017

Busi, R., Powles, S.B., Beckie, H.J. and Renton, M. (2019) Rotations and mixtures of soil-applied herbicides delay resistance. *Pest Management Science* **76**; doi: org/10.1002/ps.5534

Busi, R., Dayan, F.E., Francis, I., Goggin, D., Lerchl, J., Porri, A.,*et al.* (2020) Cinmethylin controls multiple herbicide-resistant *Lolium rigidum* and its wheat selectivity is P450-based. *Pest Management Science*; doi: org/10.1002/ps.5798

Christoffers, M.J., Berg, M.L. and Messersmith, C.G. (2002) An isoleucine to leucine mutation in acetyl-CoA carboxylase confers herbicide resistance in wild oat. *Genome* **45**, 1049–1056.

Cobb, A.H., Early, C. and Barnwell, P. (1990) Is mecoprop-resistance in chickweed due to altered auxin-sensitivity? In: Caseley, J.C., Cussans, G. and Atkin, R. (eds), *Herbicide Resistance in Weeds and Crops*. Guildford: Butterworth, pp. 435–436.

Cocker, K.M., Moss, S.R. and Coleman, J.O.D. (1999) Multiple mechanisms of resistance to fenoxaprop-*p*-ethyl in United Kingdom and other European populations of herbicide-resistant *Alopecurus myosuroides* (black-grass). *Pesticide Biochemistry and Physiology* **65**, 169–180.

Coupland, D., Lutman, P.J.W. and Health, C. (1990) Uptake, translocation and metabolism of mecoprop in a sensitive and a resistant biotype of *Stellaria media*. *Pesticide Biochemistry and Physiology* **36**, 61–67.

Cummins, I., Moss, S., Cole, D.J. and Edwards, R. (1997) Glutathione transferases in herbicide-resistant and herbicide-susceptible black-grass (*Alopecurus myosuroides*). *Pesticide Science* **51**, 244–250.

Cummins, I., Cole, D.J. and Edwards, R. (1999) A role for glutathione transferases functioning as glutathione peroxidases in resistance to multiple herbicides in black-grass. *The Plant Journal* **18**, 285–292.

Davis, G., Letarte, J., Grainger, C., Rajcan, I. and Tardif, F. (2020) Widespread resistance in pigweed species in Ontario carrot production is due to multiple PSII mutations. *Canadian Journal of Plant Science* **100**, 56–67.

Délye, C. (2005) Weed resistance to acetyl coenzyme A carboxylase inhibitors: an update [Review]. *Weed Science* **53**, 728–746.

Délye, C., Matejicek, A. and Gasquez J. (2002a) PCR-based detection of resistance to acetyl-CoA carboxylase inhibiting herbicides in black-grass (*Alopecurus myosuroides*) and ryegrass (*Lolium rigidum* Gaud.). *Pest Management Science* **58**, 474–478.

Délye, C., Wang, T. and Darmency, H. (2002b) An isoleucine–leucine substitution in chloroplastic acetyl-Co A carboxylase from green foxtail (*Setaria viridis* (L.) Beauv.) is responsible for resistance to the cyclohexanedione herbicide setoxydim. *Planta* **214**, 421–427.

Délye, C., Zhang, X.Q., Chalopin, C., Michel, S. and Powles, S.B. (2003) An isoleucine residue within the carboxyl-transferase domain of multidomain acetyl-coenzyme A carboxylase is a major determinant of sensitivity to aryloxyphenoxypropionate but not to cyclohexanedione inhibitors. *Plant Physiology* **132**, 1716–1723.

Délye, C., Jasieniuk, M. and Le Corre, V. (2013) Deciphering the evolution of herbicide resistance in weeds. *Trends in Genetics* **29**, 649–658; doi: org/10.1016/j.tig.2013.06.001

Deshpande, S. and Hall, J.C. (2000) Auxinic herbicide resistance may be modulated at the auxin-binding site in wild mustard (*Sinapis arvensis* L.): a light scattering study. *Pesticide Biochemistry and Physiology* **66**, 41–48.

Devine, M.D. (1997) Mechanisms of resistance to acetyl-coenzyme A carboxylase inhibitors: a review. *Pesticide Science* **51**, 259–264.

Devine, M.D. and Preston, C. (2000) The molecular basis of herbicide resistance. In: Cobb, A.H. and Kirkwood, R.C. (eds) *Herbicides and their Mechanisms of Action*. Sheffield/Boca Raton, FL: Sheffield Academic Press/CRC Press, pp. 72–84.

Devine, M.D. and Shukla, A. (2000) Altered target sites as a mechanism of herbicide resistance. *Crop Protection* **19**, 881–889.

Ducker, R., Zollner, P., Luemmen, P., Ries, S., Collavo, A. and Beffa, R. (2019) Glutathione transferase plays a major role in flufenacet resistance of ryegrass (*Lolium spp.*) field populations. *Pest Management Science* **75**, 11; doi: org/10.1002/ps.5425

Duke, S.O. (2019) Enhanced metabolic degradation: the last evolved glyphosate resistant mechanism of weeds? *Plant Physiology* **181**, 1401–1403; doi: org/10.1104/pp.19.01245

Edwards, R. and Cole, D.J. (1996) Glutathione transferases in wheat (*Triticum*) species with activity towards fenoxaprop-ethyl and other herbicides. *Pesticide Biochemistry and Physiology* **54**, 96–104.

Fang, J., Zhang, Y., Liu, T., Yan, B., Li, J. and Dong, L. (2019) Target-site and metabolic resistance mechanisms to penoxsulam in barnyard grass (*Echinochloa cruss-galli* (L.) P. Beauv.). *Journal of Agriculture and Food Chemistry* **67**, 8085–8095.

Food and Environment Research Agency (2013) Pesticide usage survey. www.fera.defra.gov.uk/landUseSustainability/surveys/index.cfm

Gaines, T.A., Duke, S.O., Morran, S., Rigon, C.A.G., Tranel, P.J., Kupper, A. and Dayan, F.E. (2020) Mechanisms of evolved herbicide resistance. *Journal of Biological Chemistry* **295**, 10307–10330.

Ge, X., d'Avignon, D.A., Ackerman, J.J.H. and Sammons, R.D. (2010) Rapid vacuolar sequestration: the horseweed glyphosate resistance mechanism. *Pest Management Science* **66**, 345–348.

Ge, X., d'Avignon, D.A., Ackerman, J.J.H. and Sammons, R.D. (2014) *In vivo* 31P-nuclear magnetic resonance studies of glyphosate uptake, vacuolar sequestration and tonoplast pump activity in glyphosate-resistant horseweed. *Plant Physiology* **166**, 1255–1268.

Goggin, D., Cawthray, G.R. and Powles, S.B. (2016) 2,4-D resistance in wild radish: reduced herbicide translocation via inhibition of cellular transport. *Journal of Experimental Botany* **67**, 3223–3235.

Gray, J.A., Stoltenberg, D.E. and Balke, N.E. (1996) Increased glutathione conjugation of atrazine confers resistance in a Wisconsin velvetleaf (*Abutilion theophrasti*) biotype. *Pesticide Biochemistry and Physiology* **55**, 157–171.

Gressel, J. (2002) *Molecular Biology of Weed Control*. London: Taylor & Francis.

Gruys, K.J., Biest-Taylor, N.A., Feng, P.C.C., *et al.* (1999) Resistance to glyphosate in annual ryegrass (*Lolium rigidum*) II. Biochemical and molecular analysis. *Weed Science Society of America Abstracts* **39**, 163.

Hall, L.M., Moss, S.R. and Powles, S.B. (1995) Mechanism of resistance to chlortoluron in two biotypes of the grass weed *Alopecurus myosuroides*. *Pesticide Biochemistry and Physiology* **53**, 180–192.

Hall, L.M., Moss, S.R. and Powles, S.B. (1997) Mechanisms of resistance to aryloxyphenoxypropionate herbicides in two resistant biotypes of *Alopecurus myosuroides* (blackgrass): herbicide metabolism as a cross-resistance mechanism. *Pesticide Biochemistry and Physiology* **57**, 87–98.

Harper, J.L. (1956) The evolution of weeds in relation to resistance to herbicides. *Proceedings of the third British Weed Control Conference*, 179–188.

Hatton, P.J., Dixon, D., Cole, D.J. and Edwards, R. (1996) Glutathione transferase activity and herbicide selectivity in maize and associated weed species. *Pesticide Science* **46**, 267–275.

Heap, I. (2009) *The International Survey of Herbicide Resistant Weeds*. www.weedscience.com (accessed 28 July 2009).

Heap, I. (2010) *The International Survey of Herbicide Resistant Weeds*. www.weedscience.com (accessed 1 July 2010).

Heap, I. (2020) *The International Herbicide-Resistance Database*. www.weedscience.org (accessed 8 April 2020).

Holt, J.S. and Thill, D.C. (1994) Growth and productivity of resistant plants. In: Powles, S.B. and Holtum, J.A.M. (eds), *Herbicide Resistance in Plants: Biology and Biochemistry*. Boca Raton, FL: Lewis, pp. 299–316.

Hoyos, V., Plaza, G., Gudalupe, J., Bautista, C.P., Rojano-Delgado, A.M. and De Prado, R. (2021) Confirmation of multiple resistant *Chloris radiata* population, harvested in Colombian rice fields. *Agronomy* **11**, 496; doi: org/10.3390/agronomy11030496

Ishikawa, T., Li, Z.-S., Lu, Y.-P. and Rea, P.A. (1997) The GS-X pump in plant, yeast, and animal cells: structure, function, and gene expression. *Bioscience Reports* **17**, 189–207.

Jugulam, M. and Shyam, C. (2019) Non-Target-Site Resistance to herbicides: recent developments. *Plants* **8**, 417–432; doi: org/10.3390/plants8100417

Katagi, T. and Mikami, N. (2000) Primary metabolism of agrochemicals in plants. In: Roberts, T.R. (ed.), *Metabolism of Agrochemicals in Plants*, Chichester: John Wiley, pp. 43–106.

Kemp, M.S. and Caseley, J.C. (1987) Synergistic effects of 1-aminobenzotriazole on the phytotoxicity of chlorotoluron and isoproturon in a resistant population of black-grass (*Alopecurus myosuroides*). *Proceedings of the British Crop Protection Conference – Weeds*, vol. 1. Farham: BCPC, pp. 895–899.

Kemp, M.S., Newton, L.V. and Caseley, J.C. (1988) Synergistic effects of some P450 oxidase inhibitors on the phytotoxicity of chlorotoluron in a resistant population of black-grass (*Alopecurus myosuroides*). *Proceedings of the EWRS Symposium*. Wageningen: European Weed Research Society, pp. 121–126.

Lasat, M.M., DiTomaso, J.M., Hart, J.J. and Kochian, L.V. (2006) Evidence for vacuolar sequestration of paraquat in roots of a paraquat-resistant *Hordeum glaucum* biotype. *Physiologia plantarum*; doi: org/10.1111/j.1399-3054.997tb05410.x

LeClere, S., Wu, C., Westra, P. and Sammons, R.D. (2018) Cross-resistance to dicamba, 2,4-D and fluroxypyr in *Kochia scoparia* is endowed by a mutation in an AUX/IAA gene. *Proceedings of the National Academy of Science* **115**, E2911–E2920.

Lee, L.J. and Ngim, J. (2000) A first report of glyphosate resistant goosegrass (*Elusine indica*) in Malaysia. *Pest Management Science* **56**, 336–339.

Li, J., Mu, J., Bai, J., Fu, F., Zou, T., An, F., Zhang, J., *et al.* (2013) Paraquat resistant 1, a golgi-localised putative transporter protein, is involved in intracellular transport of paraquat. *Plant Physiology* **162**, 470–483.

Li, J., Gao, X., Li, M. and Fang, F. (2019) Resistance evolution and mechanisms to ALS-inhibiting herbicides on *Capsella bursa-pastoris* populations from China. *Pesticide Biochemistry and Physiology* **159**, 17–21.

Lopez-Overjero, R.F., Takano, H.K., Nicolai, M., Ferreira, A., Melo, S.C., Cavenaghi, A.L., *et al.* (2017) Frequency and dispersal of glyphosate-resistant sourgrass (*Digitaria insularis*) populations across brasilian agricultural production areas. *Weed Science* **65**, 285–294; doi: org/10.1017/wsc.2016.31

Lu, H., Yu, Q., Owen, M.J. and Powles, S.B. (2018) A novel psbA mutation (Phe-274-Val) confers resistance to PSII herbicides in wild radish (*Raphanus raphanistrum*). *Pest Management Science* **75**, 144–151; doi: org/10.1002/ps.507

Lu, H., Yu, Q., Han, H., Owen, M.J. and Powles, S.B. (2019) Non-target-site resistance to PDS-inhibiting herbicides in a wild radish (*Raphanus raphanistrum*) population. *Pest Management Science*; doi: org/10.1002/ps.5733

Lutman, P.J. and Lovegrove A.W. (1985) Variations to the tolerance of *Galium aparine* (cleavers) and *Stellaria media* (chickweed) to mecoprop. *British Crop Protection Conference, Weeds* **2**, 411–418.

Lygin, A.V., Kaundun, S.S., Morris, J.A., McIndoe, E., Hamilton, A.R. and Riechers, D.E. (2018) Metabolic pathways of topramezone in multiple-resistant waterhemp (*Amaranthus tuberculatus*) differs from naturally tolerant maize. *Frontiers in Plant Science* **9**, article 1644; doi: 10.3389/fpls.2018.01644

Maeda, H., Murata, K., Sakuma, N., Takei, S., Yamazaki, A., Karim, M.R., *et al.* (2019) A rice gene that confers broad-spectrum resistance to beta-triketone herbicide. *Science* **365**, 393–396; doi: org/10.1126/science. aax0379

Marrs, K. (1996) The functions and regulation of glutathione *S*-transferases in plants. *Annual Review of Plant Physiology and Plant Molecular Biology* **47**, 127–158.

Martinola, E., Grill, E., Tommasini, R., Kreuz, K. and Amrhein, N. (1993) ATP-dependent glutathione *S*-conjugate 'export' pump in the vacuolar membrane of plants. *Nature* **364**, 247–249.

Mellado-Sánchez, M., McDiarmid, F., Cardoso, V., Kanyuka, K. and MacGregor, D.R. (2020) Virus-mediated transient expression techniques enable gene function studies in blackgrass. *Plant Physiology*; doi: org/10.1104/pp.20.00205

Mendes, R.R., Adegas, F.S., Takano, H.K., Silva, V.F.V., Machado, F.G. and Oliveira, R.S. (2019) Multiple resistance to glyphosate and imazethapyr in *Bidens subalternans*. *Ciencia e Agrotechnologia* **43**, e009919; doi: org/10.1590/1413-70542011943009919

Menendez, J. and De Prado, R. (1997) Detoxification of chlorotoluron in a chlorotoluron resistant biotype of *Alopecurus myosuroides*. Comparison between cell cultures and whole plants. *Physiologia Plantarum* **99**, 97–104.

Milner, L.J., Reade, J.P.H. and Cobb, A.H. (1999) An investigation of glutathione *S*-transferase activity in *Alopecurus myosuroides* Huds. (black-grass) in the field. *Proceedings of the 1999 Brighton Conference – Weeds*, vol. 1. Farnham: BCPC, pp. 173–178.

Milner, L.J.M., Reade, J.P.H and Cobb, A.H. (2001) Developmental changes in glutathione *S*-transferase activity in herbicide-resistant populations of *Alopecurus myosuroides* Huds. (black-grass) in the field. *Pest Management Science* **57**, 1100–1106.

Moss, S.R. (1987) Competition between black-grass (*Alopecurus myosuroides*) and winter wheat. *Proceedings of the British Crop Protection Conference – Weeds*, vol. 2. Farnham: BCPC, pp. 365–374.

Moss, S.R. and Cussans, G.W. (1985) Variability in the susceptibility of *Alopecurus myosuroides* (black-grass) to chlorotoluron and isoproturon. *Aspects of Applied Biology* **9**, 91–98.

Moss, S.R., Clarke, J., Blair, A.M., Culley, T., Read, M. and Turner, M. (1999) The occurrence of herbicide-resistant grass-weeds in the United Kingdom and a new system for designating resistance in screening assays. *Proceedings of the 1999 Brighton Conference – Weeds*, vol. 1. Farnham: BCPC, pp. 179–184.

Moss, S.R., Ulber, L. and den Hoad, I. (2019) A herbicide resistance risk matrix. *Crop Protection* **115**, 13–19; doi: org/10.1016/j.cropro.2018.09.005

Mougin, C.P., Corio-Costet, M.F. and Werck-Reichhart D. (2001) Plant and fungal cytochrome P-450s: their role in pesticide transformation. In: Hall, J.C. *et al.* (eds), *Pesticide Biotransformations in Plants and Microorganisms: Similarities and Divergences*. Washington, DC: American Chemical Society, pp. 166–181.

Murphy, B.P. and Tranel, P.J. (2019) Target-site mutations conferring herbicide resistance. *Plants* **8**, 382; doi: org/10.3390/plants8100382

Nakka, S., Thompson, C.R., Peyterson, D.E. and Jugulam, M. (2017) Target site-based and non-target site-based resistance to ALS-inhibitors in *Amaranthus palmeri*. *Weed Science* **65**, 681–689.

Oettmeier, W. (1999) Herbicide resistance and supersensitivity in photosystem II. *Cellular and Molecular Life Sciences* **55**, 1255–1277.

Pan, L., Yu, Q., Han, H., Mao, L., Nyporko, A., Fan, L., *et al.* (2019) Aldo-keto reductase metabolises glyphosate and confers glyphosate-resistance in *Echinochloa colonum*. *Plant Physiology* **181**, 1–16; doi: org/10.1104/pp.1900979

Panozzo, S., Milani, A., Scarabel, L., Balogh, A., Dancza, I. and Sattin, M. (2017) Occurrence of different resistance mechanisms to acetolactate inhibitors in European *Sorghum halepense*. *Journal of Agriculture and Food Chemistry* **65**, 7320–7327.

Peniuk, M.G., Romano, M.L. and Hall, J.C. (1992) Absorption, translocation and metabolism are not the basis for differential selectivity of wild mustard (*Sinapis arvensis* L.) to auxinic herbicides. *Weed Science* **32**, 165.

Perotti, V.E., Larran, A.S., Palmieri, V.E., Martnatto, A.K., Alvarez, C.E., Tuesca, D. and Permingeat, H.R. (2018) A novel triple amino acid substitution in the EPSPS found in a high-level glyphosate-resistant *Amaranthus hybridus* population from Argentina. *Pest Management Science* **75**; doi: org/10.1002/ps.5303

Piasecki, C., Yang, Y., Benemann, D.P., Kremer, F.S., Galli, V., Millwood, R.J. *et al.* (2019) Transcriptomic analysis identifies new non-target site glyphosate-resistance genes in *Conyza bonariensis*. *Plants*, **8**, article 157; doi: org/10.3390/plants8060157

Piasecki, C., Carvalho, I.R., Avila, L.A., Agostinetto., D. and Vargas, L. (2020) Glyphosate and saflufenacil: elucidating their combined action on the control of glyphosate-resistant *Conyza bonariensis*. *Agriculture* **10**, 236–254; doi: 10.3390/agriculture10060236

Popay, A.I., Bourdot, G.W., Harrington, K.C. and Rahman, A. (1990) Herbicide resistance in weeds in New Zealand. In: Caseley, J.C., Cussans, G. and Atkins, R. (eds), *Herbicide Resistance in Weeds and Crops*. Guildford: Butterworth, pp. 470–471.

Rangani, G., Salas-Perez, R.A., Aponte, R.A., Knapp, M., Craig, I.R., Mietzner, T., *et al.* (2019). A novel single-site mutation in the catalytic domain of protoporphyrinogen oxidas (PPO) confers resistance to PPO- inhibiting herbicides. *Frontiers of Plant Science*; doi: org/10.3389/fpls.2019.00568

Reade, J.P.H. and Cobb, A.H. (1999) Purification, characterisation and comparison of glutathione *S*-transferases from black-grass (*Alopecurus myosuroides* Huds) biotypes. *Pesticide Science* **55**, 993–999.

Reade, J.P.H. and Cobb, A.H. (2002) New, quick tests for herbicide resistance in black-grass (*Alopecurus myosuroides* Huds) based on increased glutathione *S*-transferase activity and abundance. *Pest Management Science* **58**, 26–32.

Reade, J.P.H., Hull, M.R. and Cobb, A.H. (1997) A role for glutathione-*S*-transferase in herbicide resistance in black-grass (*Alopecurus myosuroides*). *Proceedings of the 1997 Brighton Crop Protection Conference – Weeds*, vol. 2. Farham: BCPC, pp. 777–782.

Reade, J.P.H., Belfield, J.L. and Cobb, A.H. (1999) Rapid tests for herbicide resistance in blackgrass based on elevated glutathione *S*-transferase activity and abundance. *Proceedings of the 1999 Brighton Crop Protection Conference – Weeds*, vol. 2. Farnham: BCPC, pp. 185–190.

Refatti, J.P. and seven others (2019) High CO_2 concentration and temperature increase resistance to cyhalofop-butyl in multiple-resistant *Echinochloa colona*. *Frontiers in Plant Science* **10**, article 529; doi: org/10.3389/fpls.2019.00529

Rigon, C.A.G., Gaines, T.A., Kupper, A. and Dayan, F.E. (2020) Metabolism-based herbicide resistance, the major threat among the non-target site resistance mechanisms. *Outlooks on Pest Management* **31**, 162–168; doi: org/10.1564/v31_aug_04

Roux, F., Matejicek, A. and Reboud, X. (2005) Response of *Arabidopsis thaliana* to 22 ALS inhibitors: baseline toxicity and cross-resistance of *csr*1-1 and *csr*1-2 resistant mutants. *Weed Research* **45**, 220–227.

Ryan, G.F. (1970) Resistance of common groundsel to simazine and atrazine. *Weed Science* **18**, 614–616.

Sathasivan, K., Haughn, G.W. and Murai, N. (1990) Nucleotide sequence of a mutant acetolactate synthase gene from an imidazolinone-resistant *Arabidopsis thaliana* var. Columbia. *Nucleic Acids Research* **18**, 2188.

Sathasivan, K., Haughn, G.W. and Murai, N. (1991) Molecular basis of imidazolinone herbicide resistance in *Arabidopsis thaliana* var. Columbia. *Plant Physiology* **97**, 1044–1050.

Schultz, B. and Segobye, K. (2016) 2,4-D transport and herbicide resistance in weeds. *Journal of Experimental Botany* **67**, 3177–3179.

Shaaltiel, Y. and Gressel, J. (1986) Multienzyme oxygen radical detoxifying system correlated with paraquat resistance in *Conyza bonariensis*. *Pesticide Biochemistry and Physiology*, **26**, 22–28.

Sharples, C.R. (1996) An investigation of herbicide resistance in black-grass using safeners and synergists. PhD Thesis, The Nottingham Trent University.

Sharples, C.R., Hull, M.R. and Cobb, A.H. (1995) The effect of herbicide safeners on chlorotoluron susceptible and resistant black-grass (*Alopecurus myosuroides*). *Proceedings of the 1995 Brighton Crop Protection Conference – Weeds*, vol. 1. Farnham: BCPC, pp. 359–360.

Sharples, C.R., Hull, M.R. and Cobb, A.H. (1997) Growth and photosynthetic characteristics of two biotypes of the weed black-grass (*Alopecurus myosuroides* Huds) resistant and susceptible to the herbicide chlorotoluron. *Annals of Botany* **79**, 455–461.

Shyam, C., Borgato, E.A., Peterson, D.E., Dille, J.A. and Jugulam, M. (2021) Predominance of metabolic resistance in a six-way-resistant Palmer Amaranth (*Amaranthus palmeri*) population. *Frontiers in Plant Science* **11**, 614618; doi: org:10.3389/fpls.2020.614618

Singh, S.S., Kirkwood, R.C. and Marshall, G. (1998) Effect of ABT on the activity and rate of degradation of isoproturon in susceptible and resistant biotypes of *Phalaris minor* and in wheat. *Pesticide Science* **53**, 123–132.

Smeda, R.J. and Vaughn, K.C. (1994) Resistance to dinitroaniline herbicides. In: Powles, S.B. and Holtum, J. (eds), *Herbicide Resistance in Plants: Biochemistry and Physiology*. Boca Raton, FL: CRC Press, pp. 215–228.

Suzukawa, A.K., Bobadilla, L.K., Mallory-Smith, C. and Brunharo, C.A.C.G. (2021) Non-target site resistance in *Lolium* spp. Globally: a review. *Frontiers in Plant Science* **11**, article 609209; doi: org/10.3389/fpls.2020.609209

Takano, H.K., Melo, M.S.C., Ovejero, R.F.L., Westra, P.H., Gaines, T.A. and Dayan, F.E. (2020a) Trp2027Cys mutation evolves in *Digitaria insularis* with cross-resistance to ACCase inhibitors. *Pesticide Biochemistry and Physiology* **164**, 1–6; doi: org/10.1016/j.pestbp.2019.12.011

Takano, H.K., Ovejero, R.F.L., Belchoir, G.G., Maymone, G.P.L. and Dayan, F.E. (2020b) ACCase-inhibiting herbicides: mechanism of action, resistance evolution and stewardship. *Scientia Agricola* **78**, e20190102; doi: org/10.1590/1678-992X-2019-0102

Tal, A. and Rubin, B. (2004) Molecular characteristics and inheritance of resistance to ACCase-inhibiting herbicides in *Lolium rigidum*. *Pest Management Science* **60**, 1013–1018.

Tal, A., Romano, M.L., Stephenson, G.R., Schwann, A.L. and Hall, J.C. (1993) Glutathione conjugation: a detoxification pathway for fenoxaprop-ethyl in barley, crabgrass, oat and wheat. *Pesticide Biochemistry and Physiology* **46**, 190–199.

Tal, J.A., Hall J.C. and Stephenson G.R. (1995) Nonenzymatic conjugation of fenoxaprop-ethyl with glutathione and cysteine in several grass species. *Weed Research* **35**, 133–139.

Tan, M.-K., Preston, C. and Wang, G.-X. (2007) Molecular basis of multiple resistance to ACCase-inhibiting and ALS-inhibiting herbicides in *Lolium rigidum*. *Weed Research* **47**, 534–541.

Torra, J., Rojano-Delgado, A.M., Rey-Caballero., Royo-Esnal, A., Salas, M.L. and De Prado, R. (2017) Enhanced 2,4-D metabolism in two resistant populations of *Papaver rhoeas* from Spain. *Frontiers of Plant Science* **8**, September (2017); doi: org/10.3389/fpls.2017.01584

Vaughn, K., Vaughan, M. and Gossett, B. (1990) A biotype of goosegrass (*Eleusine indica*) with an intermediate level of dinitroaniline herbicide resistance. *Weed Technology* **4**, 157–162.

Vaughn, K.C. (2002) Cellulose biosynthesis inhibitor herbicides. In: Böger, P., Wakabayashi, K. and Harai, K. *Herbicide Classes in Development*. Berlin: Springer, pp. 139–150.

Vaughn, K.C, Marks, M.D. and Weeks, D.P. (1987) A dinitroaniline-resistant mutant of *Eleusine indica* exhibits cross-resistance and supersensitivity to antimicrotubule herbicides and drugs. *Plant Physiology* **83**, 956–964.

Vila-Aiub, M.M. (2019) Fitness of herbicide-resistant weeds: current knowledge and implications for management. *Plants* **8**, 469–480.doi.org/10.3390/plants8110469

Wakelin, A.M. and Preston, C. (2006) A target-site mutation is present in a glyphosate-resistant *Lolium rigidum* population. *Weed Research* **46**, 432–440.

Webb, S.R. and Hall, J.C. (1999) Indole-3-acetic acid binding characteristics in susceptible and resistant biotypes of wild mustard (*Sinapis arvensis* L.). *Weed Science Society of America Abstracts* **33**, 196.

Wiersma, A.T., Gaines, T.A., Preston, C., Hamilton, J.P., Giacomini, D., Buell, C.R., *et al.* (2015) Gene amplification of 5-enol-pyruvylshikimate-3-phosphate synthase in glyphosate-resistant *Kochia scoparia*. *Planta* **241**, 463–474.

Yamamoto, W., Zeng, L.H. and Baird, W.V. (1998) Alpha-tubulin mis-sense mutations correlate with anti-microtubulin drug resistance in *Eleusine indica*. *Plant Cell* **10**, 297–308.

Yanniccari, M. and Gigon, R. (2020) Cross-resistance to acetyl-CoA carboxylase-inhibiting herbicides conferred by a target-site mutation in perennial ryegrass (*Lolium perenne*) from Argentina. *Weed Science* **68**, 116–124.

Yuan, S.J., Tranel, P.J. and Stewart, C.N. (2006) Non-target-site herbicide resistance: a family business. *Trends in Plant Science* **12**, 6–13.

Zagnitko, O., Jelenska, J., Tevzadze, G., Haselkorn, R. and Garnicki, P. (2001) An isoleucine/leucine residue in the carboxyl-transferase domain of acetyl-CoA carboxylase is critical for interaction with aryloxyphenoxypropionate and cyclohexanedione inhibitors. *Proceedings of the National Academy of Sciences, USA* **98**, 6617–6622.

Chapter 14
Herbicide-tolerant Crops

DNA neither knows nor cares. DNA just is. And we dance to its music.

Richard Dawkins (1998)

The capacity to blunder slightly is the real marvel of DNA. Without this special attribute, we would still be anaerobic bacteria and there would be no music.

Lewis Thomas (1913–1993)

14.1   Introduction

The development and use of genetically modified, herbicide-tolerant (GM-HT) crops represents the single greatest change in weed control technology since the commercial arrival of selective herbicides in the mid-1940s. Herbicides that up until now have only been used when a crop was not present (or at least when a crop had yet to emerge) may now be used to perform selective weed control in emerged crops. In addition, herbicides that were not permitted in certain crops, for reasons of crop damage, may now be used safely in those crops. In practical terms, this offers growers simpler and more effective chemical weed control strategies that incorporate lower numbers of sprays and therefore less energy inputs. GM-HT crops, by allowing more efficient weed control may, however, be more likely to reduce biodiversity in crops and, through gene escape and the repeated use of a single mode of action herbicide, give rise to the spread of herbicide resistance in weeds. GM-HT crops continue to be grown commercially in an increasing number of countries and at increasing hectareage. The debate regarding their use, and the use of genetic manipulation technologies in general, continues unabated, especially in Europe. In addition to GM-HT crops, a number of crops have also been produced by non-GM techniques that exhibit herbicide tolerances which are of use in the field. These crops have many of the same benefits and drawbacks as GM-HT crops, but are arguably more acceptable to the general public as they do not contain genes sourced from other organisms. In this chapter the term genetically modified (GM) is used to denote the transfer of genes from one organism to another or the alteration of gene expression. Although the use of mutagens can cause the modification of

Herbicides and Plant Physiology, Third Edition. Andrew H. Cobb.
© 2022 John Wiley & Sons Ltd. Published 2022 by John Wiley & Sons Ltd.

genes and resultant protein sequences, they have been used in crop improvement for many years and this is not included in definition of genetic modification within the main body of this chapter and is covered separately, in Section 14.12.

14.2   History of genetically modified, herbicide-tolerant crops

Herbicide tolerance is the trait that has received the most commercialisation of all GM crops to date. Table 14.1 indicates that 14 species have so far been commercially modified for herbicide tolerance, with 347 events, out of a total of 526 events, i.e. 66%, to nine herbicides (Table 14.2). One of the first herbicide-tolerant crops grown commercially was soybean (USA, 1996) followed by cotton (USA, 1997), maize (USA, 1998) and oilseed rape (USA, 1999) (Duke, 2005). The soybean was modified to be resistant to the broad-spectrum herbicide glyphosate (Roundup™) and is known as Roundup-Ready

Table 14.1   List of species that have been genetically modified (GM) for herbicide tolerance.

Species	Number of events
Alfalfa (*Medicago sativa*)	4
Argentine canola (*Brassica napus*)	34
Carnation (*Dianthus caryophyllous*)	3
Cotton (*Gossypium hirsutum* L.)	45
Creeping bentgrass (*Agrostis stolonifera*)	1
Flax (*Linum usitatissimum* L.)	1
Maize (*Zea mays* L.)	210
Polish Canola (*Brassica rapa*)	4
Potato (*Solanum tuberosum* L.)	4
Rice (*Oryza sativa* L.)	3
Soybean (*Glycine max* L.)	33
Sugar beet (*Beta vulgaris*)	3
Tobacco (*Nicotiana tabacum* L.)	1
Wheat (*Triticum aestivum*)	1
	Total 347

Source: modified from GM Approval Database, ISAAA; www.isaaa.org/gmapprovaldatabase (accessed 13 April 2020).

Table 14.2   List of herbicides used in commercialised GM-HT crops.

2,4-D
Bromoxynil
Dicamba
Glufosinate
Glyphosate
Imazamox
Isoxaflutole
Mesotrione
Sulphonylureas

soybean (Padgette *et al.*, 1995). The gene conferring resistance to glyphosate, encoding *Agrobacterium tumefaciens* EPSP synthase, has been introduced into a number of commercially important crops including oilseed rape, maize and cotton. It has also been used to transform a number of other crops, including sugar beet and wheat, although these have yet to be grown commercially (CaJacob *et al.*, 2004). GM-mediated tolerance to the broad-spectrum herbicide glufosinate has also been commercialised in a number of similar crops: maize (1997), oilseed rape (1999), soybean (1999) and cotton (2003), using a gene from the bacterium *Streptomyces hygroscopicus* (Lyndon and Duke, 1999). These crops are commercialised under the Liberty Link™ trade name. GM-mediated tolerance to the herbicide bromoxynil has also been commercialised in oilseed rape under the trade name BXN™ since 1995 (Pallett *et al.*, 1997).

Many other crops have been genetically modified in the laboratory to possess resistance to herbicides, but have yet to be commercialised. It is worth noting that, by 2003, 54% of soybean crops grown worldwide were genetically modified and that, by 2004, 85% of US-grown soybean was genetically modified (Economic Research Service, 2007). Evans (2010) reported that 14 million farmers in 25 countries planted 134 million hectares of GM crops in 2009. GM crops have been more poorly received in Europe than in Asia and North and South America, mainly owing to public and consumer pressure against the use of GM technology. To date, the UK Government has not permitted the commercial introduction of GM-HT crops, although field-trials of GM traits have been undertaken at Research Institutes. While public perception in the UK of such crops remains negative, their commercialisation will be delayed or postponed.

Within 10 years of the adoption of GM-HT soybean, the take-up of this new technology was over 90% of the crop grown in the USA, soon to be matched by the same take-up in the cotton and maize crops. GM-HT sugar beet was commercialised in 2007 and by 2010 the adoption rate in the USA was 95%. By 2018, the percentage of the global top four crops grown as GM was between 29 and 78% (Table 14.3). Consequently, the use of glyphosate has increased dramatically, at the expense of other herbicides, and glyphosate has become the most used and most profitable agrochemical in the world (Duke, 2017).

The UK has not permitted any commercial work with GM crops. A 3 year Farm-Scale Evaluation of GM-HT oil-seed rape (canola), maize and sugar beet was undertaken during 2000–2002 (Firbank *et al.*, 2003) to establish effects of these crops on biodiversity. The main (and rather obvious) conclusion drawn was that where weed populations were reduced by herbicide treatment, biodiversity was reduced, and where more weeds were present biodiversity was increased. The GM-HT sugar beet and oil-seed rape crops had better weed control than using conventional control methods, and this was the reason for their success. It follows that a requirement for better weed control and increased biodiversity are currently incompatible.

Table 14.3 Hectares grown of the top four GM crops in 2018 (ISAAA, 2018).

Crop	Hectares grown as GM (millions)	Percentage of global GM crop
Canola	10.1	29
Cotton	24.9	76
Maize	58.9	31
Soybean	96.3	78

The uptake of GM-HT crops in the world, in addition to weed control and environmental impact, is also due to financial savings, resulting from the need for fewer spray applications, resulting in lower agrochemical costs, lower fuel and labour costs and less soil compaction owing to a reduced use of machinery and reduced tillage. Brookes and Barfoot (2018a, b) have estimated that the economic benefit for all GM-HT crops in 2016 was US$8.44 billion.

14.3   How genetically modified crops are produced

The genetic information of an organism, encoded by the sequence of nucleotide bases within its DNA, determines the proteins that it can manufacture and, through these, the physical and biological nature of the organism. Protein synthesis can be subdivided into two distinct processes: transcription and translation (Figure 14.1). Transcription is the copying of the information from a segment of DNA, often a gene encoding for a single protein, to a strand of mRNA. Translation, which takes place in association with ribosomes, involves the 'reading' of the nucleotide sequence in the mRNA molecule and the construction of the amino acid chain which forms the primary structure of the protein. Subsequent to this, the amino acid chain will fold and cross-link and may undergo post-translational modification to form the functional protein.

GM crops usually have one or more genes added to their genetic make-up and so produce one or more extra proteins. It is these proteins that cause the crop to have different properties from those of the unmodified crop. In some instances the modification may involve overexpressing one of the genes, in which case it will produce more of one of its proteins, or underexpressing one of its genes, in which case the plant will produce less or none of one of its proteins.

Since the early 1980s scientists have developed a number of ways by which genes from other organisms may be added to the genetic material of the plant. The use of the microorganism *A. tumefaciens* has proved useful in modifying a number of commercially important crops (Komari *et al.*, 2004). *A. tumefaciens* is a common soil bacterium that is a plant pathogen, causing galls to be produced on infected plants. It accomplishes this in nature by transferring some of its genes to plant cells. These genes are transferred in a plasmid, a circular DNA molecule, and in the case of *A. tumefaciens* the plasmid causing gall formation is termed the Ti (tumour inducing) plasmid. Using this naturally occurring system of genetic modification, scientists have successfully transformed a number of crop species (Komari *et al.*, 2004). The gene of interest is first inserted into the Ti plasmid of *A. tumefaciens* and the transformed bacteria are used to infect pieces of plant tissue. Successful infection will result in the Ti plasmid (and the added gene) being transferred to the plant genome. Owing to totipotency, the plant tissue can be regenerated to a whole plant, in which each cell will contain the added gene. In order for this to be successful, the genes in the Ti plasmid that are responsible for tumour initiation are first removed. The introduced genes are integrated into the nuclear DNA of the plant in the same way as when *A. tumefaciens* infects plants in the field.

Although this method has proved very successful for a number of dicotyledonous plant species, until recently it had not been so for monocotyledons, as *A. tumefaciens* would not infect them. However, techniques have since been developed that enable agrobacterium-mediated transformation of rice (Tyagi and Mohanty, 2000), maize (Ishida *et al.*, 1996),

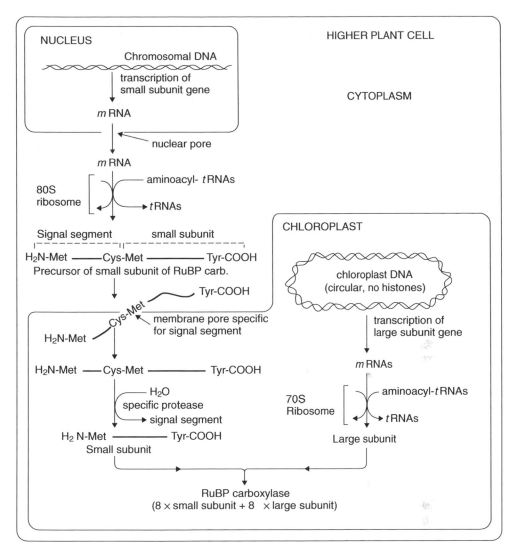

Figure 14.1 Diagrammatic representation of protein synthesis and its cellular location. The example given is for RuBisCo. Goodwin, T.W. and Mercer, E.I. (1983) Introduction to Plant Biochemistry, 2nd edn. Oxford: Pergamon Press. Reproduced with permission of Elsevier.

barley and wheat (Cheng *et al.*, 1997). It has also proved problematic with soybean, as regeneration from cell cultures is often not possible (McCabe *et al.*, 1988). An alternative method of introducing genes has been used for these species. This involves coating tungsten particles with DNA containing the gene of interest and using a microprojectile gun to shoot the particles and DNA into plant cells. This method, often referred to as the 'shotgun' method, has allowed transformation of plant species where *A. tumefaciens* is ineffective as a vector. DNA that has entered the cell is incorporated into the nuclear DNA and is then passed on as the cell divides.

In addition to these methods, DNA can also be introduced to protoplasts (plant cells that have had their cell walls enzymatically removed) using electroporation or chemical methods to open up the plasma membrane and allow DNA to enter the cell. Alternatively, microinjection can be used. The protoplast will grow back its cell wall and can be regenerated to a plant whose cells all contain the added gene.

In addition to the gene conferring the required trait, a number of other genes and pieces of genetic information are added. These include a promoter (a molecular switch to ensure that the required gene will be transcribed in the transformed plant) and a stop sequence that ensures transcription stops at the end of the required gene. In addition, a marker gene is also added as a way of screening cell cultures to determine if transformation has been successful. For many GM crops the marker gene has been one which confers antibiotic resistance to the successfully transformed plant tissue, allowing easy selection for transformed tissue by culturing it *in vitro* in the presence of antibiotics. A commonly used gene is *NPT II*, which produces the enzyme neomycin phosphotransferase, giving resistance to the aminoglycoside antibiotics (Komari *et al.*, 2004). The gene encoding for the enzyme hygromycin phosphotransferase has also been used as a marker (Bilang *et al.*, 1991).

There has been some concern regarding the use of antibiotic resistance genes, especially where they remain in the GM plant when it is at the stage of farm-scale evaluation and commercial production. Alternative methods (for instance, incorporation of the *pmi* gene encoding phosphomannose isomerise, ensuring that on a mannose-only culture medium only transformed tissue survives) have received limited use and may prove more publicly and environmentally acceptable (Joersbo, 2001). The herbicide tolerance gene *bar* has also been used as a marker gene (Vasil, 1996) and, where herbicide tolerance is the trait being introduced, the presence of a marker gene would appear unnecessary, as a simple screen in the presence of the herbicide to which tolerance is introduced should successfully select for transformed tissue by killing untransformed cells. The process of producing a GM-HT crop is summarised in Figure 14.2

14.4 Genetically engineered herbicide tolerance to glyphosate

Glyphosate is a non-selective herbicide that kills plants by inhibiting their ability to synthesise the aromatic amino acids phenylalanine, tyrosine and tryptophan (see Chapter 9). It accomplishes this by blocking the action of the enzyme EPSP synthase, the penultimate step in the shikimate pathway, vital in the biosynthesis of these three amino acids. Interestingly, glyphosate competes with one substrate (phosphoenol pyruvate, PEP) but forms a very stable herbicide–enzyme complex with the other, resulting in a 'full stop' to the metabolic pathway. This also subsequently reduces the ability of the plant to synthesise a number of vital metabolites including hormones, flavonoids and lignins. GM crops that are tolerant to glyphosate are produced by the insertion of a gene coding for a glyphosate-insensitive EPSP synthase that is obtained from a soil bacterium *Agrobacterium* CP4. In this way, plant EPSP synthase is still inhibited, but the bacterial EPSP synthase is unaffected, allowing the plant to still synthesise aromatic amino acids and the other essential metabolites resulting from the shikimate pathway (Figure 14.3). For an example of this process, see Vande Berg *et al.* (2008). The kinetic properties of plant and bacterial wild-type and mutated EPSP synthases are shown in Table 14.4.

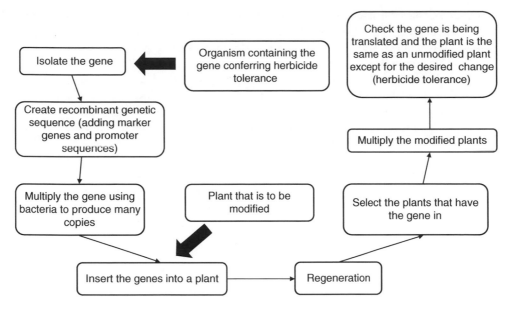

Figure 14.2   An overview of the steps taken in the production of a genetically modified plant. From Bruce and Bruce (1998).

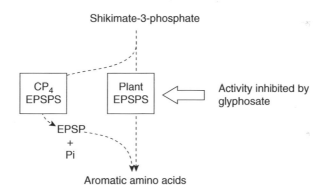

Figure 14.3   Biosynthesis of aromatic amino acids indicating how CP4 EPSPS allows synthesis in the presence of glyphosate. EPSP, 5-Enoylpyruvyl shikimic acid 3-phosphate; EPSPS, EPSP synthase.

Table 14.4   Kinetic properties for selected EPSP synthases. Ki/Km (PEP) is a measure of the selectivity of EPSPS for PEP over glyphosate. A higher value indicates a greater tolerance to glyphosate while the enzyme still possesses EPSPS activity.

Enzyme source	Km (PEP) (μM)	Ki (glyphosate) (μM)	Ki/Km
Petunia (wild type)	5.0	0.4	0.08
G101A	210	2000	9.5
T102I/P106S	10.6	58	5.5
P106S	17	1	0.06
Agrobacterium sp. CP4	12	2720	227

Source: Dill, G.M. (2005) Glyphosate-resistant crops: history, status and future. *Pest Management Science* **61**, 219–224.

Crops that have been modified by insertion of the EPSP synthase CP4 gene are commercialised under the trade name Roundup-Ready™ and represent the largest proportion of the GM-HT crop market worldwide. CP4 is a mutated form of EPSP synthase that has an amino acid replacement (Gly100 → Ala) which results in extremely high tolerance to glyphosate (Padgette *et al.*, 1996). The low mammalian toxicity, fast degradation in soil and translocation within plants make glyphosate a popular and effective herbicide. The use of glyphosate-tolerant crops has reduced the number of herbicide applications necessary for a number of crops in addition to reducing fuel costs and land compaction owing to reduced machinery use. In addition, these GM crops increase the flexibility of timing of weed control, as glyphosate can be used on weeds at growth stages in excess of those that conventional herbicides can effectively control. Such practices have allowed for the increase in no-tillage and conservation tillage practices alongside GM-HT crops in the USA and Canada.

The effectiveness of glyphosate as a herbicide means that weed control through cultivation is not always regarded as necessary. The high level of weed control achieved, however, will clearly impact on biodiversity, as indicated in some of the data collected in the UK during the 3 year Farm-Scale Evaluations. In addition, the overuse of glyphosate as the sole method of weed control has created an unacceptable selection pressure for the survival of naturally occurring populations of glyphosate-resistant weeds. This is a very real concern. As such, GM glyphosate-tolerant crops should be considered alongside GM crops that are tolerant to other herbicides and non-GM crops in rotation in cropping systems. Studies such as those by Westra *et al.* (2008) will also prove important in assessing the effects of changing land management practices under glyphosate-tolerant cropping systems on weed population dynamics.

An alternative method for conferring tolerance to glyphosate involves the insertion of the *gox* gene from *Ochrobactrum anthropi* into crop plants (CaJacob *et al.*, 2004; Reddy *et al.*, 2004). Glyphosate oxidoreductase (GOX) is an enzyme that catalyses the breakdown of glyphosate to AMPA (aminomethylphosphonic acid) and glyoxylic acid (Figure 14.4).

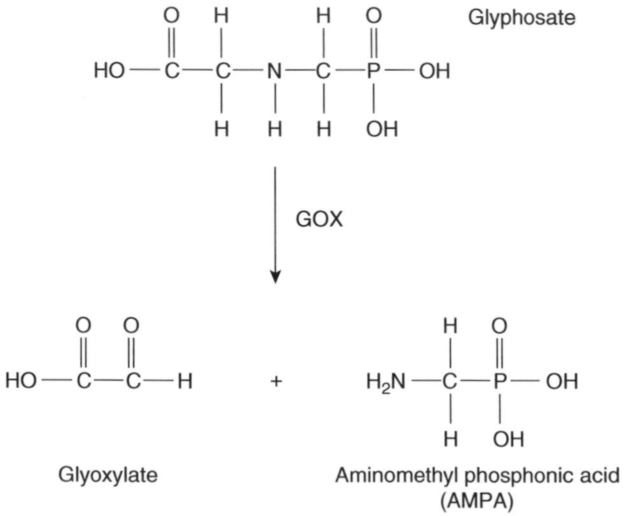

Figure 14.4 The detoxification of glyphosate by glyphosate oxidoreductase (GOX).

In some oilseed rape lines, both the CP4 EPSP synthase and *gox* genes are inserted and expressed, giving the resultant plant two methods of avoiding damage by glyphosate. The reasons for this gene stacking have not been made public and the build-up of AMPA (owing to the presence of GOX) has been implicated in phytotoxic symptoms in some GM oilseed rape lines (Reddy *et al.*, 2004).

As a result of its global success in treating GM-HT crops, and the fact that it is no longer under patent to Monsanto, glyphosate has become the most commonly used herbicide in the world. Furthermore, glyphosate alone now contributes over 50% of the global herbicide market and will become the first agrochemical to reach US$10 billion in annual sales.

14.5   Genetically modified herbicide tolerance to glufosinate

Glufosinate is a non-selective herbicide that kills plants by inhibiting the enzyme glutamine synthase (Chapter 9). This is a key enzyme in the biosynthesis of the amino acid glutamine and also plays a vital role in nitrogen metabolism. Inhibition of this enzyme by glufosinate results in a decrease in glutamine/glutamate pool size and an accumulation of phosphoglycolate, glycolate and glyoxylate. Glyoxylate inhibits the activity of ribulose 1,5 bisphosphate carboxylase-oxygenase (RuBisCo), the key enzyme in photosynthetic carbon reduction. So glufosinate effectively inhibits the ability of a plant to photosynthetically produce carbohydrates and other products.

The bacterium *S. hygroscopicus* possesses the *bar* gene that encodes the enzyme phosphinothricine acetyl transferase (PAT) that detoxifies glufosinate by acetylation (Lyndon and Duke, 1999). An alternative and very similar gene, *pat*, from *Streptomyces viridichromogenes*, also gives rise to the enzyme PAT and has also been used to modify plants (Lyndon and Duke, 1999; Figure 14.5). Further details on how glufosinate-degrading bacteria are identified and isolated from soils can be found in Hsiao *et al.* (2007).

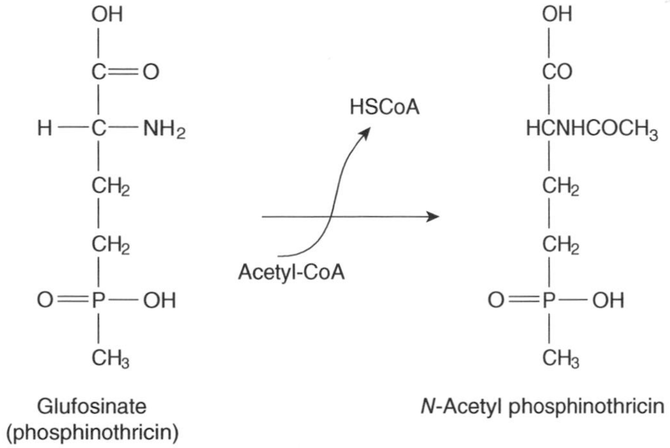

Figure 14.5   The detoxification of glufosinate by phosphinothricin acetyl transferase (PAT).

Although commercially exploited far less than tolerance to glyphosate, oilseed rape, maize, soybean and cotton have all been commercially grown with inclusion of the *pat* or *bar* gene under the trade name Liberty Link™.

14.6 Genetically modified herbicide tolerance to bromoxynil

Bromoxynil is a contact herbicide that kills plants by inhibiting photosynthesis at Photosystem II. It is used for selective weed control in a number of crops including wheat, barley and oats, but is not suitable for use in oilseed rape. GM-HT oilseed rape has been produced using the *oxy* gene from *Klebsiella ozaenae* that encodes a nitrilase enzyme (Pallett *et al.*, 1997). The *K. ozaenae* strain containing this gene, isolated from bromoxynil-contaminated soil, can utilise bromoxynil as its sole carbon source, indicating that the strain contains enzymes that can catabolise bromoxynil. It has subsequently been shown that the nitrilase is responsible for the breakdown of bromoxynil to the non-phytotoxic 3,5-dibromo-4-hydroxybromobenzoic acid (Figure 14.6). This method of producing herbicide tolerance has only had limited commercialisation including oilseed rape in Canada (under the trade name BXN). Owing to the poor performance of bromoxynil against grass weeds and certain broad-leaved weeds, it is unlikely that this method will ever prove as popular as those used for creating tolerance to glyphosate or glufosinate.

14.7 Genetically modified herbicide tolerance to sulphonylureas

The identification of genes encoding for acetolactate synthase (ALS) that contain point mutations has allowed the genetic modification of a number of crops so that they possess tolerance to sulphonylurea herbicides. The gene introduced is often from *Arabidopsis* or from tobacco. The most common mutation resulting in tolerance is Pro197 → Ser (Haughn *et al.*, 1988). Many of the commercial sulphonylurea-tolerant crops have been produced by non-GM means and are discussed in Section 14.12.

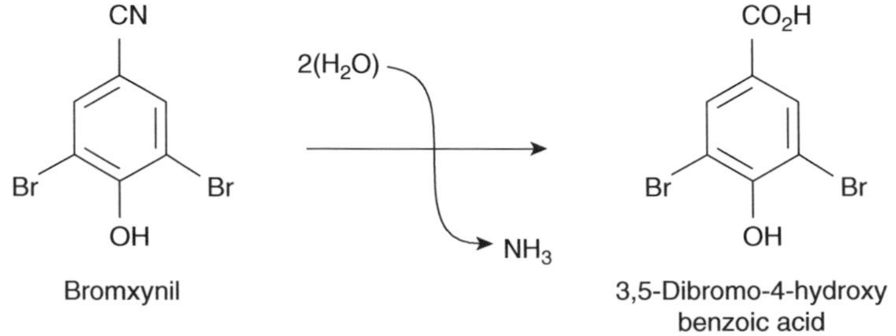

Figure 14.6 The detoxification of bromoxynil by bromoxynil nitrilase.

14.8   Genetically modified herbicide tolerance to 2,4-D

Unlike other herbicides discussed in this chapter, the primary site of action of 2,4-D is complex (Chapter 7). Consequently, attempts to date at transferring herbicide tolerance to crops have focused on enhanced breakdown rather than an altered target site. 2.4-D is broken down quickly in soils by a number of microorganisms. One such organism, the bacterium *Alcaligenes eutrophus*, has been shown to be able to utilise 2,4-D as its sole source of carbon (Llewellyn and Last, 1996). It can do this because it possesses a number of genes (called *tfd* genes) that encode for enzymes that metabolise 2,4-D (Llewellyn and Last, 1996, 1999). Of the six identified *tfd* genes, the first (*tfdA*) is of most importance with regard to engineering crop tolerance. 2,4-D (which is phytotoxic) is metabolised to 2,4-DCP (which is not phytotoxic; Figure 14.7), and *tfdA* encodes the enzyme 2,4-D dioxygenase, which carries out the first step in this process.

14.9   Genetically modified herbicide tolerance to fops and dims

Although resistance in weeds to the fops and dims readily occurs and is attributed to an altered target site, enhanced metabolism and possibly altered membrane electrochemical properties, no crops with tolerance to this group of herbicides have yet been produced by GM methods, probably owing to the numbers of existing weeds that are resistant to ACCase inhibitors. A few have been produced by non-GM means and are discussed in Section 14.12.

14.10   Genetically modified herbicide tolerance to phytoene desaturase inhibitors

A point mutation in the gene encoding phytoene desaturase (the *pds* gene) results in tolerance to herbicides acting at this target site. Mutations that cause Ser, Cyst or Pro substitution for Arg at position 195 all produce this effect. A herbicide-tolerant phytoene

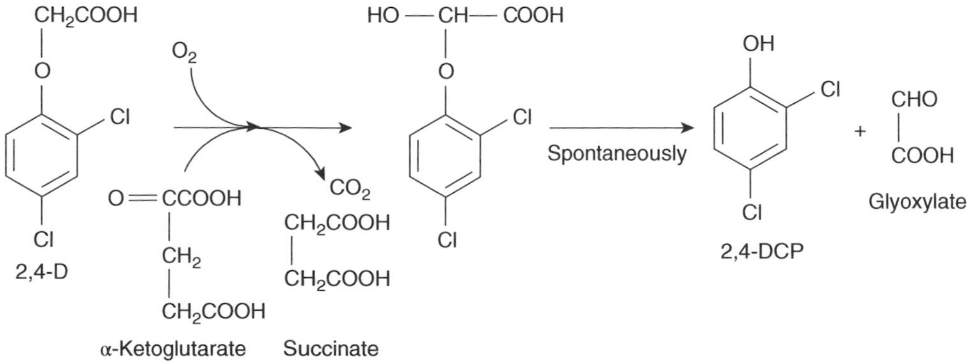

Figure 14.7   The detoxification of 2,4-D by 2,4-D dioxygenase.

desaturase from *Erwinia uredovora* (encoded by the gene *crtI*) has been used to genetically engineer tobacco to be tolerant to norflurazon and fluridone (Misawa *et al.*, 1993).

14.11 Herbicide tolerance owing to genetic engineering of enhanced metabolism

Specific examples of increases in herbicide metabolism as a means of conferring tolerance have already been presented for glufosinate, bromoxynil and the GOX system for glyphosate tolerance. These all involve highly specific enzyme–herbicide interactions and so are not likely to impart a great degree of cross-tolerance to other herbicides. An alternative approach is to engineer the cytochrome P450 and/or glutathione *S*-transferase (GST) enzyme systems that have been implicated in the metabolism of many herbicides and other xenobiotics. In this way, multiple tolerance may be produced within a crop. Certain herbicide safeners, by increasing P450 and GST activities, already aid the crop in metabolising herbicides so that it is are not damaged by them (Chapter 4). An interesting study by Inui *et al.* (1999) introduced human and human/yeast-fused genes encoding for cytochrome P450s into potato plants using standard genetic engineering techniques. The resultant transgenic potato plants showed tolerance to the herbicides chlorotoluron, atrazine and pyriminobac-methyl. Although this method of producing GM-HT crops might been seen as advantageous owing to the multiple tolerance that is obtained, it should be considered that there may also be a number of drawbacks. As less is known regarding herbicide metabolism, the herbicides that can be used in such crops and the control of such crops if they occur as volunteers will be a lot less predictable. Additionally, the P450 and GST enzyme systems appear to be involved in stress responses as well as protection from xenobiotics, and therefore disruption of these systems may alter the plant's abilities to cope with environmental stressors. Finally, it is highly probable that the consumer, who has already shown some distrust of GM crops, will reject crops that contain human genes.

14.12 Herbicide tolerance through means other than genetic modification

Crop tolerance to herbicides does not necessarily need to be produced by the introduction of genes from other species. A number of methods have been employed, with some success, to produce herbicide-tolerant crops by non-GM means. These include the screening of cultivars for tolerance to herbicides during standard breeding programmes, as well as screening for tolerance in cell cultures and in seed germination assays. In addition to these methods, mutations have been induced, using chemicals and/or exposure to X-rays, to increase the probability of such screening methods successfully isolating herbicide-tolerant lines. Pollen mutagenesis and crossing with wild species that demonstrate herbicide tolerance have also proved successful (Somers, 1996).

In some cases, tolerance to herbicides varies greatly between cultivars of the same species, as is the case for chlorotoluron tolerance in wheat. In this case, the herbicide can only be used on cultivars listed on the product label, as unacceptable levels of phytotoxicity are observed in sensitive cultivars. This could form the basis of breeding programmes to increase tolerance to chlorotoluron, although in reality yield, quality factors and disease resistance are traits considered more important by crop breeders. It is also advantageous that the level of tolerance in herbicide-tolerant crops is far greater than is commonly found in naturally occurring cultivars. Selection of ryegrass lines showing resistance to glyphosate did not produce plants with tolerance levels anywhere near those exhibited by genetically engineered crops.

The use of mutagens, often in cell culture, can address this and has been used to produce a number of herbicide-tolerant crops. By causing mutations, the probability of selection of a herbicide-tolerant line is increased, but compared with the more certain outcomes of genetic engineering these methods are still somewhat 'hit and miss'.

Glyphosate tolerance owing to overexpression of the gene encoding for EPSP synthase in *Petunia* suggested that the low levels of tolerance obtained using this method would not be commercially viable at present. Similar observations have been made in *Salmonella typhimurium* (Ser101 → Pro), *Escherichia coli* (Gly96 → Ala) and can be compared with the mutation in the CP4 EPSP synthase gene (Gly100 → Ala) that has been used for genetic engineering of glyphosate tolerance, as detailed above, especially in Table 14.4. Overexpression of the target site for glufosinate, glutamine synthase, in transgenic tobacco suggested that this method might be of some use in producing glufosinate-tolerant crops, although the genetic engineering methods outlined above have proved more successful to date (Eckes *et al.*, 1989).

Tolerance to triazines in brassica species has been produced simply by cultivar screening and conventional breeding techniques. The physiological basis of this resistance is point mutations in the target site of the triazines, that is the D1 protein. A mutation on the *psbA* gene results in an amino acid replacement (Ser264 → Gly) (Hall *et al.*, 1996) that has a greater tolerance to atrazine. A similar replacement (Ser264 → Thr) results in a greater degree of cross-tolerance to other triazines. In addition, glutathione *S*-transferases has been implicated in resistance to triazines in the weed species *Abutilon theophrasti* (Anderson and Gronwald, 1991; Gray *et al.*, 1996). The level of this resistance is a lot lower than that imparted by *psbA* gene mutations and has not been exploited in crop tolerance studies.

Tolerance to sulphonylurea (SU) herbicides has been obtained by screening cell cultures in order to select for SU-tolerant mutants in the presence of chlorsulfuron. Using these techniques, SU-tolerant tobacco, sugar beet, oilseed rape, soybean and a host of other crops have been obtained (Saari and Mauvais, 1996; Shaner *et al.*, 1996). The basis of this tolerance is point mutations in the gene encoding for ALS. In *Arabidopsis*, a single mutation results in Pro197 → Ser and increased tolerance to SU herbicides. In tobacco, Pro196 → Ser produces similar results. Substitution of almost any amino acid for Ala (position 117), Pro (position 192) or Try (position 586) results in increased SU tolerance. These multiple sites of mutation that result in viable yet SU-tolerant ALS variants may go some way towards explaining why resistance in weeds to this family of herbicides appeared so soon after their introduction in the early 1980s and has risen to such a high level today.

Tolerance to imidazolinones, which act at the same target site as SUs, has been obtained in maize (pollen mutagenesis), wheat (seed mutagenesis), oilseed rape (microspore mutagenesis) and maize (tissue culture in the presence of imazaquin). The latter, which selected for cells containing an altered ALS target site, produced maize plants that were up to 100 times as tolerant as susceptible wild-type maize. Mutations causing Ser653 → Asn give rise to resistance that is specific to imidazolinones. The resulting ALS is still susceptible to SUs and triazolopyrimidines.

Limited research into increasing tolerance to fops and dims in ryegrass (by sexual hybridisation with resistant species), maize and wheat (by selection in tissue cultures) has produced successes, although these have yet to be commercialised (Somers, 1996).

Soybean is one of the biggest monoculture crops in the world, with 122.44 million hectares harvested in 2019, being widely used for food and animal feed (Ali *et al.*, 2020). Weeds, however, are responsible for an average yield reduction of 37%, although many broad-leaf weeds are controlled by bentazone. Since bentazone may damage crops after application, the search is now on to discover bentazone-tolerant cultivars in GM-free countries, including the EU, to increase the global yield of this important crop. Ali *et al.* (2020) report a screen of 338 soybean cultivars in the USA under field conditions, and several have been found to have increased tolerance to bentazone, mainly owing to increased rates of bentazone metabolism.

14.13   Gene editing

By the end of the twentieth century it was well known that bacteria were resistant to viruses and bacteriophages, which was especially important in the production of yoghurts and cheeses. Sequencing the genome of *Streptococcus thermophilus* led to the discovery that stretches of DNA conferred resistance against bacteriophage attack. These clustered, regularly interspaced short palindromic repeats, or CRISPRs, were fragments of DNA acquired from bacteriophages, and so bacteria with these sequences then became resistant to the bacteriophages. Subsequently, scientists led by Dr Jennifer Doudna *et al.* in 2012, at the University of California, Berkeley, showed that CRISPR can be used to precisely target any part of a genome with the aid of a cutting enzyme, termed Cas9. They demonstrated that Cas9 could be directed to almost any DNA sequence using a single guide RNA molecule. Cas9 then edits the DNA by making a double-stranded break, to either disrupt a gene or to insert a desired replacement sequence. In this way, 'repair or remove' becomes possible. The technique itself is relatively cheap and quick, and easy to use, and has become a major breakthrough in the manipulation of DNA. This approach, termed gene editing (GE), has massive potential in biomedicine, plant breeding and agriculture, and the annual market for this technology is estimated at tens of billions of dollars, and growing. In agriculture, it could be used to breed improved crop resistance to pests, weeds and diseases, and to enhance food quality. It is no surprise that the number of publications using CRISPR/Cas9 has increased exponentially in recent years and that the technology is becoming increasingly refined with time. As an example, Ming *et al.* (2020) have demonstrated a new system for GE that uses CRISPR–Cas12b for engineering the rice genome. Cas12b recognises a different DNA targeting sequence, allowing for larger cuts in the DNA. The authors suggest that it provides greater efficiency for gene activation and the potential for broader targeting sites for gene repression. Second-generation GE crops now feature improved nutritional and industrial traits, and the perceived risks of using this new technology are being addressed. The reader may wish to refer to the reviews by Bortesi and Fischer (2015), Songstad *et al.* (2017), Wolt *et al.* (2016) and Es *et al.* (2019) for further details.

While the production of genetically modified organisms (GMOs) involves the transfer of a gene from one organism to another, GE allows scientists to change the DNA sequence of an organism without the insertion of foreign DNA. We are now seeing many examples of the use of this new technology.

14.14   Economic, environmental and human health benefits from the adoption of GM technology

After 25 years of GM crop production, much data has been collected on the benefits from their global adoption, as summarised below. Data are taken from Brookes and Barfoot (2018a, b), the ISAAA Pocket of Knowledge, Documented Benefits of GM Crops, updated April 2020, and Smyth (2020).

14.14.1   Economic

In 2016, direct global farm income benefit was US$18.2 billion. Since 1996, cumulative benefits have increased by US$186.1 billion. Over that time, planting GM crops has reduced chemical inputs by an estimated 37%, increased yields by 22% and improved farm profits by 68%. Up to 2014, it has been estimated that there have been grown an additional 158 million tonnes of soybean and 332 million tonnes of maize.

14.14.2   Environmental

The environmental impact of reducing herbicide and insecticide inputs, as measured by the Environmental Impact Quotient, has been 18.4%. GM technology has also resulted in less fuel use, less soil tillage and a reduction in greenhouse gas emissions, equivalent to removing an estimated 16.7 million cars from roads.

14.14.3   Human health benefits

These have recently been documented and quantified by Smyth (2020) as (a) reductions in pesticide poisonings in developing countries resulting from reduced applications and less exposure to insecticides, (b) lower rates of suicide in India since the adoption of GM cotton, (c) reductions in cancer rates from fungal infections following the introduction of GM maize, since there are fewer mycotoxins, thricotecens and fumonisins, which are both toxic and carcinogenic to humans, (d) nutritional benefits resulting from biofortified GM crops and (e) mental health benefits for farmers, knowing that it is less likely that their crop will fail from pests, weeds and diseases, and be more resilient to drought and other climatic changes. Finally, there is no evidence to suggest that food from GM crops is unsafe for human and animal consumption.

Kniss (2017) has made a detailed analysis of herbicide use in the USA over a 25 year period. He found that herbicide use intensity, as measured by the number of applications made to each field, had increased in maize, cotton, rice and wheat, whether they were GM or non-GM. As herbicide use increased, the associated chronic or acute toxicity declined. In the years of his study (2014/2015), glyphosate accounted for 26% of maize, 43% of soybean and 45% of cotton herbicide applications. Owing to its relatively low chronic toxicity, however, glyphosate contributed only 0.1, 0.3 and 3.5% of the chronic toxicity hazard in those crops. It follows that if glyphosate was banned in the EU or elsewhere, it would be replaced by more relatively toxic herbicides which would have more negative impacts on both herbicide applicators and the environment. His analysis suggests that GM-HT with glyphosate has had a positive effect regarding herbicide use intensity and lowering toxicity to mammals.

14.15   Gene stacking

Gene stacking is the process of combining two or more genes of interest into a single plant. The combined traits are termed stacked traits. A very successful example is the generation of crop plants that have been transformed with genes for both HT and insect resistance. In the latter case, genes that encode *Bacillus thuringiensis* proteins have been most effective in controlling insect pests in maize, cotton and eggplant (brinjal). This technology offers resistance to major above- and below-ground pests, including stem and stalk borers in maize, bollworm in cotton and caterpillars in soybean. Of the 77.7 million hectares of GM crops grown with stacked traits in 2017, the USA and Brazil contributed 41% each (i.e. 32 million ha), followed by Argentina (7.6 million ha), Canada (1.5 million ha), South Africa (1.3 million ha), Paraguay (1.2 million ha) and smaller areas in the Philippines, Australia, Uruguay, Mexico, Colombia, Vietnam, Honduras, Chile and Costa Rica.

Soybean cultivars with the Roundup Ready (RR) trait, i.e. glyphosate resistant, alone account for 30% of the soybean market in the USA. The addition of the Xtend trait to the stack, i.e. dicamba-resistant (DR), will account for at least 50% of the soybean market in 2020. The Liberty Link (LL) cultivar was introduced in 2009 and confers resistance to glufosinate. It has about 20% of the soybean market, as it also controls glyphosate-resistant weeds, especially *Amaranthus* spp. Soybean cultivars with RR, LL and DR traits are also now available. Monsanto introduced dicamba-tolerant cotton in 2015 and soybean in 2016 (XtendiMax), so that weeds that had become resistant to glyphosate could be controlled by this auxin-type herbicide. Corteva have also commercialised cultivars with RR, LL and 2,4-D resistance. The addition of resistance to isoxaflutole, as an additional mode of action, is also available. Although increasing the stack in crop cultivars to control a wider range of herbicide-resistant weeds is becoming more widespread, one wonders if this strategy will be sustainable in the coming decades. Will it simply lead to an even more widespread occurrence of herbicide-resistant weeds?

One unforeseen outcome has been the reports by farmers in the USA that dicamba has caused extensive damage to organic and conventional peach crops and fruits, apparently across several million hectares, owing to spray drift and volatilisation. They have taken Monsanto, Corteva and BASF to court and damages of US$265 million were awarded in February 2020. Bayer, the parent company of Monsanto, has challenged this ruling and BASF is providing training for those applying dicamba to address this concern. The Environmental Protection Agency (EPA), the organisation in the USA responsible for the approval of agrochemicals, is investigating the cause of the crop damage. In June 2020 a decision by the US Court of Appeals has ruled that, since the EPA did not estimate the amount of damage that would result from dicamba use, the registration of dicamba was invalidated for use in GM soybean and cotton crops. The current registration of the use of dicamba in this context was due to expire in December 2020. This decision will no doubt be appealed by Bayer, BASF and Corteva Agriscience, and perhaps the dicamba formulation will be amended to reduced volatility and spray drift. Time will tell if EPA approval will be extended beyond this time. In the summer of 2020, Bayer agreed to pay US$400 million to settle the allegations that dicamba caused damage to crops.

Soybean, maize and cotton cultivars are the main focus for herbicide-resistant crops with stacked traits. Market acceptance is thought to be the reason for why this technology has not been developed in wheat and rice. Future stacks are likely to include traits with engineered metabolic pathways for nutritional enhancements that include β-carotene, vitamins A, C and E, and folic acid, but do not compromise yield (Naqvi *et al.*, 2010).

While no separate and additional regulatory approval is needed for the introduction of stacked traits in the USA and Canada, authorities in Japan and the EU regard stacks as separate events and so applications for registration must pass through a more detailed regulatory process. This includes risk assessments that could arise from the combined genes in the stack, especially in relation to effects on non-target organisms and possible environmental risks. The regulation of GM crops resulting from various GM techniques in Argentina has been recently reviewed by Whelan *et al.* (2020).

14.16   Will the rise of glyphosate be inevitably followed by a fall?

Three recent developments suggest that the rise of glyphosate might indeed be followed by a fall, namely glyphosate-resistant weeds, the suggestion that it may be a human carcinogen and public perception of GM crops.

First, the widespread use and relative supremacy of glyphosate, not only in GM-HT crops, but also in the public and domestic sectors, has created a huge selection pressure and so there are now over 43 weed species that are known to have evolved resistance to it. This is an inevitable and predictable response to its global use, especially in fields where glyphosate use has been largely exclusive, year-on-year. Consequently, farmers now have to adopt improved weed control management strategies with increased costs. Weed resistance is especially evident in the genera *Lolium*, *Amaranthus* and *Conyza*, and so the benefits of using GM-HT with glyphosate are being slowly eroded. It is not difficult, therefore, to envisage a decline in glyphosate sales and use in the coming decades.

Second, in 2016, the International Agency for Research on Cancer assessed hazard data on glyphosate and characterised it as a category 2A carcinogen, i.e. likely to cause cancer in humans. The response was both strong and partisan. The agrochemical industry has strongly opposed this new categorisation, based on extensive scientific evidence over 40 years, while the environmental and public lobbies have been equally vociferous, proposing that it should be banned with immediate effect. This ruling, however, has resulted in county councils in the UK re-assessing their use of glyphosate, and several European countries, including France, Denmark and the Netherlands, proposing to ban the use of glyphosate altogether, with inevitable negative consequences to the farmers in those countries. Anxieties also developed in Germany with the suggestion that there were residues of glyphosate in breast milk. A subsequent study commissioned by the Federal Institute for

Risk Assessment in Germany tested 114 samples of breast milk and found no detectable residues of glyphosate (www.bfr.bund.de/en/press_information/2015/16/the_national_breastfeeding_committee_and_the_bfr_recommend_that_mothers_continue_to_breastfeed_194547.html).

Glyphosate was not classified as a carcinogen by the European Chemicals Agency in 2017 as it was not thought to fulfil the criteria for such categorisation. They proposed that glyphosate should therefore be authorised for sale and use for a further 15 years. However, the environmental lobby groups Greenpeace and the Soil Association consider that the European Chemicals Agency rejected evidence of cancer in laboratory animals.

The EU approval for the sales and use of glyphosate lapsed in 2015 and there was considerable controversy whether its approval should be continued for a further 15 years. A petition to prevent glyphosate use was signed by 1.3 million people in a European Citizens Initiative, while the glyphosate producers strongly presented their case for continuation. A binding vote, taken by the EU Commissioner's Appeals Committee in November 2017, decided that glyphosate was awarded a 5 year renewal for approval of sales and use in the EU. Many observers consider that the decision was unduly influenced by public opinion and politics. It is hoped that an unbiased and independent review is initiated for when renewal is next applied for in 2022.

Meanwhile, in August 2018, Monsanto was ordered to pay US$289 million in compensation to a former grounds-keeper who was diagnosed with non-Hodgkin's lymphoma in 2014, after having sprayed glyphosate for many years. An appeal by Bayer in October 2018 reduced the sum to US$78.5 million, and Bayer has since (April 2019) requested an appeals court to quash the entire award. Bayer will seek a re-trial if the decision is not overturned, based on their findings that glyphosate-based weedkillers do not cause cancer. Indeed, in 2018 a two-decade-long definitive analysis of data from nearly 45,000 farmworkers who had applied glyphosate in their careers was conducted by the National Institutes of Health, which showed no association with non-Hodgkin's lymphoma or increased overall cancer risk. In the meantime, Bayer has not only acquired Monsanto's agrochemical portfolio, but also an estimated 13,400 lawsuits filed by persons in the USA claiming that glyphosate has caused cancer to themselves and/or to their families. While the scientific evidence, gained over four decades, supports the newly merged company, it will still need to defend glyphosate as safe, and such a legal defence will be costly. Indeed, in June 2020, Bayer agreed to pay US$10.9 billion to settle glyphosate cancer claims.

At the time of writing (summer 2020 to spring 2021) it is clear that continuing EU intransigence on the use of GM and GE technologies and accompanying bureaucratic administrative processes, will continue to prevent EU farmers from adopting these technologies. It is also clear that other countries are continuing to adopt GM and GE owing to the perceived benefits involved. Furthermore, it is beyond doubt that the scientific consensus is that the technology to generate GM crops is safe. Constructive dialogue is essential to solve this impasse.

The US EPA has since reviewed glyphosate and reported in 2020 that there are no risks of concern to human health when glyphosate is used in accordance with label recommendations. The EPA scientists concluded that:

- glyphosate is unlikely to be carcinogenic in humans;
- glyphosate has no risks to human health;
- children are not more sensitive to glyphosate, nor women of child-bearing age; and
- glyphosate is not an endocrine disruptor.

Furthermore, the EU Assessment Group on Glyphosate has concluded in June 2021 that a classification of glyphosate as a carcinogen is not justified.

A further possibility is that if pure glyphosate alone is safe, then is it possible that other molecules added to the formulation that constitutes the commercialised products are responsible for the genotoxic effects reported? These co-formulants are referred to in Chapter 3 as adjuvants but, at the present time, herbicide manufacturers are not required to disclose details on the co-formulants as this information is considered to be business confidential. If this suggestion is to be tested, then much work needs to be done to establish beyond doubt the biological activity of formulation components.

This author recommends a helpful article by David Cox in the Guardian Newspaper (Saturday 9th March, 2019), entitled 'The Roundup row: is the world's most popular weed-killer carcinogenic?', that presents a balanced and informed view.

Finally, in July 2018, the European Court of Justice ruled that genetically modified crops are subject to the EU GMO Directive and that gene-edited crops must abide by the same regulations. This decision has been received with dismay within the European plant science community, who see gene-editing as a faster, cheaper and more precise technology than existing conventional or GM plant breeding processes.

The ruling does not appear to be founded on science and will inevitably have the effect of preventing the introduction of new GM crop varieties in the EU, hence placing EU farmers at a competitive disadvantage in global agricultural markets. It is possible that the UK Government will depart from the EU ruling post-Brexit, so that the economic and agricultural challenges faced by UK farmers and growers can be addressed using gene-editing technology. The EU position, however, is shared by the organic farming sector, the Soil Association and other anti-GM non-government organisations.

14.17   Why is there so much opposition to GM technology?

In 2003, Gaskell *et al.* published a survey regarding public attitudes to biotechnology in Europe over the period 1991 to 2002. They found a wide distinction in attitudes between medical and agri-food biotechnology. The former was widely supported, as it is perceived to combat disease in humans, but the latter attracted much less support. Most anxieties regarding GM food centred on perceptions of food adulteration, infection and monstrosities (Table 14.5).

Table 14.5   Menacing images of food biotechnology in Europe (1996–2002).

	Statement	Correct answer?	Percentate correct in	
			1996	2002
1.	Ordinary tomatoes do not contain genes, but GM tomatoes do	No	22	26
2.	By eating a genetically modified food, a person's genes also become modified	No	15	15
3.	GM animals are always bigger than ordinary ones	No	27	26

Source: Gaskell, G., Allum, N., Bauer, M.W., Jackson, J., Howard, S. and Lindsey, N. (2003) Ambivalent GM nation? Public attitudes to biotechnology in the UK 1991–2002. Life Sciences in European Society Report, London School of Economics and Political Science.

Public opposition to GMOs remains strong, notwithstanding many studies that show that GM crops can make a positive contribution to sustainable agriculture. Blancke *et al.* (2015) suggest that the discrepancy between public opinion and scientific evidence is due to intuitive reasoning being used to form a judgement on GM, which includes folk biology and disgust. Anti-GMO activists then exploit this reasoning to promote their various causes, hence the strong appeal of anti-GM messages to intuitions and emotions. The authors give three examples and one conclusion:

1 As DNA is the essence of an organism, introducing genes from another species is not natural. It is against our religious beliefs.
2 Nature is as intended and we should not meddle with Nature. Scientists are playing God.
3 Changing the DNA contaminates the organism, leading to disease, sterility and the contamination of our food.
4 Therefore, GMOs are both dangerous and immoral.

Public perception is invariably fickle and stubbornly ill-informed. Tell the public that there might be one part per billion of a pesticide residue in their beer and they will take offence. The resultant headlines in the popular press can easily be imagined, expressing outrage that their beer is poisoned and that, as a consequence, we are all doomed. The media might overlook the fact, however, that their beer might contain 50 million times more of a proven carcinogen, called alcohol!

Without confidence in experts, it is suggested that the public do not correctly perceive risks and uncertainties, and are swayed by exaggerated and sensational claims of those who oppose new technology, for whatever reason. Or, to quote Charles Darwin (1809–1882):

> Ignorance more frequently begets confidence than does knowledge. It is those who know little, and not those who know much, who so positively assert that this or that problem will never be solved by science.

More recently, Fernbach *et al.* (2019) reported that in the USA extreme opposition to GM food was associated with low levels of objective knowledge and high levels of self-assessed knowledge. This pattern also applied to medical applications of genetic engineering technology.

It is difficult for some to understand the anti-GM stance in some EU organisations when there are so many malnourished persons in the world, since GM crops address food quality and yield, in addition to adaptations to climate change, such as tolerance of salinity and water stress. Such activism is considered to be misplaced and not supported by evidence. Indeed, it is ironic that so many GM and GE discoveries were made, and continue to be made, in Europe, but are not exploited there. The scientific evidence is clear: crops developed using GM technology pose no more risk to humans, animals or the environment than equivalent crops developed using conventional breeding techniques (Herman and Price, 2013). For the substantial equivalence of GM and non-GM crops see www.isaaa.org/kc Pocket K no. 56, March 2008.

McPhetres *et al.* (2019), in surveys on public attitudes towards GM foods, concluded that teaching people about the science of GM foods will increase more positive attitudes and lead to an increased willingness to eat them. The European public need to be armed with evidence, rather than emotion and anecdote, and scientists have a responsibility to provide such objective evidence in a clear and measured fashion. Only then will both politicians and the public be able to arrive at informed decisions. In conclusion, facts rather than uninformed opinions are the way forward, cutting through the subjective opinions of pressure groups, and allowing for open discussion, public education and debate based on evidence.

Finally, it is relevant to note that millions of people worldwide with diabetes now owe their lives, and an extension of them, to genetically modified bacteria. The human gene for insulin biosynthesis has been inserted into the DNA of the *E. coli* bacterium. The bacteria are then grown in vats and the insulin harvested. This has ensured a consistent supply of pure insulin for human use. More recently, the covid-19 pandemic of 2020/2021 has claimed millions of lives throughout the world, disrupting modern lifestyles and national economies. Fortunately, novel mRNA vaccines have now been developed that generate antibodies to viral surface proteins, which prime the immune system. It is anticipated that these treatments will save millions of lives, which will overcome any public outcry. Indeed, over 85% of the over-60s population in the UK alone in August 2021 have now accepted two covid-19 vaccinations, which may be one of the biggest ever utilisations of gene technology to date.

14.18   Future prospects

It is beyond doubt that molecular plant scientists will continue to explore and develop new technologies for the manipulation of the crop genome. Indeed, the further and widespread growth of GM-HT, GE and gene stacking is most likely in the coming decades, with the effect of changing global cropping systems, to the short-term benefit of the global population. However, will more reliance on GM-HT inevitably lead to a decline in biodiversity, and an increased incidence of weed cultivars that express multiple-resistance to more herbicide target sites? The Luddites, however, as persons opposed to change, for a multitude of reasons, will continue to challenge these developments and accuse molecular scientists of 'playing God'. A wider debate with the general public will need to continue and improve, especially in the media.

While techniques such as CRISPR–Cas9, and its variants, can now be used to edit genes, a new emerging concept in 2021 is Gene Drive. This suggests that genes can be altered to by-pass the laws of heredity. Thus, selected genes can be passed on to a following generation, thereby over-ruling Natural Selection. A consquence of this may be extinctions. As an example, mosquitoes can harbour the malaria parasite, but only the female offspring can bite and transmit the disease to humans. It follows that by genetically engineering all-male populations of mosquitoes that do not bite humans, the disease may not continue. A new debate is now urgently needed on the regulation and governance of gene-drive research.

As for glyphosate, regulatory organisations will continue to rule on its status, hopefully driven by evidence rather than by emotion and politics. Perhaps a new generation of herbicides will emerge in the years ahead, based on new target sites, to challenge the current eminence of glyphosate and so improve on the existing armoury of herbicides. Is there a 'new glyphosate' out there, currently under development?

In the meantime, it is likely that plant breeders will focus on developing new cultivars with enhanced traits to compete more effectively with weeds. Examples may include an increased early vigour to ensure more rapid ground cover via canopy development, and an enhanced competition for resources.

What is certain, to this observer, is that 'nature abhors a vacuum' (an idiom attributed to the ancient Greek philosopher Aristotle), and so the weed population will continue to evolve and adapt to different selection pressures. The onus will remain on weed scientists to develop more effective means to control weeds (not forgetting pests and diseases) by a range of methods, perhaps relying less on only chemical control in the future.

References

Ali, L., Jo, H., Song, J.T. and Lee, J-D. (2020) The prospect of bentazone-tolerant soybean for conventional cultivation. *Agronomy* **10**, 1650; doi: org/10.3390/agronomy10111650

Anderson, M.P. and Gronwald, J.W. (1991) Atrazine resistance in a velvetleaf (*Abutilon theophrasti*) biotype due to enhanced glutathione *S*-transferase activity. *Plant Physiology* **96**, 104–109.

Bilang, R., Iida, S., Peterhans, A., Potrykus, I. and Paszkowski, J. (1991.) The 3′-terminal region of a hygromycin-B-resistance gene is important for its activity in *Escherichia coli* and *Nicotiana tabacum*. *Gene* **100**, 247–250.

Blancke, S., van Breusegem, F., de Jaeger, G., Braeckman, J. and van Montagu, M. (2015) Fatal attraction: the intuitive appeal of GMO opposition. *Trends in Plant Science* **20**, 414–418.

Bortesi, L. and Fischer, R. (2015) The CRISPR–Cas9 system for plant gene editing and beyond. *Biotechnology Advances* **33**, 41–52.

Brookes, G. and Barfoot, P. (2018a) Environmental impacts of genetically modified (GM) crop use 1996–2016: impacts on pesticide use and carbon emissions. *GM Crops and Food* **9**, 109–139.

Brookes, G. and Barfoot, P. (2018b) *GM Crops: Global Socio-economic and Environmental Impacts 1996–2016*. PG Economics Ltd, UK.

Bruce, D. and Bruce, A. (eds) (1998) *Engineering Genesis: the Ethics of Genetic Engineering in Non-human Species*. London: Earthscan.

CaJacob, C.A., Feng, P.C.C., Heck, G.R., Alibhai, M.F., Sammons, R.D. and Padgette, S.R. (2004) Engineering resistance to herbicides. In: Christou, P. and Klee, H. (eds), *Handbook of Plant Biotechnology*, vol. 1. Chichester: John Wiley and Sons, pp. 353–372.

Cheng, M., Fry, J.E., Pang, S., Zou, H., Hironaka, C.M., Duncan, D., *et al.* (1997) Genetic transformation of wheat mediated by *Agrobacterium tumefaciens*. *Plant Physiology* **115**, 971–980.

Dill, G.M. (2005) Glyphosate-resistant crops: history, status and future. *Pest Management Science* **61**, 219–224.

Duke, S.O. (2005) Taking stock of herbicide-resistant crops ten years after introduction. *Pest Management Science* **61**, 211–218.

Duke, S.O. (2017) The history and current status of glyphosate. *Pest Management Science* **74**, June 2017; doi: org/10.1002/ps.4652

Eckes, P., Vijtewaal, B. and Donn, G. (1989) Synthetic gene confers resistance to the broad spectrum herbicide L-phosphinothricin in plants. *Journal of Cell Biochemistry* suppl. 13D.

Economic Research Service (2007) Adoption of genetically engineered crops in the US, USDA, Washington, DC. http://www.ers.usda.gov/Data/BiotechCrops/ (accessed July 2009).

Es, I., Gavahian, M., Marti-Quijal, F.J., Lorenzo, J.M., Khaneghah, A.M., Tsatsanis, C., *et al.* (2019) The application of the CRISPR–Cas9 genome editing machinery in food and agricultural science: current status, future perspectives and associated challenges. *Biotechnology Advances*; doi: org/10.1016/j.biotechadv.2019.02.006

Evans, J. (2010) Food security is all in the genes. *Chemistry and Industry* **9**, 14–15.

Fernbach, P.M., Light, N., Scott, S.E., Inbar, Y. and Roizin, P. (2019) Extreme opponents of genetically modified foods know the least but think they know the most. *Nature, Human Behaviour* **3**, 251–256.

Firbank, L.G., Heard, M.S., Woiwod, I.P., Hawes, C., Haughton, A.J., Champion, G.T. *et al.* (2003) An introduction to the farm-scale evaluations of genetically modified herbicide-tolerant crops. *Journal of Applied Ecology* **40**, 2–16.

Gaskell, G., Allum, N., Bauer, M.W., Jackson, J., Howard, S. and Lindsey, N. (2003) Ambivalent GM nation? Public attitudes to biotechnology in the UK 1991–2002. Life Sciences in European Society Report, London School of Economics and Political Science.

Goodwin, T.W. and Mercer, E.I. (1983) *Introduction to Plant Biochemistry*, 2nd edn. Oxford: Pergamon Press.

Gray, J.A., Balke, N.E. and Stoltenberg, D.E. (1996) Increased glutathione conjugation of atrazine confers resistance in a Wisconsin velvetleaf (*Abutilon theophrasti*) biotype. *Pesticide Biochemistry and Physiology* **55**, 157–171.

Hall, J.C., Donnelly-Vanderloo, M.J. and Hume, D.J. (1996) Triazine-resistant crops: the agronomic impact and physiological consequences of chloroplast mutation. In: Duke, S.O. (ed.), *Herbicide Resistant Crops: Agricultural, Environmental, Economic, Regulatory and Technical Aspects*. Boca Raton, FL: CRC Press, pp. 107–126.

Haughn, G.W., Smith, J., Mazur, B. and Somerville C. (1988) Transformation with mutant *Arabidopsis* acetolactate synthase gene renders tobacco resistant to sulfonylurea herbicides. *Molecular and General Genetics* **211**, 266–271.

Herman, R.A. and Price, W.D. (2013) Unintended compositional changes in genetically modified (GM) crops: 20 years of research. *Journal of Agricultural and Food Chemistry* **61**, 11695–11701; doi: org/10.1021/jf00135r

Hsiao, C.L., Young, C.C. and Wang, C.Y. (2007) Screening and identification of glufosinate-degrading bacteria from glufosinate-treated soils. *Weed Science* **55**, 631–637.

Inui, H., Ueyama, Y., Shiota, N., Ohkawa, Y. and Ohkawa, H. (1999) Herbicide metabolism and cross-tolerance in transgenic potato plants expressing human CYP1A1. *Pesticide Biochemistry and Physiology* **64**, 33–46.

Ishida, Y., Saito, H., Ohta, S., Hiei, Y., Komari, T., Kumashiro, T. (1996) High efficiency transformation of maize (*Zea mays* L.) mediated by *Agrobacterium tumefaciens*. *Nature Biotechnology* **14**, 745–750.

Jinek, M., Chylinski, K., Fonfara, I., Hauer, M., Doudna, J.A. and Charpentier, E. (2012) A programmable dual-RNA-guided DNA endonuclease in adaptive bacterial immunity. *Science* **337**, 816–821.

Joersbo, M. (2001) Advances in the selection of transgenic plants using non-antibiotic marker genes. *Physiologia Plantarum* **111**, 269–272.

Knioo, A.R. (2017) Long-term trends in the intensity and relative toxicity of herbicide use. *Nature Communications* **8**, article number 14865; doi: org/10.1038/ncomms14865

Komari T., Ishida, Y. and Hiei, Y. (2004) Plant transformation technology: agrobacterium-mediated transformation. In: Christou, P. and Klee, H. (eds), *Handbook of Plant Biotechnology*, vol. 1. Chichester: John Wiley and Sons, pp. 233–261.

Llewellyn, D. and Last, D. (1996) Genetic engineering of crops for tolerance to 2,4-D. In: Duke, S.O. (ed.), *Herbicide Resistant Crops: Agricultural, Environmental, Economic, Regulatory and Technical Aspects*. Boca Raton, FL: CRC Press, pp. 159–174.

Llewellyn, D. and Last, D. (1999) A detoxification gene in transgenic *Nicotinia tabacum* confers 2,4-D tolerance. *Weed Science* **47**, 401–404.

Lyndon, J. and Duke, S.O. (1999) Inhibitors of glutamine biosynthesis. In: Singh, B.K. (ed.), *Plant Amino Acids: Biochemistry and Biotechnology*. New York: Marcel Dekker, pp. 445–464.

McCabe, D.E., Swain, W.F., Martinell, B.J. and Christou, P. (1988) Stable transformation of soybean (*Glycine max*) by particle acceleration. *Bio/Technology* **6**, 923–926.

McPhetres, J., Ritjens, B.T., Weinstein, N. and Brisson, J.A. (2019) Modifying attitudes about modified foods: increased knowledge leads to more positive attitudes. *Journal of Environmental Psychology* **64**, 21–29; doi: org/10.1016/j.envp.2019.04.012

Ming, M., Ren, Q., Pan, C., He, Y., Zhang, Y. and Liu, S. (2020) CRISPR–Cas12b enables efficient plant genome engineering. *Nature Plants* **6**, 202–208.

Miswawa, N., Yamano, S., Linden, H., Lucas, M., Ikenaga, H. and Sandmann, G. (1993) Functional expression of the *Erwinia uredovora* carotenoid biosynthesis gene *crt1* in transgenic plants showing an increase of beta-carotenoid biosynthesis activity and resistance to the bleaching herbicide norflurazon. *The Plant Journal* **4**, 833–840.

Naqvi, S., Farre, G., Sanahuja, G., Capell, T., Zhu, C. and Cristou, P. (2010) When more is better: multigene engineering in plants. *Trends in Plant Science* **15**, 48–56; doi: org/10.10.16/j.tplants.2009.09.010

Padgette, S.R., Kolacz, K.H., Delanny, X., Re, D., Bradley, J., LaVallee, B.J. *et al.* (1995) Development and characterisation of a glyphosate-tolerant soybean line. *Crop Science* **35**, 1451–1461.

Padgette, S.R., Re, D.B., Barry, G.F., Eichholtz, D.E., Delanny, X., Fuchs, R.L., *et al.* (1996) New weed control opportunities: development of soybeans with Roundup Ready™ gene. In: Duke, S.O. (ed.), *Herbicide-resistant Crops: Agricultural, Environmental, Economic, Regulatory and Technical Aspects.* Boca Raton, FL: CRC Press, pp. 53–84.

Pallett, K.E., Veerasekaran, P., Freyssinet, M., Pelissier, B., Leroux, B. and Freyssinet, G. (1997) Herbicide tolerance in transgenic plants expressing bacterial detoxification genes. The case of bromoxynil. In: Hatzios, K.K. (ed.), *Regulation of Enzymic Systems Detoxifying Xenobiotics in Plants.* Dordrecht: Kluwer, pp. 337–350.

Reddy, K.N., Duke, S.O. and Rimando A.M. (2004) Aminomethylphosphonic acid, a metabolite of glyphosate, causes injury in glyphosate-treated, glyphosate-resistant soybean. *Journal of Agriculture and Food Chemistry* **52**, 5139–5143.

Saari, L.L. and Mauvais, C.J. (1996) Sulfonylurea herbicide-resistant crops. In: Duke, S.O. (ed.), *Herbicide Resistant Crops: Agricultural, Environmental, Economic, Regulatory and Technical Aspects.* Boca Raton, FL: CRC Press, pp. 127–142.

Shaner, D.L., Bascombe, N.F. and Smith W. (1996) Imidazolinone-resistant crops: selection, characterisation and management. In: Duke, S.O. (ed.), *Herbicide Resistant Crops: Agricultural, Environmental, Economic, Regulatory and Technical Aspects.* Boca Raton, FL: CRC Press, pp. 143–157.

Smyth, S.J. (2020) The health benefits from GM crops. *Plant Biotechnology Journal* **18**, 887–888; doi: org/10.1111/pbi.13261

Somers, D.A. (1996) Aryloxyphenoxypropionate and cyclohexanedione resistant crops, In: Duke, S.O. (ed.), *Herbicide Resistant Crops: Agricultural, Environmental, Economic, Regulatory and Technical Aspects.* Boca Raton, FL: CRC Press, pp. 175–188.

Songstad, D.D., Petolino, J.F., Voytas, D.F. and Reichert, N.A. (2017) Genome editing in plants. *Critical Reviews in Plant Science* **36**, 1–23; doi: org/10.1080/07352689.2017.1281663

Tyagi, A.K. and Mohanty, A. (2000) Rice transformation for crop improvement and functional genomics. *Plant Science* **158**, 1–18.

Vande Berg, B.J., Hammer, P.E., Chun, B.L., Shouten, L.C., Carr, B., Guo, R., *et al.* (2008) Characterization and plant expression of a glyphosate-tolerant enolpyruvylshikimate phosphate synthase. *Pest Management Science* **64**(4), 340–345.

Vasil, I.K. (1996) Phosphinothricin-resistant crops. In: Duke, S.O. (ed.), *Herbicide-resistant Crops: Agricultural, Environmental, Economic, Regulatory and Technical Aspects.* Boca Raton, FL: CRC Press, pp. 85–91.

Westra, P. Wilson, R.G., Miller, S.D., Stahlman, P., Wicks, G.W., Chapman, P., *et al.* (2008) Weed population dynamics after six years under glyphosate- and conventional herbicide-based weed control strategies. *Crop Science* **48**(3), 1170–1177.

Whelan, A.I., Gutti, P. and Lema, M.A. (2020) Gene editing regulation and innovation economics. *Frontiers in Bioengineering and Biotechnology* 8, 303; doi: org/10.3389/fbioe.2020.00303n

Wolt, J.D., Wang, K. and Yang, B. (2016) The regulatory status of genome-edited crops. *Plant Biotechnology Journal* **14**, 510–518.

Chapter 15
Further Targets for Herbicide Development

What a marvellous cooperative arrangement – plants and animals each inhaling each other's exhalations, a kind of planet-wide mutual mouth-to-stoma resuscitation, the entire elegant cycle powered by a star 150 million kilometres away.

Carl Sagan (1934–1996)

15.1   Introduction

This text is not intended as an exhaustive, nor exhausting, list of herbicides and their proposed sites of action, but instead focuses on the main physiological functions of plants that are sensitive to herbicidal inhibition. The inhibition of photosynthetic electron flow and pigments, growth regulation by auxins, lipid biosynthesis and amino acid biosynthesis have been well exploited agrochemically, and much imitative chemistry is still evident in the literature. Nevertheless, it is the conference sessions on potential or novel herbicidal targets that always attract the best and the most inquisitive audiences. This is because the rational design of new herbicides is regularly attempted, but invariably lacks success, owing to our incomplete knowledge of plant biochemistry in particular and plant growth in general. It is likely that our increasing understanding of the plant genome will remedy this situation in the years ahead.

Many unsuccessful attempts to design herbicides using biochemical reasoning have been published. In each case a strategic reaction in a key pathway is identified as a suitable target for manipulation, and novel inhibitors are designed on the basis of existing knowledge of the structures of the substrates and the reaction mechanism. The reader is referred to Chapter 2 for potential and patented targets that may lead to new herbicides in the next decade. However, such molecular designs are seldom, if ever, successfully transmitted to the field and commercialised. For example, Kerr and Whitaker (1985) chose phosphoglycollate phosphatase, a key enzyme in photorespiratory carbon oxidation, as a suitable target for herbicide attack. They reasoned that the inhibition of this enzyme would lead to an accumulation of phosphoglycollate, which had been reported to inhibit triose phosphate synthesis and hence photosynthetic carbon reduction. This would also have the effect of preventing photorespiratory carbon recycling that dissipates excess ATP and reducing

Herbicides and Plant Physiology, Third Edition. Andrew H. Cobb.
© 2022 John Wiley & Sons Ltd. Published 2022 by John Wiley & Sons Ltd.

power generated under high-intensity light conditions. Detailed studies by these workers found that this enzyme could indeed be inhibited *in vitro*, but not in intact leaves, and concluded that 'the effects of an extremely complex chemical on an equally complex biochemical target may be just too subtle for our present understanding'. Indeed, the process of photorespiration has to date defied all attempts at chemical manipulation. It follows that if we are not to rely solely on the routine and empirical screening of new chemicals, then only an increased knowledge of plant physiology is sure to generate the new leads, ideas and areas for future agrochemical development. The following sections offer a brief, personal choice of some potential areas of plant physiology and biochemistry that deserve further consideration.

15.2   Protein turnover

15.2.1   *Introduction*

Proteins are constructed from amino acids linked together by peptide bonds. The breakage of these bonds, termed proteolysis, is performed by proteolytic enzymes that are widely referred to as proteases or peptidases. The terms are synonymous, although proteases will be used in this text. Protease research generates thousands of publications each year and the reader is referred to the website of the International Proteolysis Society (www.protease. org) for an overview. The reason for this interest is the realisation in recent years that proteases now appear to be involved in almost all cellular processes in plant growth and development (Figure 15.1).

The protein content of plant cells is in a state of constant flux, balanced by rates of synthesis and degradation. Alterations in protein content are essential for plant growth and development, and for the most appropriate responses to environmental stimuli. In actively growing leaves the total soluble proteins may turn over every week, while the lifetime of individual proteins is often less than that of a cell. Thus, ACC synthase has a half-life of an hour and the D1 protein may turn over in 2 hours at high light intensities. This raises the important question: why should a plant expend considerable energy synthesising a protein, only to degrade it within minutes? Clearly, proteolysis regulates metabolism. Protein degradation will release a pool of amino acids to enable the synthesis of new proteins on a different developmental pathway, or simply remove a rate-limiting enzyme. Indeed, many key enzymes are rapidly turned over, indicating metabolic control via proteolysis, controlling their abundance.

Protein mobilisation is therefore essential to the plant cell to:

- supply amino acids for new protein synthesis;
- mobilise storage proteins during germination;
- modify proteins post-translation to their mature, active/inactive forms;
- degrade damaged or misfolded proteins;
- remove enzymes no longer required in metabolism; and
- remove regulatory proteins, such as transcription factors, that are no longer needed.

The general principles governing proteolysis in plants have become better understood in recent times and may offer potential new targets for herbicide development. Unravelling

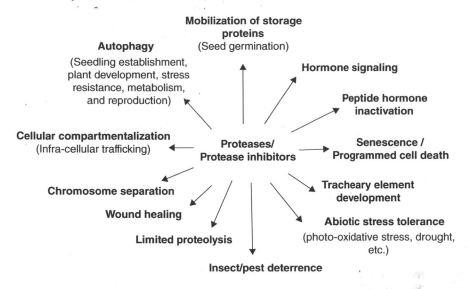

Figure 15.1 Biological significance of protease and protease inhibitor interactions in plants. Source: Rustgi, S., Boex-Fontvielle, E., Reinbothe, C., von Wettstein, D. and Reinbothe, S. (2018) The complex world of plant protease inhibitors: insights into a Kunitz-type cysteine protease inhibitor of *Arabidopsis thaliana*. *Communicative and Integrative Biology*, e1368599; doi: org/10.1080/19420889.2017. 1368599. Reproduced with permission of Taylor & Francis.

the *Arabidopsis* genome has identified about 1300 genes involved in the ubiquitin/proteosome pathway and over 800 proteases! Understanding their roles may allow the design of new herbicides to control the accumulation of regulatory proteins and prevent cell division as normal developmental responses to environmental change.

15.2.2 Proteases

Proteolysis, or the hydrolytic cleavage of peptide bonds, is highly selective and involves a large family of protease enzymes. In essence, hydrolysis is achieved by nucleophilic attack at the carbonyl carbon, supported by the donation of a proton to the NH group of the peptide bond. The proteases themselves are either endoproteases (cleaving internal peptide bonds) or exoproteases, which progressively cleave peptide bonds from either the *C*-terminus (carboxypeptidases) or the *N*-terminus (aminopeptidases).

The endoproteases are further characterised according to the catalytic mechanism. In serine, threonine and cysteine proteases, the hydroxyl or sulphydryl groups of the amino acids at the active site act as the nucleophile during catalysis. On the other hand, aspartic, glutamic and metallo-proteases rely on an activated water molecule as the nucleophile. Further classification in the *MEROPS* database (merops.sanger.ac.uk) is according to sequence similarity. There are 888 proteases in the *Arabidopsis* genome. The serine proteases are the most numerous (45%), followed by cysteine- (25%), metallo- (15%), aspartate- (11%) and threonine proteases (4%). The observation that there are so many implies considerable structural and functional diversity (Stael *et al.*, 2019). As a consequence, they are involved in the regulation of probably all plant functions. They constitute about 2% of the gene products in all organisms. It is estimated that 14% of the 500 human proteases are under investigation as drug targets.

The following is a brief summary of protease function:

1 *Cysteine proteases* – the phytocalpains (calcium-dependent cysteine proteases) regulate numerous development processes, including embryonic pattern formulation, shoot apical meristem formation, cell fate specification in the endosperm and leaf epidermis, and regulation of the balance between cell differentiation and proliferation. Papain-like enzymes are also cysteine proteases involved in protein degradation during seed germination, leaf senescence and programmed cell death. They also contribute to plant resistance against pathogens and insects. The legumains are involved in the processing of seed storage proteins and tissue senescence.

2 *Serine proteases* – this is the largest class of plant proteases, over half being carboxypeptidases and subtilases. The serine carboxypeptidases function in protein turnover and nitrogen mobilisation. Apparently, their abundance and diversity are due to differences in substrate specificity. They have been implicated in programmed cell death, brassinosteroid signalling and seed development. They may also have a function in the formation of plant secondary metabolites. The functions of the subtilases remains uncertain. The study of *Arabidopsis* mutants suggests that they may act as highly specific regulators of plant development, such as the regulation of stomatal density, cuticle formation and embryo development.

3 *Aspartic proteases* – the phytepsins are located in the vacuole of storage tissues (seeds and tubers) and have been implicated in the breakdown of storage proteins (e.g. patatins in potatoes) during germination and sprout growth. They have roles in organ senescence, cell death, defence against pathogens and insect herbivores. Intriguingly, the nepenthesins are found in the digestive fluids of the insect-trapping organs in carnivorous pitcher plants (*Nepenthes* sp.). They appear to be remarkably stable, being able to function at 50°C over a wide range of pH and substrates.

4 *Metalloproteases* – these enzymes rely on a divalent cation for activity, commonly Zn, Co or Mn. They are very structurally diverse and few have been characterised to date. The leucine aminopeptidases may be important in the regulation of protein half-life by the cleavage of the *N*-terminal amino acid exposing or removing destabilising residues.

5 *Threonine proteases* – threonine proteases have a threonine residue at their active site and are catalytic subunits of the proteasome. The *N*-terminal threonine acts as a nucleophile for catalysis.

Figure 15.1 provides an overview of the biological importance of proteases and protease inhibitors in plants, after Rustgi *et al.* (2018).

Most chloroplast proteins are encoded in the nuclear genome and translated in the cytoplasm as large precursor proteins. Import into the chloroplast is governed via a transit peptide, which is then proteolytically removed in the chloroplast stroma by the stromal processing peptidase, a soluble metallo-protease. When the gene for this enzyme was antisensed in *Arabidopsis*, it was shown that altered protein import was vital for chloroplast biogenesis, photosynthesis and consequently plant growth. Van der Hoorn and Rivas (2018) have reviewed the mode of action of plant proteases.

15.2.3   *Programmed cell death*

Programmed cell death (PCD) is a sequence of potentially interruptible events that lead to the controlled and organised (suicidal) destruction of the cell. It is crucial for defence responses to restrict the spread of pathogens and for the correct development of the multi-cellular plant body. There are two distinct forms of PCD, apoptosis and necrosis. The former is characterised by cell shrinkage, nuclear condensation and fragmentation, leading to the breakdown of the cell. Apoptopic cell destruction is driven in animals by a family of cysteine proteases, termed caspases. Caspase activation is via the release of cytochrome c in the mitochondria, which drives the assembly of an apoptosome, a cytoplasmic complex. Thus, a loss of mitochondrial integrity promotes the process. DNA fragmentation, the accumulation of active oxygen species and organelle breakdown are common features of the PCD process. On the other hand, necrosis is regarded as an uncontrolled form of cell death in which the cell is unable to initiate apoptopic pathways. Necrosis is characterised by swelling, rather than shrinkage, resulting from a loss of osmotic regulation (Threape *et al.*, 2008). Plants lack caspases, but have the similarly functioning plant metacaspases. The papain-like cysteine proteases (PLCPs) are the essential regulators of PCD in plants. PLCPs are also involved in seed germination, leaf senescence, the mediation of responses to abiotic stresses and plant immunity (Liu *et al.*, 2018). How they are involved in the perception of stresses and signals remains to be understood.

Once the proteases have been activated, endogenous inhibitors can act as safety valves by modulating protease activity. These inhibitors, termed serpins, can inhibit both serine and cysteine proteases in either a reversible or an irreversible fashion (Stael *et al.*, 2019).

15.2.4   *The ubiquitin–proteosome pathway*

While found in all eukaryotes, genome data reveals that the ubiquitin–proteosome pathway is more elaborate in plants. The extent to which plants rely on this pathway for regulating their life-cycle via signalling processes is becoming increasingly evident (Sadanandon *et al.*, 2012). In *Arabidopsis*, more than 1600 genes (encoding for about 6% of the total proteins in this species), encode components of the ubiquitin–proteosome pathway. About 90% of them code for subunits of the E3 ubiquitin ligases, which confer substrate specificity.

A central mechanism in the degradation of cytoplasmic and nuclear proteins utilises the small protein ubiquitin. This globular protein contains 76 amino acids (molecular mass 8.5 kDa). Its structure is highly conserved and very stable in a range of conditions. Its function is to mark proteins for degradation, which it does via two key features: a *C*-terminal tail (Arg–Gly–Gly) and a lysine residue at position 48. Proteins destined for breakdown are conjugated to ubiquitin in a ligase reaction involving ATP at the carboxy terminal tail. The target protein may then be modified by the binding of more ubiquitins, resulting in a poly-ubiquitin chain to be degraded at the proteosome.

Proteosomes are large protein complexes (molecular mass 2000 kDa) found in the nucleus and cytoplasm of all plant cells, where they function to degrade proteins. They are composed of a core particle and two regulatory particles (Figure 15.2). The core particle (CP) is constructed of 14 different proteins assembled in groups of seven, each group forming a ring. Four rings are arranged in stacks. There is a regulatory particle (RP) at each end

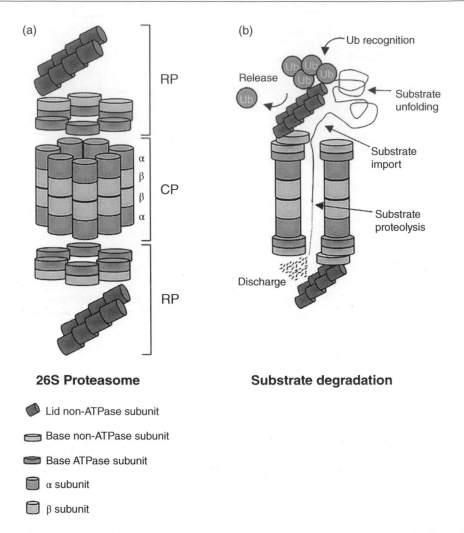

Figure 15.2 Structure of the 26S proteosome. A. Relative arrangement of proteosome subunits depicting the association of heptameric rings within the core protease (CP) and subunit distribution of the Lid and Base sub-complexes within the regulatory particle (RP). B. Diagram representing known and predicted activities required for substrate degradation. Source: Sadanandom, A., Bailey, M., Ewan, R., Lee, J. and Nelis, S. (2012)The ubiquitin–proteosome system: central modifier of plant signalling. *The New Phytologist* **196**, 13–28. Reproduced with permission of John Wiley & Sons. *See color inset section for the color representation of this figure.*

of the CP made from 14 different proteins, some of which are ATPases and others that recognise poly-ubiquitin. The target protein is unfolded by the ATPases in the RP and the ubiquitins are released for reuse. The protein moves into the central cavity of the CP where specific proteases on the interior surface degrade the protein releasing peptides of about eight amino acids long.

A good example of how this pathway may control a physiological process involves the role of the cyclins in the cell cycle (Chapter 10). This family of proteins is involved in the cell cycle, where their concentrations are crucial. They accumulate during G_1, S and G_2

phases, and when a critical concentration is reached, they initiate mitosis by binding to and activating the cdc2 protein kinase. During metaphase the cyclins are rapidly degraded, reducing the pool of active cdc2 protein kinase, thereby preventing newly formed daughter cells from entering into another phase of mitosis. Once the cyclin pool declines, degradation ceases and cyclin concentrations being to rise again, initiating another round of the cell cycle. Clearly, interference with cyclin synthesis and degradation will have a profound effect on the cell cycle and is a potential target for herbicide development.

Even though the chloroplasts do not contain proteosomes, the ubiquitin–proteosome pathway is implicated in chloroplast processes. The E3 ligase SP1 is located at the outer chloroplast envelope membrane where it controls the ubiquitination of the translocon protein machinery which itself controls the import of nuclear-encoded proteins into the chloroplast. *Arabidopsis* plants overexpressing SP1 have enhanced tolerance to stressors that generate reactive oxygen species. Note also that the cytosolic E3 ligase PUB4 tags impaired or damaged chloroplasts that are removed in a process termed chlorophagy (Mendoza *et al.*, 2020).

Why is the ubiquitin–proteosome pathway so abundant in plants? Why does the plant devote so much resource and energy to the turnover of proteins? Reasons may include the sessile growth habit of plants, which require additional regulation in response to biotic and abiotic stresses, genome duplication during evolution, roles in immunity and the need for annual plants to confine their life-cycle to short growing seasons. By controlling the protein content of the cell, including essential enzymes such as kinases, phosphatases and proteases, the ubiquitin–proteosome pathway allows plant cells to maintain responses to internal and external signals (Miricescu *et al.*, 2018; and Table 15.1).

The proteosome inhibitor bortezomib (Figure 15.3) was approved for use in 2003 as an anticancer drug that treats multiple myeloma and mantle cell lymphoma. In these cancers, proteins that are involved in killing cancer cells are broken down rapidly in the proteosome, and bortezomib slows this process. The boron atom binds to the catalytic site of the 26S proteosome and may trigger PCD in the cancer cells.

Table 15.1 Proteosome subunits with known roles in tolerance to environmental stress in *Arabidopsis*.

Regulatory subunit	Biological function
RPN 10	Salt, sucrose, heat, nitrogen, UV radiation, abscisic acid signalling
RPN 1a	Salt, heat and oxidative stress
RPN 8a	Disease resistance
RPN 6	Drought stress tolerance
RPN 12a	Heat, drought stress, disease resistance
RPT 2a	Heat, zinc deficiency, disease resistance
RPT 5a	Zinc deficiency, high boron stress tolerance
Core protease subunit	
PAE 1/2 alpha 5	Anti-virus response
ARS 5 alpha 6	Arsenic tolerance
PBA 1 beta 1	Pathogen resistance and stress-induced PCD
PBB 1/2 beta 2	Pathogen resistance and salt stress
PBE 1/2 beta 5	Pathogen resistance and sat stress

Source: Xu, F.-Q. and Xue, H.-W. (2019) The ubiquitin–proteosome system in plant responses to environments. *Plant, Cell and Environment* **42**, 2931–2944; doi: org/10.1011/pce.13633. Reproduced with permission of John Wiley & Sons.

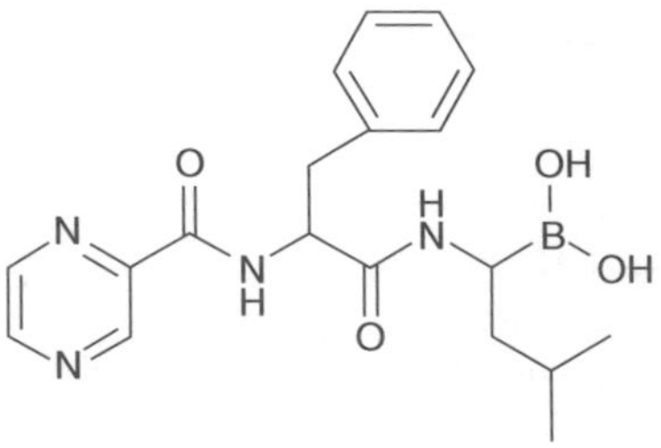

Figure 15.3   Structure of bortezomib.

In recent years, proteolysis targeting chimeras (PROTACs) have been developed that allow proteins associated with disease to be degraded. A PROTAC is a small molecule with two binding regions; one end binds to a E3 ligase enzyme and the other to a disease-causing protein to be degraded. The linker region in between directs the molecule to the targeted protein. This technology is already being used in trials for the control of prostate and breast cancers, and Bayer proposes to extend this approach to agricultural applications (O'Driscoll, 2019).

15.2.5   Small plant peptides

Many secreted peptides can act as regulators of signalling events and cell-to-cell communication in plants. Most small signalling peptides are derived from larger, inactive, precursor proteins, with the *N*-terminal signal sequence directing the protein to a secretory pathway. They are classified into two groups, the cysteine-rich peptides (CRP) and the post-translationally modified peptides (de Coninck and de Smet, 2016).

The CRPs contain six to 12 cysteine residues and have known roles in stomatal patterning and density; pollen tube germination, guidance and burst; and gamete activation and seed development. They are located in both male and female gametophytes, and are thought to have evolved from antimicrobial peptides, such as the defensins (an α-helix and triple-stranded β subunit, stabilised by four disulphide bridges) and the cyclotides (cyclic peptides with three disulphide bridges). RALF (rapid alkalinisation factor) peptides are small CRPs of about 5 kDa that affect cell and organ growth via calcium responses, activation of MAPK signalling and pH modulation.

The post-translationally modified peptides have up to 20 amino acids altered by modifications such as tyrosine sulphation, proline hydroxylation and hydroxyproline glycosylation. There are 32 CLE (clavate 3/embryo surrounding proteins) in *Arabidopsis*, each containing 12–15 amino acids. They function in the maintenance of root, shoot, floral meristems, lateral root emergence and vascular development. About half are involved in signalling pathways and plant–environment interactions, including responses to stress. They are located predominantly in vegetative tissues. CEP5s (*C*-terminally encoded peptide) have 15 amino acids and are involved in the negative regulation of lateral root initiation. RGF/CLE/GLV (root growth factor/CLE-like/Golven) peptides have roles in root gravitropism, maintenance of the root apical meristem and root and shoot development.

Little is currently known about how these small peptides function in signalling processes. Future studies on peptide-receptor kinase interactions and identifying which processes are involved downstream are urgently awaited.

Although plant scientists have long known that abscisic acid (ABA) regulates stomatal movement in guard cells to prevent water loss from the leaf, no mobile signals have been identified to trigger ABA accumulation in leaves. Takahashi *et al.* (2018) have recently demonstrated that the CLE25 peptide transmits water deficiency signals from roots to shoots in *Arabidopsis* via BAM (barley-any-meristem) receptors in the leaves. The CLE25 is expressed in vascular tissues and in roots in response to dehydration stress.

In a further 2018 publication, Toh *et al.* screened chemical libraries to identify compounds that affected stomatal movement in *Commelina benghalensis*. They found nine compounds that suppressed light-induced stomatal opening (Figure 15.4) and two that induced stomatal opening in the dark. On further investigation of the molecular mechanisms involved, they suggested that two of the compounds that suppressed light-induced stomatal opening acted by inhibiting the blue light-induced activation of an ATPase at the plasma membrane, rather than affecting the ABA-signalling pathway. The compounds involved are likely to be studied further and their effects optimised to influence drought-stress in the field.

Finally, Xu *et al.* (2021) have reported that γ amino butyric acid (GABA) accumulates in plants under stress, and modulates stomatal opening by acting as a negative regulator of an anion transporter located at the guard cell tonoplast membrane. Consequently, stomatal opening is reduced and water use efficiency is improved. It will be interesting to observe whether GABA derivatives may be synthesized and screened for activity in this process.

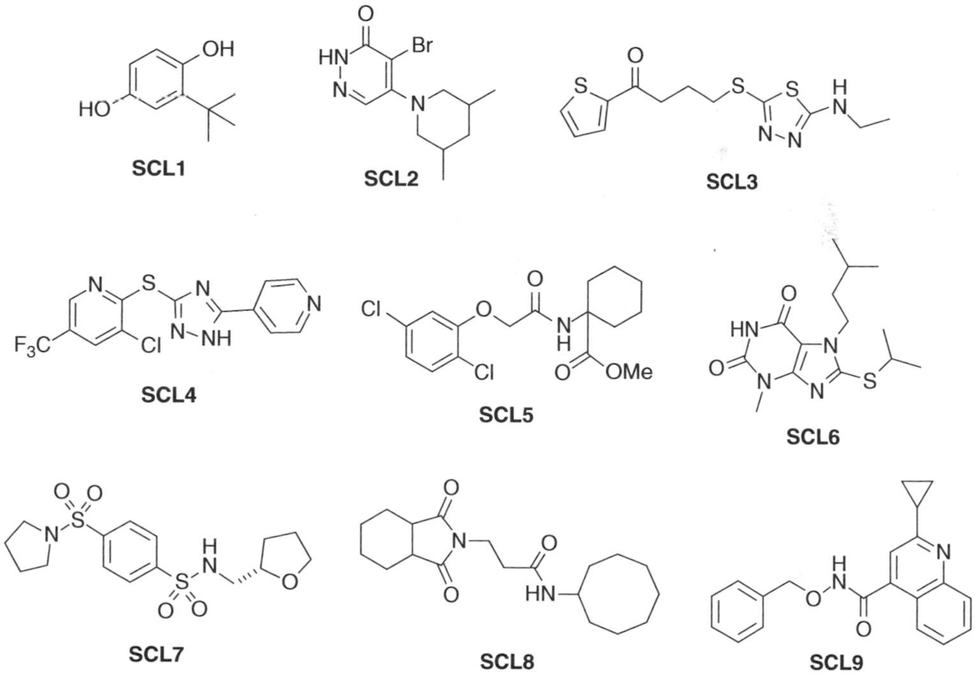

Figure 15.4 Structures of the nine stomatal-closing (SC) compounds which suppress light-induced stomatal opening. Source: Toh, S., Inoue, S., Toda, Y., Yuki, T., Suzuki, K., Hamamoto, S., *et al.* (2018) Identification and characterisation of compounds that affect stomatal movement. *Plant and Cell Physiology* **59**, 1568–1580.

15.3 The promotion of ageing in weeds?

Leaf senescence is an important part of the life-cycle of plants. It is a complex process that results in the reallocation of nutrients and resources to other parts of the plant, so increasing the chances of plant viability in the next generation. How it is regulated, however, is poorly understood. Chen *et al.* (2016) have demonstrated that in *Arabidopsis*, histone deacetylase 9 (HDAC9) and acetyltransferases are involved in genome expression in senescence. Acetylation is often associated with active transcription. Histone deacetylases act as transcriptional repressors, removing the acetyl groups and inducing histone contraction. They consider that this HDAC may therefore promote the onset of plant aging. It is the case that members of the HDAC family act as negative regulators of gene expression in diverse processes and environmental stress signalling. HDAC9 modulates ABA-dependent drought stress signalling and represses the expression of ABA catabolism-related genes in *Arabidopsis thaliana* (Baek *et al.*, 2020). Perhaps this enzyme could be studied further as a target for herbicide development with a view to manipulating its activity in weeds.

15.4 Herbicide leads at the apicoplast

Strange but true. The parasitic protozoans involved in malaria, *Plasmodium* spp., share an evolutionary relationship with plants! A feature of their evolutionary past is the retention in all apicomplexan parasites of a non-photosynthetic plastid, termed the apicoplast. It is thought that they originated from endosymbiotic photosynthetic ancestors, perhaps similar to the modern dinoflagellate zooxanthellae (McFadden, 2011). The apicoplast houses a range of plastid-like metabolic pathways that are assumed to be essential for parasite life. McFadden envisages the apicoplast as a bacterium inside the parasite, complete with a circular genome and a stroma. Discovering what the apicoplast is metabolically capable of may yield new herbicide targets.

Corral *et al.* (2017) have demonstrated that four antimalarial drugs effective in humans (namely sulfadiazine, sulfadoxine, pyrimethamine and cycloguanil) are phytotoxic in *Arabidopsis*. The modes of action for many antimalarials, however, remain unknown. Ciprofloxacin, a quinolone-based antibiotic, was more active than glyphosate and glufosinate in a plate assay. The target of ciprofloxacin in *Arabidopsis* is a DNA gyrase, a type 2 topoisomerase that may be targeted to chloroplasts and mitochondria, as a disruptor of organelle replication (Evans-Roberts *et al.*, 2016). DNA gyrase is a key enzyme in plants that catalyses the ATP-dependent super-coiling of closed circular DNA, essential for the replication and transcription of circular DNA found in chloroplasts. It may therefore be attractive as a novel target site in the development of new herbicides.

More recently, Corral *et al.* (2018) have discovered a novel antimalarial compound, MMV 007978 (Figure 15.5), that is also herbicidal. In this case, the thiophenyl group has been subjected to 22 variations and the resulting, novel herbicide action may be related to an inhibition of seed germination and cell division. Since these molecules appear to be non-toxic to organisms other than plants and apicomplexans, they show promise as herbicides. The reader is also referred to the paper of Khan *et al.* (2018) that reviews progress with DNA gyrase inhibitors.

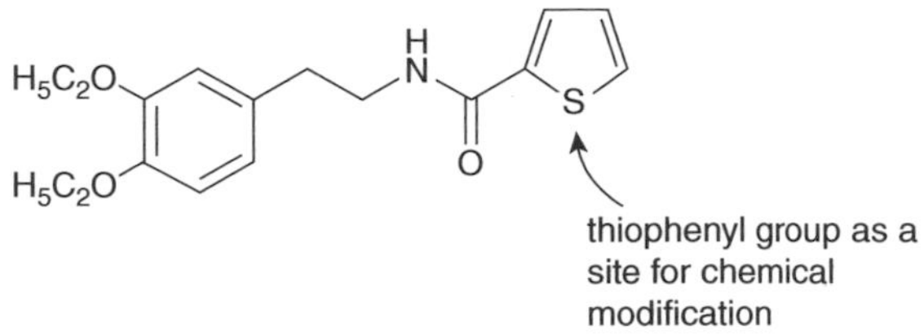

thiophenyl group as a
site for chemical
modification

Figure 15.5 Structure of MMVOO7978.

15.5   Control of seed germination and dormancy

The estimated more than 100,000 viable weed seeds per square metre of arable soil represent a massive potential for competition with both existing and future crops. These numbers, combined with long-term viability in the soil, present real problems for crop growth and yield, requiring innovative weed control strategies. It has long been understood that many exogenous chemicals can stimulate seed germination, including potassium nitrate and thiourea (Mayer and Poljakoff-Mayber, 1989). Potassium nitrate in combination with red light is effective in many weed species, implicating a central role of phytochrome (e.g. Hilton, 1984). The germination of many seeds in the dark is stimulated by thiourea. In both instances, concentrations of about 100 mM to be appear optimal. Changes in seasonal conditions can often overcome physiological dormancy.

Dormancy can also be broken by chemicals in smoke and smoke products, the main one being karrikinolide, KAR1, chemically termed a butanolide (Figure 15.6). It might be possible, therefore, to attempt a two-stage strategy for weed control, thus initially breaking weed dormancy followed by applications of herbicides. Reynolds *et al.* (2014) describe this approach to the control of an important weed in Australia, *Chrysanthemoides monolif era* ssp. monilifera (L.), Norl commonly known as boneseed. Six karrikins have been discovered in smoke (Figure 15.6), although KAR1–4 are thought to be the most active (Nelson *et al.*, 2012). The burning of sugars and cellulose in the laboratory will generate karrikins, while the burning of dried grass is sufficient in the field. As yet, there is no evidence that karrikins occur naturally in the plant.

In *Arabidopsis*, karrikin perception is via a hydrolase, termed karrikin-insensitive 2 (KAI2). This assumes that the enzyme hydrolyses a ligand *in vivo* to release an active molecule, although the nature of the ligand remains to be identified (Waters *et al.*, 2014). These authors report close parallels between karrikin signalling and strigolactone signalling as the perception systems employ similar mechanisms. The KAI2 protein appears to be present in all angiosperms, implying a fundamentally important function in plant physiology.

In September 2008 a new group of natural plant hormones was announced, the strigol actones, thought to be involved in the control of the branching process. Mutant pea seedlings that exhibited uncontrolled branching were found not to contain these hormones and the effect could be overcome by the addition of strigolactones (Gomez-Roldan *et al.*, 2008; Umehara *et al.*, 2008).

These terpenoid lactones are biosynthetically derived from carotenoids (Figure 15.6). Inhibition of shoot branching can have a profound effect on plant architecture and apical dominance. Thus, cut flowers or ornamental potted plants with either more or less branching may have ornamental appeal, while crop yield might be manipulated.

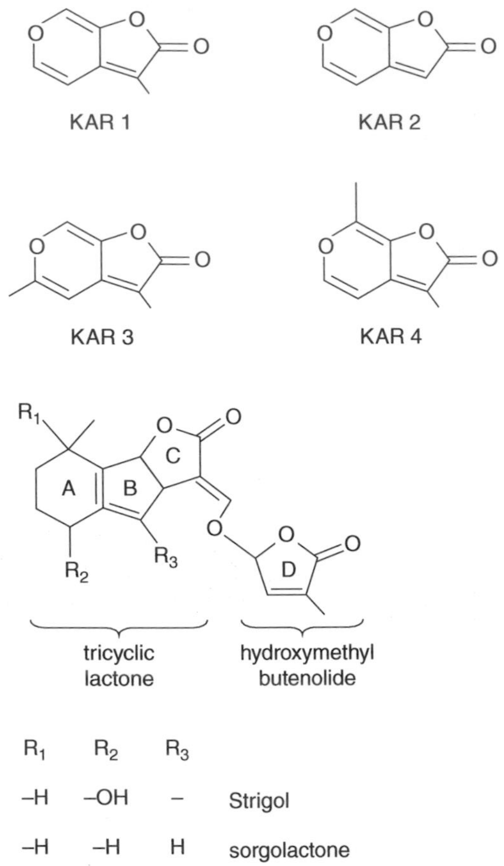

Figure 15.6   Chemical structures of active karrikins and strigolactones.

The strigolactones are synthesised in the roots and are involved in the germination of parasitic plants at concentrations as low as 5 ppm. They also promote the recognition of symbiotic fungi and inhibit branching in plants (Umehara *et al.*, 2015). Infestations of the parasitic plant *Striga hermonthica* cause significant crop losses in sub-Saharan Africa, equating to US\$7–10 billion, and the Striga seedbank is vast and long lived. Kountche *et al.* (2019) describe the application of strigolactone analogues in the absence of crop plants to stimulate germination of this parasitic weed, resulting in a reduction between 55 and 65% of Striga emergence in pearl millet and sorghum fields. This 'suicidal germination' strategy may increase crop production and yields in the long term.

Noting the chemical similarities shown in Figure 15.6, and the array of physiological processes that involve both karrikins and strigolactones, these lactones may become lead chemicals in the generation of new herbicides and plant growth regulators.

15.6   Natural products as leads for new herbicides

Secondary metabolites are synthesised from primary metabolism, i.e. photosynthetic carbon reduction, and their pathways often display great complexity (Figure 15.7). For more biochemical details, the reader is referred to the volumes edited by Wink (2010).

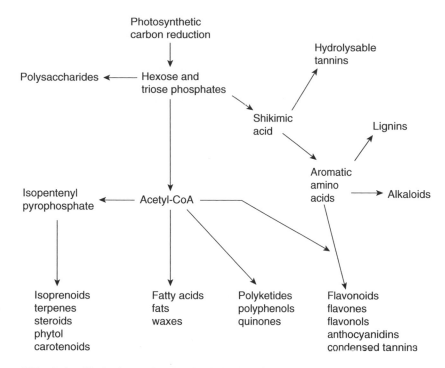

Figure 15.7 A simplified scheme showing the derivation of the main groups of plant secondary metabolites. Source: modified from Duke, S.O., Dayan, F.E. and Rimando, A.M. (2000) Natural products and herbicide discovery. In: Cobb, A.H. and Kirkwood, R.C. (eds) *Herbicides and their Mechanisms of Action*. Sheffield: Sheffield Academic Press, chapter 5.

Chemical analysis shows that the pathways of secondary metabolite biosynthesis are not only found in plants, but also in fungi and some bacteria, and some key enzymes are found in animals! One explanation for this is that these key molecules evolved in prokaryotes which became endosymbionts as proto-bacteria and cyanobacteria, progenitors of mitochondria and chloroplasts, respectively. Termed, horizontal gene transfer, these genomes have evolved and endogenous plant enzymes have transformed these novel metabolites. Similarly, mycorrhizal root associations, found in about 80% of all plants, produce secondary metabolites important in defence against herbivores and pathogens. As a final example, recent research has demonstrated that a key gene (Arc) in memory formation in the human brain encodes a repurposed retrotransposon protein which looks and behaves like a retrovirus, moving RNA between cells in a virus capsid. Evolutionary analysis has shown that the Arc protein is distantly related to a class of retrotransposons, but it has evolved to mediate a novel form of communication between neurons (Pasturzyn *et al.*, 2018; Pasturzyn, 2019). Thus, it is likely that horizontal gene transfer introduced genes coding for the biosynthesis of some secondary metabolites which, following millions of years of evolution, were metabolised by host enzymes to generate the numbers of secondary metabolites observed today.

Yet what relevance is this to a text on herbicides? The answer is that such chemical diversity has produced molecules of such amazing diversity that one family of novel diterpenoids, termed taxols and taxanes, are potent chemotherapeutic agents, active against a range of human cancers and commonly used to treat breast and ovarian cancers. They were originally discovered in the bark of the Pacific Yew Tree (*Taxus brevifolia*). A further example is the recent discovery of miyabeacin by Ward *et al.* (2020) in willow (*Salix* spp.). In laboratory tests, this cyclodimer has been found to kill cancer cells, including those resistant to other drugs. It is active against neuroblastoma and several breast, throat and ovarian cancer cell lines.

The screening of secondary metabolites against important weeds and key enzymes has proven value, and continues to be a useful route for the discovery of lead structures to new herbicides. This conclusion is confirmed by the recent report by Choi *et al.* (2017) that extracts from an actinomycete soil bacterium, obtained from a forest floor in Korea and termed G-0299, demonstrated strong phytotoxic activity to the weed *Digitaria ciliaris* (southern crabgrass).

In the plant kingdom there is an incredible diversity of secondary metabolites. At least 10^5 active plant metabolites are now known and the reader is directed to the NAPRALERT website (www.napralert.org) for a comprehensive database, covering the literature for natural products.

Why do so many natural molecules exist and what is their physiological function in plants? They were first thought to exist as metabolic waste products but, since the metabolic cost to the plant is high, there must be clear benefits to the plant for producing them. It is now thought that secondary metabolism has evolved with the co-evolution of microbial and insect parasites and animal herbivores, largely as defence and protection mechanisms. Since specialist organisms have evolved that have succeeded in overcoming plant poisons and chemical defences, new metabolic pathways and chemical modifications have evolved and continue to evolve accordingly. The earliest defence molecules were the resins, lignins, condensed tannins and flavonoids. The evolution of the angiosperms and herbaceous plants was marked by the proliferation of further secondary metabolites as products of the acetate–mevalonate pathway.

During steady-state photosynthesis, at least 20% of the carbon fixed in green plants is directed, via triose phosphates, towards synthesis of an impressive array of biologically important end products (Figure 15.7). These include lignins, alkaloids, tannins, isoprenoids and a wide range of phenolic compounds, including the 15-carbon flavonoids. Indeed, it is estimated that about 2% of all carbon fixed, equivalent to 10^9 tonnes each year, is converted to flavonoids alone. The precise end products formed depend on the metabolic needs of the plant at any given time, which are highly dependent on the prevailing environmental conditions and growth stage. Furthermore, plants under stress and those with plentiful supplies of nitrogen are preferentially attacked by insects, and the plant responds by increasing the biosynthesis of defence molecules. During pathogen attack, the concentrations of flavonoids and related compounds greatly increase at the site of infection, to concentrations that are toxic to pathogens in *in vitro* assays. Wounding and feeding by herbivores can induce the biosynthesis of toxic coumarins and tannins, and phenolic acids as precursors to lignins and suberins in wound healing. The accumulation of flavonols, especially kaempferol at wound sites, may also prevent microbial infection. Recent research indicates that stresses such as infection and exposure to ultraviolet light that are perceived in one part of the plant can be communicated to the rest of the plant and elicit systemic effects. This communication is thought to be mediated by salicylic acid, which shares part of its biosynthetic pathway with the flavonoids.

Essential oils are natural products mainly composed of volatile organic compounds belonging mainly to the phenylpropanoid and terpenoid families. The terpenes are classified by the number of isoprene subunits and the phenylpropanoids are generated from L-phenylalanine. They show promise as herbicides, and can also possess antimicrobial, antibacterial and insecticidal properties. To become commercialised it will be necessary to

Table 15.2 Physiological properties of some essential oils.

Essential oil	Physiological action
Carvone	Inhibition of potato sprout growth
Camphor, citral	Altered water status
Cineole, thymol, menthol, geraniol, camphor	Lipid peroxidation
Pinene, citronellol	Induction of reactive oxygen species
Pinene, farnesene, engenol	Inhibition of photosynthesis
Juglone, pulegone, eucalyptol, limonene	Inhibition of respiration
Limonene, citral, carracrol, menthone	Microtubule disruption

Source: modified from Werrie, P.Y., Durenne, B., Delaplace, P. and Fauconnier, M.-L. (2020) *Phytotoxicity* of essential oils: opportunities and constraints for the development of biopesticides. A review. *Foods* **9**, 1291; doi: org/10.3390/foods9091291

overcome their potential phytotoxicity to crops, and they have to be comparable to existing herbicides in terms of concentrations and costs. Their transient effects and formulations will require optimisation and their physico-chemical properties explored. Most studies have so far been in the laboratory and the glasshouse, so effective formulations will need to be developed for activity in the field, with reduced volatility. Verdeguer *et al.* (2020) have reviewed the main compounds in essential oils, their botanical species of origin, their biological properties and their herbicidal activity (Table 15.2).

It is now recognised that serotonin (5-hydroxytryptamine) is ubiquitous, playing diverse roles in all forms of life. In mammals, it is an important neurotransmitter, helping to regulate mood, appetite, sleep and memory; in insects it is used to seek out resources and food, while in plants it is involved in growth and development, photosynthesis, reproduction and the signalling of responses to stress. Of course, plants lack a nervous system, but we know that their physiology is closely in tune with their ever-changing environment.

The serotonin found in plants (Figure 15.8A), termed phyto-serotonin, is derived from tryptophan and shares structural similarity to auxin (Figure 15.8C). How it functions in plants and its mechanisms of action, however, are little understood. Lu *et al.* (2018) report that when rice is infected by insects, serotonin biosynthesis occurs, conferring resistance to planthoppers and stem borers, implying an important role in defence for serotonin. Salicylic acid (Figure 15.8B) was also produced by insect attack in rice, but suppressing serotonin biosynthesis increased the resistance to these pests.

That serotonin is an ancient molecule, present in distant and distinctive organisms, raises to question of what its function is in these organisms. Billions of years of evolution appear to have been sufficient to develop additional functions from taxa to taxa. Indeed, it is a good example of functional diversity in evolution, although its precise roles in plants are far from being understood. The reader is referred to the review by Erland and Saxena (2017) for further details.

Melatonin (*N*-acetyl-5-methoxytryptamine) is also an indoleamine (Figure 15.8D), produced in the chloroplasts and mitochondria of roots and leaves. It is considered to be a master plant growth regulator, as it stimulates plant growth and development by acting as a signalling molecule associated with defence mechanisms against biotic and abiotic stresses such as cold, drought, salinity and nutritional deficiencies. The number of publications related to the physiological and genetic effects of melatonin has rapidly increased in the last decade, and its status as a plant growth regulator is being debated (Nawaz *et al.*, 2021).

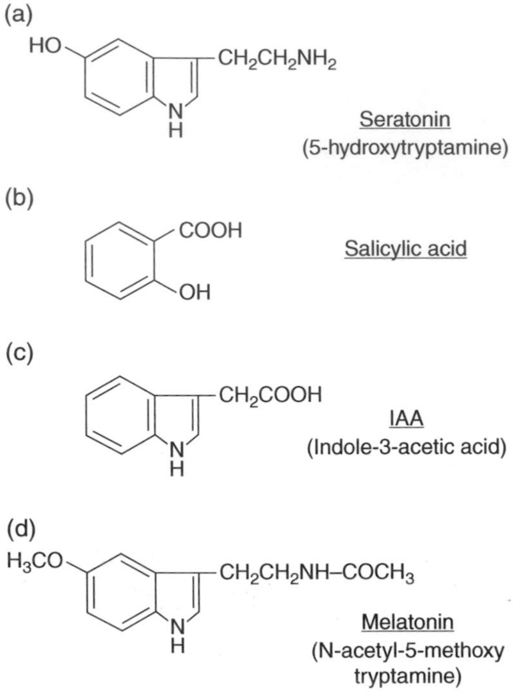

Figure 15.8 Structure of Seratonin (a), salicylic acid (b), IAA (c) and melatonin (d).

Swain (1977) has estimated that perhaps as many as 400,000 secondary products are synthesised in the Plant Kingdom, and this surely represents a major source of diverse chemical structures as herbicidal leads. For example, over 700 natural amino acids have been reported in the literature, a significant number of which show promising activity as natural inhibitors of pyridoxal phosphate-dependent enzymes, the syntheses of glutamine and glucosamine, and proteases and peptidases (Jung, 1989). One is therefore forced to conclude that, since only a small percentage of plant species have been analysed for natural product chemistry, further investigations will surely prove rewarding.

References

Baek, D., Shin, G., Kim, M.C., Shen, M., Lee, S.Y. and Yun, D-J. (2020) Histone deacetylase 9 with ABA contributes to abscisic acid homeostasis in drought stress response. *Frontiers of Plant Science*; doi: org/10.3389/fpls.2020.00143

Chen, X., Lu, L., Mayer, K.S., Scalf, M., Qian, S., Lomax, A., *et al.* (2016) POWERDRESS interacts with histone deacetylase 9 to promote aging in *Arabidopsis*. *eLife*, 5: e17214; doi: 10.7554/eLife.17214

Choi, J.-S., Kim, Y.S., Kim, J.D., Kim, H.J., Ko, Y-K, Park, K.W. and Moon, S-S. (2017) Herbicidal characteristics of soil bacteria Actinomycetes G-0299 to southern crabgrass. *Weed and Turfgrass Science* **6**, 212–221.

de Coninck, B., and de Smet, I. (2016) Plant peptides – taking them to the next level. *Journal of Experimental Botany* **67**, 4791–4795.

Corral, M.G., Leroux, J., Stubbs, K.A. and Mylne, J.S. (2017) Herbicidal properties of antimalarial drugs. *Nature, Scientific Reports* **7**, 45871; doi: 10.1038/srep45871.

Corral, M.G., Haywood, J., Stehl, L.H., Stubbs, K.A., Murcha, M.W. and Mylne, J.S. (2018) A herbicide structure–activity analysis of the antimalarial lead compound MMV007978 against *Arabidopsis thaliana*. *Pest Management Science* **74**, 1558–1563.

Duke, S.O., Dayan, F.E. and Rimando, A.M. (2000) Natural products and herbicide discovery. In: Cobb, A.H. and Kirkwood, R.C. (eds) *Herbicides and their Mechanisms of Action*. Sheffield: Sheffield Academic Press, chapter 5.

Erland, L.A.E. and Saxena, P.K. (2017) Beyond a neurotransmitter: the role of serotonin in plants. *Neurotransmitter*, 4 e1538; doi: 10.14800/nt.1538

Evans-Roberts, K.M., Mitchenall, L., Wall, M.K., Leroux, J., Mylne, J.S. and Maxwell, A. (2016) DNAGyrase is the target for the quinolone drug ciprofloxacin in *Arabidopsis thaliana*. *Journal of Biological Chemistry* **291**, 3136–3144.

Gomez-Roldan, V., Fermas, S., Brewer, P.B., Puech-Pages, V., Dun, E.A., Pillot, J.-B., *et al.* (2008) Strigolactone inhibition of shoot branching. *Nature* **455**, 189–194.

Hilton, J.R. (1984) The influence of light and potassium nitrate on the dormancy and germination of *Avena fatua* L. (wild oat) seed and its ecological significance. *The New Phytologist* **96**, 31–34.

van der Hoorn, R.A.L. and Rivas, S. (2018) Unravelling the mode of action of plant proteases. *The New Phytologist* **218**, 879–881.

Jung, M.J. (1989) Natural amino acids as enzyme inhibitors. In: Copping, L.G., Dalziel, J. and Dodge, A.D. (eds) *Prospects for Amino Acid Biosynthesis Inhibitors in Crop Protection and Pharmaceutical Chemistry*. British Crop Protection Council Monograph No. 42. Farnham: BCPC, pp. 15–22.

Kerr, M.W. and Whitaker, D.P. (1985) Energy losses – photorespiration. *Annual Proceedings of the Phytochemical Society of Europe* **26**, 45–57.

Khan, T., Sankhe, K., Suvarna, V., Sherje, A., Patel, K.N. and Dravyakar, B. (2018) DNA gyrase inhibitors: progress and synthesis of potent compounds as antibacterial agents. *Biomedicine and Pharmacotherapy* **103**, 923–938.

Kountche, B.A., Jamil, M., Yonli, D., Minimassom, P., Nikiema, P., Blanco-Ania, D., *et al.* (2019) Suicidal germination as a control strategy for *Striga hermonthica* (Benth.) in smallholder farms of sub-Saharan Africa. *Plants, People, Planet* **1**; doi: org/10.1002/ppp3.32.

Liu, H., Hu, M., Wang, Q., Cheng, L. and Zhang,Z. (2018) Role of papain-like cysteine proteases in plant development. *Frontiers of Plant Science*, article 01717.

Lu, H.P., Luo, T., Fu, H.-W., Wang, L., Tan, Y.-Y., Huang, J.-Z. *et al.* (2018) Resistance of rice to insect pests mediated by suppression of serotonin biosynthesis. *Nature, Plants* **4**, 338–344.

Mayer, A.M. and Poljakoff-Mayber, A. (1989) *The Germination of Seeds*, 4th edn. Oxford: Pergamon Press.

McFadden, G. (2011) The apicoplast. *Protoplasma* **248**, 641–650.

Mendoza, F., Berry, C., Prestigiacomo, L. and van Hoewyk, D. (2020) Proteosome inhibition rapidly exacerbates photoinhibition and impededes recovery during high-light stress in *Chlamydomonas reingardtii*. *BMC Plant Biology* **20**, article 22; doi: org/10.1186/s12870-020-2236-6

Miricescu, A., Goslin, K. and Graciet, E. (2018) Ubiquitinylation in plants: signalling hub for the integration of environmental signals. *Journal of Experimental Botany*, **69**, 4511–4527.

Nawaz, K., Chaudhary, R., Sarwar, A., Ahmad, B., Gul, A., Hano, C., *et al.* (2021) Melatonin as a master regulator in plant growth, development and stress alleviator for sustainable agricultural production: current status and future perspectives. *Sustainability* **13**, 294; doi: org/10.3390/su13010294

Nelson, D.C., and four others (2012) Regulation of seed germination and seedling growth by chemical signals from burning vegetation. *Annual Review of Plant Biology* **63**, 107–130.

O'Driscoll, C. (2019) Game changer for Agchems. Reporting on the Bayer Future of Farming Dialogue Event, September 2019. *Chemistry and Industry* **10**, 8.

Pastuzyn, E.D. (2019) Are our memories formed by an ancient virus? *The Biologist* **65**, 14–17.

Pastuzyn, E.D., Day, C.E.,Kearns, R.B., Kyrke-Smith, M., Taibi, A.V., McCormick, J., *et al.* (2018) The neuronal gene Arc encodes a repurposed retrotransposon Gag Protein that mediates intercellular RNA transfer. *Cell* **172**, 275–288.

Reape, T.J., Molony, E.M. and Case, P.F. (2008) Programmed cell death in plants: distinguishing between different modes. *Journal of Experimental Botany* **59**, 435–444.

Reynolds, C.J., Long, R.L., Flematti, G.R., Cherry, H. and Turner, S.R. (2014) Karrikins promote germination of physiologically dormant seeds of *Chrysanthemoides monilifera* ssp. monilifera (boneseed). *Weed Research* **54**, 48–57.

Rustgi, S., Boex-Fontvielle, E., Reinbothe, C., von Wettstein, D. and Reinbothe, S. (2018) The complex world of plant protease inhibitors: insights into a Kunitz-type cysteine protease inhibitor of *Arabidopsis thaliana. Communicative and Integrative Biology*, e1368599; doi: org/10.1080/1942 0889.2017.1368599

Sadanandom, A., Bailey, M., Ewan, R., Lee, J. and Nelis, S. (2012)The ubiquitin–proteosome system: central modifier of plant signalling. *The New Phytologist* **196**, 13–28.

Stael, S., van Breusegem, F., Gevaert, K. and Nowack, M. (2019) Plant proteases and programmed cell death. *Journal of Experimental Botany* **70**, 1991–1995.

Swain, T. (1977) Secondary compounds as protective agents. *Annual Review of Plant Physiology* **28**, 479–501.

Takahashi, F., Suzuki, T., Osakabe, Y., Betsuyaku, S., Kondo, Y., Dohmae, N., *et al.* (2018) A small peptide movia abscisic acid in long distance signalling. *Nature* **556**, 235–238.

Threape, T.J., Molony, E.M. and McCabe, P.F. (2008) Programmed cell death in plants: distinguishing between different modes. *Journal of Experimental Botany* **59**, 435–444

Toh, S., Inoue, S., Toda, Y., Yuki, T., Suzuki, K., Hamamoto, S., *et al.* (2018) Identification and characterisation of compounds that affect stomatal movement. *Plant and Cell Physiology* **59**, 1568–1580.

Umehara, M., Hanada, A., Yoshida, S., Akiyama, K., Arite, T., Takeda-Kamiya, N., *et al.* (2008) Inhibition of shoot branching by new terpenoid plant hormones. *Nature* **455**, 195–200.

Umehara, M., Cao, M., Akiyama, K., Akatsu, T., Seto, Y., Hanada, A., *et al.* (2015) Structural requirements of strigolactones for shoot branching inhibition in rice and *Arabidopsis. Plant and Cell Physiology* **56**, 1059–1072.

Verdeguer, M., Sanchez-Moreiras, A.M. and Araniti, F. (2020) Phytotoxic effects and mechanisms of action of essential oils and terpenoids. *Plants* **9**, 1571; doi: org/10.3390/plants9111571

Wakabayashi, K. and Böger P. (2004) Phytotoxic sites of action for molecular design of new herbicides (part 2): amino acid, lipid and cell wall biosynthesis, and other targets for future herbicides. *Weed Biology and Management* **4**, 59–70.

Ward, J.L., Wu, Y., Harflett, C., Onafuye, H., Corol, D., Lomax, C., *et al.* (2020) Miyabeacin: a new cyclodimer presents a potential role for willow in cancer therapy. *Nature Scientific Reports*; doi: org/10.1038/s41598-020-63349-1

Waters, M.T., Scaffidi, A., Sun, Y.K., Flematti, G.R. and Smith, S.M. (2014) The karrikin response system in *Arabidopsis. The Plant Journal*; doi: org/10.1111/tpj.12430.

Werrie, P.Y., Durenne, B., Delaplace, P. and Fauconnier, M.-L. (2020) Phytotoxicity of essential oils: opportunities and constraints for the development of biopesticides. A review. *Foods* **9**, 1291; doi: org/10.3390/foods9091291

Wink, M. (ed.) (2010) *Biochemistry of Plant Secondary Metabolism*. Annual Plant Reviews 40, 2nd edn. Chichester: Wiley-Blackwell.

Xu, B., Long, Y., Zhu, F.Q., Sai, N., Chirkova, L., Betts, A., *et al.* (2021) GABA signalling modulates stomatal opening to enhance plant water use efficiency and drought resilience. *Nature Communications* **12**, article 1952; doi: org/10.1038/s41467-021-21694-3

Xu, F.-Q. and Xue, H.-W. (2019) The ubiquitin–proteosome system in plant responses to environments. *Plant, Cell and Environment* **42**, 2931–2944; doi: org/10.1011/pce.13633

Yan Y. Liu, Q., Zang, X., Yuan, S., Bat-Erdene, U., Nguyen, C., *et al.* (2018) Resistance-gene-directed discovery of a natural product herbicide with a new mode of action. *Nature* **559**, 415–418.

Ziemert, N., Alanjary, M. and Weber, T. (2016) The evolution of gene mining in microbes – a review. *Natural Products Reports* **33**, 988–1005.

Glossary

The following provides a short explanation of some relevant terms that are not defined in the text. For further information of a botanical nature the reader may find *The Penguin Dictionary of Plant Science* (1999) to be of value.

absorption   The process whereby chemicals gain entry into plant tissues. This may be active (against an energy gradient) or passive (no energy expended).

acceptable daily intake   The concentration of **herbicide** or its **residue** that a human may be exposed to on a daily basis that, according to current information, does not appear to create an unacceptable risk to the well-being of the individual. Expressed as mg per kilogram body weight per day.

active ingredient   The chemical in a commercial **formulation** that is responsible for the herbicidal effect.

active site   The site at which the substrates of an enzyme are bound during catalysis.

acute toxicity   Toxicity of a single chemical after a 24 h exposure period.

adjuvant   A substance added to the formulation or spray tank to modify application characteristics, eg to enhance coverage of leaf surfaces.

adsorption   Chemical or physical attraction to a surface, e.g. soil.

allelopathy   The release of a chemical by a plant that inhibits the growth of nearby plants and so reduces competition.

annual   A plant that germinates, grows, flowers, produces seeds and dies within one year.

antagonism   Reduced activity of a herbicide in the presence of another chemical. The opposite is synergism.

apoplast   The continuum of cell walls throughout the plant. Important in the movement of water and water-soluble herbicides.

autoradiography   A technique for detecting the presence and distribution of radioactive compounds in a biological material. In essence, radioactive emissions darken a photographic plate which, when developed, reveals the location of the isotope as dark patches of silver grains.

Herbicides and Plant Physiology, Third Edition. Andrew H. Cobb.
© 2022 John Wiley & Sons Ltd. Published 2022 by John Wiley & Sons Ltd.

auxin   A natural (e.g. IAA) or synthetic (e.g. MCPA) plant **growth regulator** involved in cell division, enlargement and differentiation.

biennial   A plant that completes its life cycle in two years. In the first year vegetation growth predominates and photosynthates are stored over winter. In the following year these products are used for the generation of leaves, flowers and seeds.

bioassay   A biological assay. The assessment of the effect of a chemical on an organism by comparison with the effects of standard substances of known concentration.

biological control   The control of an unwanted organism by making use of natural predators.

biorational design   The use of biological control agents and analogues of naturally occurring biochemicals for the discovery of new agrochemicals.

biotechnology   The use of (usually) microorganisms, or their metabolites and products, in industrial processes.

biotype   Sub-group within a species that differs in some respect (for example in **herbicide** resistance) from that species.

carboxylase   An **enzyme** that catalyses the transfer or incorporation of carbon dioxide into a substrate molecule.

carotenoids   Yellow, orange, brown or red lipophilic pigments that function as accessory photosynthetic pigments.

cell cycle   Sequence of events leading to the formation of two daughter cells.

chemical synthesis   Production of potentially new active ingredients from basic chemical entities, often enhanced by combinatorial chemistry.

chemophobia   An irrational prejudice against chemicals.

chlorosis   Loss of green colour in foliage. Leaves appear typically pale or yellow in colour.

chronic toxicity   Toxicity of a single chemical after a prolonged period of exposure, usually from a day to several weeks.

clone   A population of genetically identical cells or individuals.

coleoptile   A cylindrical sheath of tissue that encloses and protects young shoots of grasses and cereals during growth to the soil surface.

combinatorial chemistry   A rapid, mechanised system for the production of a large number of potentially active ingredients from basic chemical reagents.

contact herbicide   A compound that kills on contact rather than relying on translocation for activity.

conventional breeding   A method of making target organisms express desirable traits by artificial and repeated crossing. It can take years, but development costs are low. No safety regulations.

cotyledon   Embryonic leaf in seed plants that acts either as a storage organ or in absorbing food reserves from the **endosperm**. Dicotyledonous plants (such as broadleaf weeds) have two cotyledons and monocotyledonous plants (cereals and grasses) have one.

cultivar   A cultivated variety generated by human selection and so not normally found in natural populations.

cuticle   A continuous waxy layer that covers the aerial parts of a plant to prevent excessive water loss.

cutin   A water-repellant waxy polymer that is a major component of the **cuticle**.

cytoskeleton   An intracellular scaffold of proteins. A dynamic structure that maintains cell shape and function, including transport and cell division.

desiccant   A chemical that induces rapid desiccation of plant parts.

dormancy   An inactive phase during which growth and developmental processes stop.

ED$_{50}$ The chemical dose that produces a desired effect in 50% of the test organisms exposed to the chemical.

emulsifiable concentrate   A formulation containing organic solvent and emulsifier to aid mixing with water.

environment   All the biotic and abiotic conditions in which an organism lives.

environmental chemistry   Investigation of the physical and metabolic breakdown of a product in plant, animal, soil and water systems. Identification of the compounds in these systems.

enzyme   A protein that catalyses biological reactions.

ephemeral   A plant with such a short life cycle that it may be completed many times in one growing season.

epinasty   Plant tissue movement to an external stimulus in which increased growth on one side of an organ causes bending of that organ. Commonly observed following application of auxins to young tissues.

ethylene   A gaseous plant hormone that affects the growth, development, ripening and senescence.

fatty acid   A long-chain aliphatic carboxylic acid that may be saturated or unsaturated.

field trials   An assessment of the activity of a herbicide against target weeds, pets or diseases in the field, including comparison with standard treatments.

formulation   The form in which an agrochemical is prepared for commercial use, including wettable powders, **emulsifiable concentrates**, granules, etc.

gene   The unit of inheritance.

gene editing   A method of making organisms express desirable traits by site-specific alterations of genetic materials, such as by deleting or adding sequences of DNA. Can be done in a short time, with low to medium costs.

genetic engineering   Isolation of 'useful' **genes** from a donor organism and their functional incorporation into an organism that does not normally possess them.

genetic modification   A method of making target organisms express desirable traits by modifying parts through the exchange of genetic materials, by inserting genes from other species, including micro-organisms. Needs time and development costs are high. Need to address safety regulations.

genomics   The application of biotechnology to further understand genetic structure and function.

glycosyltransferase   An **enzyme** that transfers a sugar residue from one molecule to another.

graminicide   A **herbicide** which kills grasses.

growth regulator   A substance that at very low concentrations may affect growth and differentiation in plants.

herbicide   A chemical that kills plants.

high-throughput screening   A rapid, mechanical system for assessing the biological activity of very low volumes of chemical.

Hill reaction   Transfer of electrons from water to non-physiological acceptors by thylakoids in the presence of light and with the evolution of oxygen.

induced enzyme   An **enzyme** that is synthesized in response to elevated concentrations of its substrate or a specific inducer.

isotopic tracer   A stable, radioactive isotope that can be used to label a herbicide or metabolite to monitor its fate within an intact organism.

isozyme (or isoenzyme)   One of a number of distinct proteins with the same enzyme activity but different kinetic charactertistics.

kinase   An enzyme that adds a phosphate group to a protein.

kinome   A complete set of protein kinases and phosphatases encoded in the genome.

LD_{50}   The dose required to kill 50% of test organisms, usually expressed as milligrams of chemical orally ingested per kilogram body weight.

leaching   The movement of water soluble chemicals down the soil profile.

meristem   A region containing actively dividing cells.

metabolism   The sum of the enzymatic reactions taking place in a cell, organ or organism.

metabolite   A product of **metabolism** within an organism.

microtubule   A cytoplasmic tubule composed of the protein tubulin.

morphology   The study of form and shape, particularly with respect to external structure.

mutagen   An agent that causes an increased frequency of mutation, i.e. an inherited genetic change.

necrosis   Death of a plant cell, group of cells or tissue, while the rest of the plant is still alive. A particularly useful term when dead tissues undergo typical colour changes in comparison with healthy tissues.

non-selective herbicide   A treatment to kill all vegetation.

organelle   A membrane-bound structure in the cytoplasm in which specific but essential processes take place. Examples include photosynthesis in chloroplasts and oxidative phosphorylation in mitochondria.

perennial   A plant that lives for many years, surviving as either herbaceous perennials (with underground storage organs) or woody perennials (whose aerial stems persist above ground).

persistence   The property of a compound to persist in soil and give prolonged protection to a crop or prevent plant regrowth for extended periods. Persistence is a function of dosage, chemical volatility and stability in a given environment.

pest   Any organism that damages crop growth or reduces yield.

phosphatase   An enzyme that removes a phosphate group from a protein.

photosynthate   Organic products of photosynthetic carbon reduction.

physiology   The study of the processes and functions of life.

phytotoxicity   Damage to plants.

plasticity   The ability of an organism to change its form in response to varying environmental conditions.

population   A community of potentially interbreeding organisms.

post-emergence   After a **weed** or crop has emerged.

pre-emergence   Before a **weed** or crop has emerged.

proteases   Enzymes that break down the peptide bonds of proteins.

protoplast   The part of the plant cell internal to the cell wall and bounded by the plasma membrane.

rate   Amount of active ingredient applied to a unit area.

recombinant DNA   Genetic material that contains novel **gene** sequences using techniques of **genetic engineering**.

registration   Preparation and submission of data dossiers to, and subsequent negotiations with, registration authorities with the aim of obtaining approval to market a new product.

residual   To have a profound effect over a period of time.

residue   Trace of a pesticide and its metabolites remaining on and in a crop or the environment.

rhizome   An underground stem capable of horizontal growth. Buds can play an important role in the vegetative propagation of weeds.

safener   A chemical that reduces the **phytotoxicity** of another chemical when used together.

seed dressing   A chemical seed treatment to protect against fungal or insect attack.

selective herbicide   A herbicide that will kill some species and not others.

sink   A site in a plant where a demand exists for particular substrates or photosynthate.

source   A site of production of particular substrates or **photosynthate** for movement to specific **sinks**.

stoma   A pore formed from guard cells on aerial plant surfaces that can open or close to allow gaseous exchange (pl. stomata).

symplast   The continuum of cytoplasm throughout the plant, linked by **plasmodesmata**.

synergism   The combined effects of two treatments when acting together greatly exceed the sum of their effects when each acts alone.

systemic   A chemical that is absorbed and translocated throughout the plant body.

tank mixture   Mixture of two or more pesticides in a spray tank at application.

taproot   A persistent, primary root often penetrating to considerable depth and specialized for storage.

target   The organisms, structures, tissues, cells and **enzymes** to be affected by the **herbicide**.

tiller   A shoot that develops at the base of a stem.

tissue culture   The growth of isolated plant cells or pieces of tissue under controlled conditions in a sterile growth medium.

tolerant   Capable of withstanding effects.

toxicology   Safety assessment of a new product candidate in biological systems.

translocation   Transfer of soluble molecules, such as **herbicides**, from one part of the plant to another.

transpiration   The loss of water by evaporation from a plant surface, primarily through open **stomata**.

trichome   An outgrowth from an epidermal cell that is variable in shape and function. Often termed leaf hairs.

tuber   A swollen part of a stem or root that is modified for storage.

vacuome   The collective term for all of the vacuoles within a cell.

volatility   A measure of the tendency of a substance to vapourise.

weed   A plant growing where it is not wanted, or interferes with the activities or welfare of humankind. A plant whose virtues remain to be discovered!

xenobiotic   A chemical foreign to the organism.

Index

Herbicides and Plant Physiology, Third Edition. Andrew H. Cobb.
© 2022 John Wiley & Sons Ltd. Published 2022 by John Wiley & Sons Ltd.

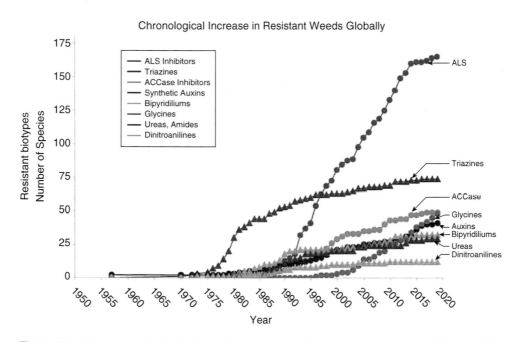

Figure 13.2 Development of herbicide resistance in weeds, from Ian Heap, accessed 18th April 2020.

Herbicides and Plant Physiology, Third Edition. Andrew H. Cobb.
© 2022 John Wiley & Sons Ltd. Published 2022 by John Wiley & Sons Ltd.

Figure 13.7 Black-grass in a field of wheat. On the left side the field is free of black-grass. On the right the black-grass population is approximately 500 plants per m². Yield will be seriously reduced in the part of this trial where black-grass has been allowed to remain. Source: photograph courtesy of J.P.H. Reade.

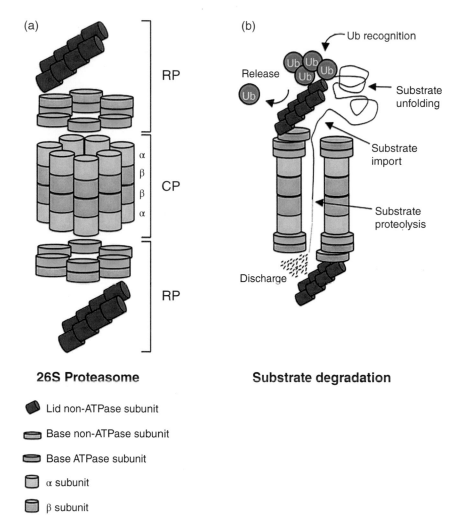

Figure 15.2 Structure of the 26S proteosome. A. Relative arrangement of proteosome subunits depicting the association of heptameric rings within the core protease (CP) and subunit distribution of the Lid and Base sub-complexes within the regulatory particle (RP). B. Diagram representing known and predicted activities required for substrate degradation. Source: Sadanandom, A., Bailey, M., Ewan, R., Lee, J. and Nelis, S. (2012) The ubiquitin–proteosome system: central modifier of plant signalling. *The New Phytologist* **196**, 13–28. Reproduced with permission of John Wiley & Sons.